Nanocomposites and Polymers with Analytical Methods

Nanocomposites and Polymers with Analytical Methods

Editor

Leroy Sidney

scitus
academics

Nanocomposites and Polymers with Analytical Methods
Edited by **Leroy Sidney**

ISBN: 978-1-68117-526-3
Library of Congress Control Number: 2016942894

www.scitusacademics.com
Box No. 4766, 616 Corporate Way,
Suite 2, Valley Cottage,
NY 10989

Printed in United States of America on Acid Free Paper

Preface

Nanocomposites are composites in which at least one of the phases shows dimensions in the nanometer range. These are high performance materials that exhibit unusual property combinations and unique design possibilities and are thought of as the materials of the 21st century. With an estimated annual growth rate of about 25% and huge demand for engineering polymers, their potential is so promising that they are useful in several applications ranging from packaging to bio-medical.

Nanocomposites and Polymers with Analytical Methods cover a wide range of polymer matrices including both thermoplastics and thermosets. The text provides both a description of the fundamental aspects and an up-to-date overview of the recent developments in manufacturing technologies, fabrication, spectroscopic characterization, properties, structure-property relationships, rheology, crystallization behavior, degradability of biodegradable nanocomposites, processing, and finally commercial applications of polymer nanocomposites. The book describes new materials and analytic methods. These materials include chapters on gold nanoparticles and Sol-Gel metal oxides, nanocomposites with carbon nanotubes, methods of evaluation by depth sensing, and other methods. The book is intended for a very wide audience including students, researchers, scientists, engineers and professionals working in the fields of composite science and technology, nanocomposites, nanotechnology, materials science, surface science, polymer science and engineering, biodegradable materials, chemistry, and solid-state physics.

Table of Contents

Chapter 1

Polymer/Carbon-Based Hybrid Aerogels: Preparation, Properties and Applications

Lizeng Zuo [1], Youfang Zhang [1], Longsheng Zhang [1], Yue-E Miao [2], Wei Fan [2,*] and Tianxi Liu [1,2,*]

[1]State Key Laboratory of Molecular Engineering of Polymers, Department of Macromolecular Science, Fudan University, 220 Handan Road, Shanghai 200433, China

[2]State Key Laboratory of Modification of Chemical Fibers and Polymer Materials, College of Materials Science and Engineering, Donghua University, 2999 North Renmin Road, Shanghai 201620, China

ABSTRACT

Aerogels are synthetic porous materials derived from sol-gel materials in which the liquid component has been replaced with gas to leave intact solid nanostructures without pore collapse. Recently, aerogels based on natural or synthetic polymers, called polymer or organic aerogels, have been widely explored due to their porous structures and unique properties, such as high specific surface area, low density, low thermal conductivity and dielectric constant. This paper gives a comprehensive review about the most recent progresses in preparation, structures and properties of polymer and their derived carbon-based aerogels, as well as their potential applications in various fields including energy storage, adsorption, thermal insulation and flame retardancy. To facilitate further research and development, the technical challenges are discussed, and several future research directions are also suggested in this review.

KEYWORDS

polymer aerogels; carbon aerogels; hybrids; porous; three-dimensional network

1. INTRODUCTION

The term aerogel is to designate a sol-gel material that the liquid component of the gel has been replaced with gas to leave an intact solid nanostructure without pore collapse in which 90%~99% is air by volume [1,2,3]. The density of aerogel ranges from 1000 $kg \cdot m^{-3}$(above the solid density) to 1 $kg \cdot m^{-3}$ (lower than the density of air), which induces dramatic changes in the properties [4]. Due to the high porosity as well as dual microscopic (*i.e.*, nanoscale skeleton) and macroscopic (*i.e.*, condensed state matter) structural features, aerogel exhibits versatile unique properties such as ultralow thermal conductivity, sonic velocity and dielectric constant, high specific surface area, as well as ultrawide adjustable density and refractive index [5,6,7,8,9,10]. Therefore, it is an attractive material for applications in chemical sensors, thermal insulations, chemical adsorbents, catalysts or catalytic carriers, and space explorations.

Evolution of the aerogels is illustrated in Figure 1a, and the number of papers published on aerogels is booming, especially in the last 20 years (Figure 1b). Aerogel was first invented by Kistler in 1931 [11], which was named as "aerogel" (air + gel) because the liquid component inside the wet gel was replaced by air without damaging the solid microstructure. This silica aerogel was fabricated using sodium silicate (that is water-glass) as the precursor of silica, exhibiting some fantastic properties. Nevertheless, aerogel did not arouse broad interests due to the high manufacturing cost, tedious and time-consuming procedures during the synthesis process. The significant innovation was the replacement of water-glass/water systems with organic precursors and the corresponding organic solvents to facilely prepare aerogels, which was represented by the usage of tetramethyorthosilicate by Teichner's group in 1968 [12] and safer tetraethylorthosilicate by Russo *et al.*, in 1986 [13], and the development of carbon dioxide (CO_2) supercritical fluid drying [14]. However, aerogel research was limited to silica and several inorganic compositions only during about 70 years after the aerogel was invented. However, at the end of the 1980s, a new type of aerogel was invented by Pekala through the usage of an organic polymer, resorcinol-formaldehyde (RF), during the sol-gel process to develop organic and carbon aerogels [15]. Indeed, it was the right beginning of organic or polymer aerogels. Afterwards, the chemical composition of organic aerogels has been progressively diversified and the family of organic aerogels steadily increases. Various organic aerogels have been prepared, including polyimide (PI)-based aerogels [16], poly(vinyl alcohol) (PVA)-based aerogels [17] and poly(vinyl chloride)-based aerogels [18]. Additionally, aerogels derived from supramolecular gels, which are formed by the self-assembly of low molecular weight organic gelators, pave the way for their developments in high-tech materials and biomaterials [19,20,21]. At the same time, it is found that the polymer that makes up an organic aerogel can be dehydrated after heating to certain temperatures in an inert atmosphere (such as nitrogen or argon), leaving behind the carbon aerogel. Unlike silica aerogel, the carbon aerogel is an electrical conductor, which is called the aero-capacitor and characterized as an "electrochemical double capacitor with high power density and high energy density" [22]. After entering the 21st century, research on aerogels was booming with new types of aerogels emerging, such as non-silica aerogels [23,24,25,26], chalcogenide aerogels [27,28,29,30,31], gradient aerogels and other aerogel composites [32,33,34,35,36,37,38,39,40]. More recently, novel aerogels such as carbon nanotube (CNT) aerogels [41,42,43,44,45,46], graphene aerogels [47,48,49,50,51], silicon aerogels and carbide or carbonitride aerogels [52,53,54,55] were continually added into the aerogel community.

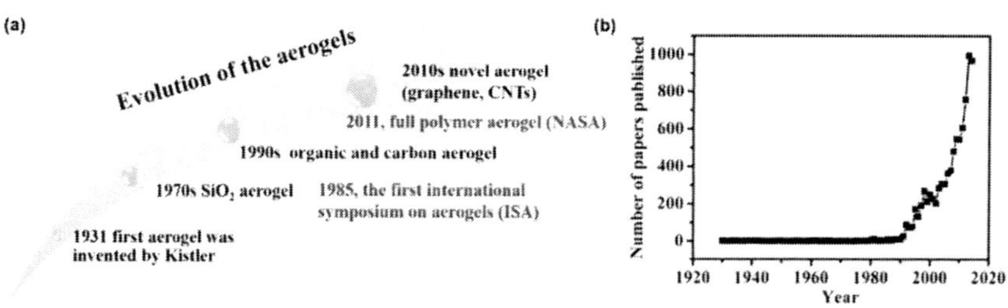

Figure 1. (**a**) Evolution of the aerogels. (**b**) The number of papers published every year during the last decades (search by aerogel from Web of Science).

Besides single-component aerogels, aerogels with multi-components were also fabricated to meet the ever-increasing desire for high performance and multi-functions. Indium-tin oxide aerogel, yttria-stabilized zirconia aerogel and binary oxide aerogel have been designed as oxide conductors or catalysts [56,57,58]. For example, ultralight metal oxide/silica aerogels combine the function of metal oxide and the rigid microstructure of the silica aerogel, which can greatly increase the efficiency for laser-X-ray conversion [59]. Moreover, organic aerogels containing various inorganic fillers such as clay, metal oxides or CNT/graphene were synthesized with improved performance and expanded applications in various fields [17,32,33,34,35,36,37,39]. In addition, the preparation of metal oxide/carbon composite aerogels is presumably a good way to improve the capacitive properties of carbon aerogels considerably [60,61,62].

Even though silica aerogels are the most widely studied, potential applications of silica aerogels in aerospace, industry, and daily life have been restricted because of their poor mechanical properties and hygroscopic nature. On the other hand, polymer aerogels including cellulose-based, resin-based, PI-based and their derived carbon aerogels exhibit diverse structures and properties, which widely broaden the application of aerogel based materials. Furthermore, hybrid aerogels with multi-components show enhanced performance and multi-functions, which are attracting more and more interest recently. Therefore, in this paper, recent advances in the field of polymer/carbon aerogels and their derivative hybrid aerogels will be discussed, with particular emphasis placed on their fabrication/synthesis methods, structures, properties and applications.

2. PREPARATION OF POLYMER/CARBON AEROGELS

The potential application of the aerogel is based on its properties, which is strongly relied on the microstructure. Therefore, it is very important to realize the microstructure control during the preparation process. Normally, the preparation process of the aerogel includes following three or four steps (Figure 2):

- Sol-gel transition (gelation): Nanoscale sol particles are crosslinked and hierarchically assembled into a wet gel spontaneously or catalyzed by catalysts via hydrolysis or condensation reactions.
- Network perfection (aging): Mechanically reinforce the tenuous solid skeleton generated during the sol-gel process.
- (Gel-aerogel transition (drying): The solvent inside the wet gel is replaced by air without serious microstructure damage.

All three steps could determine the microstructure of the aerogels and affect their properties and applications. Besides, it is worth to mention that an additional carbonization process (step 4, Figure 2) is required for fabricating carbon aerogels. In the following sections, the basic preparation procedure of aerogels including the sol-gel process, aging, drying methods and carbonization will be introduced in details.

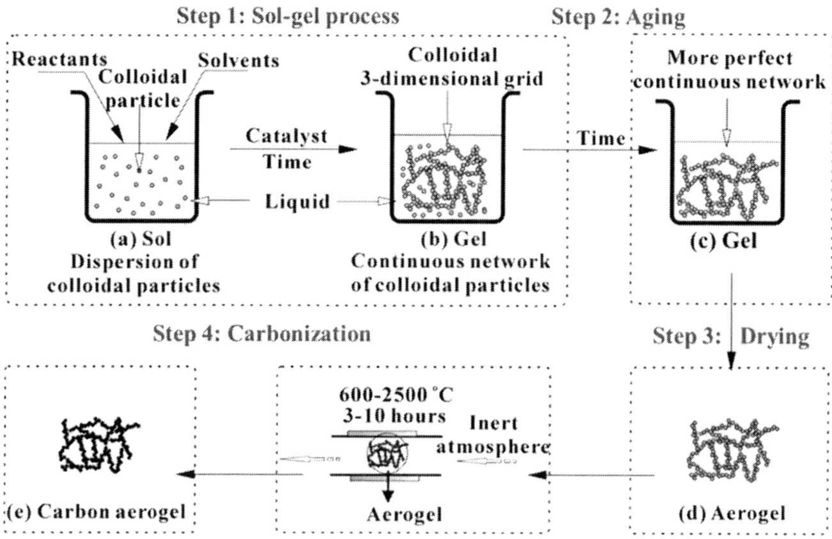

Figure 2. General preparation process of polymer or carbon aerogels.

2.1. The Sol-Gel Process

The sol-gel process designates the transition of a colloidal solution into a disordered, branched and continuous network linked by colloidal particles, which is interpenetrated by a liquid at low temperature (typically $T < 100$ °C). This gelation process can take place by the aggregation of colloidal particles via destabilizing the sol, or by hydrolysis and polycondensation of organometallic precursors, or by polycondensation of water-glass. The second route is used most frequently for the preparation of aerogels, which allows the most convenient variation of reaction parameters. Using this route, aerogels of main group oxides, transition and semimetal oxides, as well as organic compounds, can be produced.

Typically, there are two steps in the sol-gel process [63,64]. The first step is the dissolution of monomers in a solvent. Then, independent solid colloidal particles with the size of approximately 2–10 nm are formed as a result of the polymerization process, which involves the establishment of chemical bonds between the monomers. Thus, a colloidal suspension also termed a sol is obtained with the solid colloidal particles stably dispersed in the solvent. Sol can be prepared by two techniques, condensation and dispersion of particles. Condensation proceeds by nucleation growth of particles to the appropriate size, whereas dispersion involves the reduction of large particles down to the colloidal dimensions. The size and properties of the resulting particles depend on the relative rates of these two processes. Sol formation is favored when the rate of nucleation is high and the rate of crystal growth is low. Depending on the degree of crosslinking and the growth process by which they are formed, the inorganic clusters can be either colloidal or polymeric in nature and can range from 10 to 200 Å in diameter. Several factors, such as polarity of the solvent, ionic strength of the reaction medium, and temperature, can be used to manipulate the formation of the sol.

In the second step, these colloidal particles are made to link with each other as polymerization process continues or crosslinked by additional crosslinkers in the solvent, so as to build a three-dimensional (3D) open grid which is termed a gel. Sol particles and small clusters in the solution attach to the spanning cluster and extend the gel network as time goes on. Commonly, the gels transformed from a sol by the gelation process are termed colloidal gels. For preparation of aerogels, the gelation is most conveniently induced through a change in the pH of the reaction solution. The mechanical state of the gel depends very much upon the number of cross-links in the network. It is obvious that the greater the degree of cross-linking, the more rigid the structure formed. On the other hand, it is also possible to form some gels directly from linear polymers instead of a precursor solution, without the intermediate occurrence of individual particles. In this condition, the gels are termed polymeric gels.

It is obvious that the sol-gel process allows us to tailor nanostructured materials by choosing appropriate reaction parameters. In addition, the sol-gel process provides the possibility of doping gels with molecular compounds. This can be performed either chemically by using suitable alkoxide derivates, or physically by introducing additives in the porous network. For example, the integration of hydrophobic alkyl groups can improve the water resistance of aerogels and the inclusion of soot can enlarge the infrared extinction.

2.2. Aging

When a gel is maintained in its liquid form, its structure and properties continue to change long after the gel point. This process is termed aging. Generally, the aim of such aging step is to mechanically reinforce the tenuous solid skeleton generated during the sol-gel process [65,66].

During aging, four processes can take place singly or simultaneously, including polycondensation, syneresis, coarsening, and phase transformation. Polycondensation reactions continue to occur within the gel network as long as the neighboring sol particles are close enough to react, which effectively increases the connectivity of the network and its fractal dimension. Syneresis is the spontaneous shrinkage of the gel, resulting from the expulsion of liquid from the pores. This process is generally attributed to the formation of new bonds during polycondensation reactions, which increases the bridging bonds and causes contraction of the gel network. As a result, larger particles grow at the expense of smaller ones and necks between the particles are filled up, thus leading to a narrow particle size distribution with larger mean particle diameter and decreased specific surface area but without shrinkage. This phenomenon can be further enhanced by increasing pH, temperature or pressure, resulting in coarsening of the gel structure.

2.3. Gel-Aerogel Transition (Drying)

For aerogels, the drying procedure is a very important step in order to preserve the highly porous structure. The gel, still embedded in the solvent, which consists mainly of alcohol with some water and catalyst, is called an alcogel or hydrogel. To remove the liquid, several extraction methods, such as supercritical drying (using CO_2, acetone, or ethanol) [15,23,25,67, 68,69,70,71,72,73,74], ambient pressure drying [24,75,76,77,78,79], freeze-drying [17,35,37,38], microwave drying [80] and vacuum drying [81,82], can be employed. Normal drying at ambient conditions leads to capillary tensions as the vapor/liquid interface retreats into the porous structure. The resulting shrinkage or even cracking of the gel network proceeds until the densified structure is able to withstand these forces. Depending on the aging, the structure of the dry aerogels consists of a more or less densely packed arrangement of gel particles or network chains with the remaining porosity amounts to about 50%. Shrinkage can be reduced by prolonged aging, by aging with additional catalyst, or by using chemical additives.

2.3.1. Supercritical Drying

Supercritical drying is a traditional drying technique that has been extensively studied. This type of drying prevents the formation of a liquid-vapor meniscus that recedes during the emptying of the pores in the wet gels. Practically, supercritical drying consists of heating the wet gel in a closed container, so that the pressure and temperature exceeds the critical temperature, T_c, and critical pressure, P_c, of the solvent entrapped in the pores inside the gel. As in a supercritical fluid, liquid and vapor phase become indistinguishable, and no capillary forces occur. After releasing the fluid through the outlet valve and subsequent cooling, the aerogel can be taken out from the autoclave. It is important that enough solvent is provided in order to guarantee the supercritical conditions throughout the whole drying process, otherwise shrinking and cracking will occur.

Critical conditions are very different depending on the fluid which impregnates the wet gel, and the most commonly used is CO_2[83,84]. For example, PI aerogels with an intrinsically hydrophobic nature and a dielectric constant as low as 1.19 were prepared by a supercritical CO_2 drying procedure [85]. Kobayashi *et al.* [86] reported a structurally new type of aerogel with a 3D ordered nanofiber skeleton of liquid-crystalline nanocellulose by supercritical CO_2 drying. Furthermore, monolithic pectin aerogels were prepared with the thermal conductivity ranging from 0.016 to 0.022 $W \cdot m^{-1} \cdot K^{-1}$ via dissolution gelation coagulation and subsequent drying with supercritical CO_2 [87].

Compared to other drying methods, supercritical drying is a more effective method for preventing the shrinkage or collapse of mesopores. However, one main disadvantage for supercritical CO_2 drying is that it is a time-consuming procedure. Therefore, a less time-consuming drying method without solvent exchange, *i.e.*, supercritical acetone drying, has been reported for the good dissolvability of acetone in supercritical CO_2 [69,71]. The results show that RF aerogels prepared by these two methods have similar chemical composition and microstructures, but the aerogels dried by supercritical acetone show larger shrinkage and pore size. On the other hand, aerogels dried by supercritical ethanol show similar framework with little shrinkage compared to supercritical CO_2 or acetone drying [68].

2.3.2. Ambient Pressure Drying

Ambient pressure drying is known as a promising technique that can be applied on large scale for industrial purposes. This method relies on a passivation of the pore surface inside the gel, so as to impede further formation of new chemical bonds after condensation reactions when the gel network is compressed under the drying stresses. At the end of the solvent evaporation process, an aerogel monolith is no longer submitted to capillary stresses so that it can resume its wet size by a spring-back effect. For example, Schwan *et al.* prepared RF organic aerogels via a sol-gel process in sodium carbonate solution, followed by solvent exchange with acetone and then dried under ambient pressure condition. Carbon aerogels using watermelon as the starting materials were also prepared through a hydrothermal method, followed by ambient pressure drying. Though ambient pressure drying is a simple and energy-saving process to prepare aerogels, simple evaporation of solvent from the hydrogels under ambient conditions may cause significant shrinkage or even form solid films essentially without porosity.

2.3.3. Freeze-Drying

In general, freeze-drying technique is a simple, more economic and environmentally friendly process to obtain aerogels with good porous structures. In this method, liquid in the wet gel is first frozen and thereafter dried by sublimation under low pressures. Thus obtained materials are also termed cryogels. These freeze-dried "cryogels" show a porosity of 80% at maximum and only half the inner surface of a comparable aerogel. Special precautions for freeze-drying are long aging periods to stabilize the gel body, a solvent exchange to provide a low expansion coefficient and high sublimation pressure, and the addition of salts to achieve low freezing rates and freezing temperatures [3]. Compared to aerogels prepared with supercritical CO_2 drying, cryogels show more macroporous structures with larger shrinkage and Brunauer–Emmett–Teller (BET) value. In constrast, compared to aerogels synthesized by hot-drying or vacuum-drying, cryogels show smaller shrinkage, narrower pore size distribution and smaller BET value [88,89]. Practically, the aerogels prepared from aqueous solution or regenerated from water (e.g., cellulose, PI, PVA, pectin) are usually dried by this method.

Cellulose aerogels prepared by different drying methods were investigated [90]. The regular freeze-dried aerogels (Figure 3a) had highly porous structure consisting of microfibrillar networks, but close examination reveals that the fibrils are severely coagulated to form film-like masses, especially at lower cellulose concentrations. In contrast, the solvent-exchange drying of water-methanol-*t*-butyl alcohol was effective in preserving the network structure (Figure 3b), resulting in aerogels consisting of finer fibrils of about 50 nm wide. The nitrogen adsorption surface area of the solvent-exchange dried aerogels ranged in 160–190 $m^2 \cdot g^{-1}$, is about twice of those of regular freeze-dried materials. Rapid freeze drying by metal plate-contact technique provided partial preservation of the original network structure, giving asymmetric porous structures as shown in Figure 3c. The surface area of regular freeze-dried material (70–120 $m^2 \cdot g^{-1}$) was about half of those of solvent-exchange materials, and also was more strongly dependent on starting cellulose concentration.

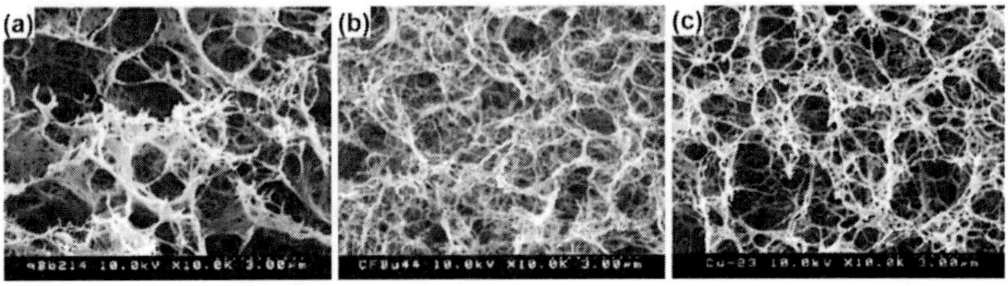

Figure 3. SEM images of 1% cellulose aerogel by (**a**) regular freeze drying; (**b**) solvent exchange drying; and (**c**) rapid freeze drying. Reproduced with permission of Ref. [90] (Copyright Elsevier 2004).

2.3.4. Other Drying Methods

Other techniques such as microwave drying and vacuum drying have also been applied to obtain aerogels with high surface area and porosity [80,81,82]. Microwave drying is a time-saving method to prepare mesoporous and 3D interconnected macroporous aerogels. Aerogels prepared by this method show similar structures to those prepared by freeze-drying, but with more small-sized macropores. Vacuum drying is an energy-conserving process in which materials are dried under a reduced pressure, which needs less heat for rapid drying [91,92]. For example, microporous PI aerogels were prepared with BET surface area up to 1454 $m^2 \cdot g^{-1}$ and a narrow pore size distribution in the range from 5 to 6 angstrom via vacuum drying [91].

Among various drying methods, supercritical drying technique is the most effective method to prevent the drying shrinkage or collapse of mesopores and obtain well-defined structures. However, the long-time solvent exchange, the involvement of massive organic solvent, high temperature and pressure used in the drying process make it a time-consuming and high-cost method, which severely limit their practical applications. On the contrary, freeze-drying technique is a much easier, more economic and environmentally friendly method. Although the direct freeze drying is simple, it is heavily dependent on freezing rate and concentration of precursors in the wet gel. Ordered porous aerogels can only be obtained by elaborately controlling the freeze-drying process. Ambient pressure drying is designed to dry the wet gel at ambient pressure, which can be applied in large-scale industrial production, whereas it may result in environment pollution for the evaporation of solvents. For microwave drying and vacuum drying, they usually produce large pores or even induce the collapse of materials. Therefore, the drying methods have a great influence on the structure, surface area, porosity, and pore volume of aerogels and proper drying methods should be considered according to the practical requirements.

2.4. Carbonization of Organic Aerogels

Carbon aerogels are mostly obtained by pyrolysis, at temperatures above 600 °C, of organic aerogels. In this way, the organic aerogels transform to an electrically conductive carbon network.

Generally, carbonization of the dried aerogels is carried out by heating the sample in flowing N_2 or Ar at a temperature between 600 and 2500 °C [93,94]. During carbonization, the aerogel loses oxygen and hydrogen functionalities, thus resulting in an enrichment in carbon and a highly pure carbon structure. Interestingly, the electrical conductivity of the samples increases significantly during this transformation, as indicated by the common appearance of a broadband, semi-metal-like infrared absorbance. Variations in the pyrolysis conditions cause significant changes in the properties of carbon aerogels. Higher pyrolysis temperatures tend to reduce the surface areas of the carbon aerogels, and their electrochemical double layer capacitance also decreases. The minor decrease in the surface area with increasing pyrolysis temperature is limited to temperature above 600 °C, whereas increasing pyrolysis temperature while under 600 °C increases the surface area [95]. However, the carbon aerogels may not be electrically conductive unless carbonized above 750 °C. The electrochemical double layer capacitance also exhibits a maximum between 800 and 900 °C [96]. At low carbonization temperatures, the macropore volume decreases and the mesopore volume increases due to the shrinkage of the material, whereas the micropore volume and surface area are enlarged due to the evolution of gases during carbonization. At higher carbonization temperatures, all these parameters tend to decrease. For example, carbon aerogels carbonized at 1200 °C are significantly denser than those carbonized at 600 °C [97]. The temperature required for complete graphitization of the carbon aerogel may exceed ≥2000 °C, but aerogels pyrolyzed at ~1050 °C contain separated graphitic structures, which has also been reported by others based on XRD patterns [98]. Therefore, for achieving better electrochemical performance, the aerogels are normally carbonized between 600 and 900 °C, whereas higher carbonization temperature (above 1000 °C) is employed to realize graphitization of the carbon aerogels for better electrical conductivity. In addition, activation of the aerogel can be carried out by its gasification with steam or CO_2 at between 800 and 900 °C for varying time periods. As a result, the surface area, pore volume, and pore size distribution of carbon aerogels are tunable in a wide spectrum related to the synthesis and processing conditions.

3. STRUCTURES AND PROPERTIES OF POLYMER/CARBON-BASED AEROGELS

3.1. Polymer Aerogels

Unlike silica aerogels being fragile and hygroscopic, polymer aerogels usually exhibit better mechanical and environmental stability for a variety of applications, especially in extreme environments such as aerospace. These organic aerogels can be produced with very little shrinkage during the processing and have thermal conductivity comparable to silica aerogels of similar density (only 14 $mW \cdot m^{-1} \cdot K^{-1}$) for aerogels with density of 0.1 $g \cdot cm^{-3}$) but are much stronger with compressive modulus ranging from 1 to 5 MPa [99]. Types of Polymer aerogels are diversified with different types, including cellulose-based, resin-based, PI-based and PVA-based aerogels, exhibiting diverse structures and properties. Depending on the polymer type and fabrication conditions, the structure of polymer aerogels can be changed from colloidal-like nanoparticles to nanofibrillar/microfibrillar networks and even to sheet-like skeletons. In addition, structural parameters such as pore size, ordering, and even pore shape (regular or irregular) will strongly affect the final macroscopic performance of the aerogel materials.

3.1.1. Cellulose-Based Aerogels

Cellulose-based aerogels are one of the "oldest" aerogels and the first organic aerogels, which were invented accompanying the birth of silica aerogels [100]. Commonly, the cellulous gels are formed by dissolution and regeneration of cellulose in an aqueous or organic solvent [101,102]. Considering the nanoscale lateral dimensions of nanocellulose, it can potentially generate porous structures with ultralow density (less than 5 $mg \cdot cm^{-3}$) and exceptionally high porosity (over 99%). The aerogels have highly porous networks consisting of fibrils ranging from tens of nanometers to tens of micrometers [103,104]. The open 3D structure can transform from nanofibrillar to microfibrillar networks and even to sheet-like skeletons by changing the initial cellulose concentration and drying methods [90,105]. In another case, cellulose aerogels are composed of "globules" attached to each other and their size decreases with the increase of cellulose concentration [106,107]. Moreover, Cai *et al.* [108] demonstrated that ultralight pure natural aerogel microspheres can be fabricated using cellulose nanofibrials directly. The aerogel microspheres are highly porous with bulk density as low as 0.0018 $g \cdot cm^{-3}$ and pore size ranging from nanometers to micrometers. Covalent crosslinking between the native nanofibrils and crosslinkers made the aerogel microspheres very stable even in a harsh environment.

The use of naturally existing substances (e.g., biomass) is an emerging approach for the production of new functional nanoporous aerogel materials. The production of aerogels from inexpensive biomass-derived precursors has recently gained a lot of interests both academically and commercially because of the range of economic/process/chemistry advantages offered by such synthetic approaches. A variety of raw materials, such as paper waste [109], bamboo leaf [110], rice straw [111], and jute fibers [112] have been utilized to fabricate cellulose aerogels. Nguyen *et al.* provided a cost effective and scalable recipe for fabricating biodegradable cellulose aerogels from paper waste. The cellulose aerogels are macroporous with a low density of about 0.04 $g{\cdot}cm^{-3}$ and thermal conductivity from 0.029 to 0.032 $W{\cdot}m^{-1}{\cdot}K^{-1}$. Cellulose aerogels were also synthesized from bamboo leaf in acidic conditions, which exhibit large specific surface area (547.2 $m^2{\cdot}g^{-1}$) and low thermal conductivity (0.024 $W{\cdot}m^{-1}{\cdot}K^{-1}$) [110]. An ultra-light (1.7 to 8.1 $mg{\cdot}cm^{-3}$) and ultra-porous (99.5% to 99.9%) aerogel has been assembled from cellulose nanofibrils that were defibrillated from rice straw cellulose at a yield of 96.8%. Furthermore, with both hydrophilic hydroxyls and hydrophobic pyranose rings, amphiphilic cellulose nanofibrils aerogels allow the uptake of both polar and nonpolar liquids, in contrast to the hydrophobic carbon-based aerogels and the hydrophilic inorganic oxide aerogels [111].

As mentioned above, nanofibrous aerogels with high continuity and an open-cell cellular structure were fabricated from cellulosic materials, including bacterial cellulose fibrils, cellulose nanocrystals and lignocellulose, *etc.* However, the inherent limits on the diversity of bulk materials, combined with the lack of precise control of the physicochemical and mechanical properties, present major challenges in the synthesis of nanofibrous aerogels that must be addressed before their extensive practical applications. In contrast, electrospun nanofibers, which combine the merits of robust mechanical strength, low density, fine flexibility, extremely high aspect ratio and ease of scalable synthesis from various materials (polymer, ceramic, metal, carbon and so on), hold great promise as an exceptional nanoscale building block for constructing macroscopic nanofibrous aerogels. Recently, Si and coworkers presented a robust methodology for creating superelastic nanofibrous aerogels with a hierarchical cellular structure that consists of bonded nanofibers, which are called "fibrous, isotropically bonded elastic reconstructed" (FIBER) nanofibrous aerogels (Figure 4) [113]. The intrinsically lamellar deposited electrospun nanofibers are reconstructed into 3D elastic bulk aerogels with tunable densities and desirable shapes on a large scale. The resulting FIBER nanofibrous aerogels exhibit the integrated properties of extremely low density (minimum of 0.12 $mg{\cdot}cm^{-3}$), super recyclable compressibility and multifunctionality in terms of combining thermal insulation, sound absorption, emulsion separation and elasticity-responsive electric conduction, all originating from the synergistic effect of hierarchical cellular fibrous networks and well-bonded nanofibers. In dramatic contrast to the brittle nature of traditional colloidal aerogels, the FIBER nanofibrous aerogels show superelastic mechanical properties, which can bear a compressive strain (ε) as high as 80% and can recover their original volume after the release of the stress (σ). The maximum σ was 10.6 kPa for the first elastic regime and 36.1 kPa at 80% strain. These values were significantly higher than those of other fibrous aerogels with similar densities. In addition, no significant decrease in the stiffness or strength was observed for FIBER nanofibrous aerogels by 1000 cyclic compressions. The successful synthesis of such fascinating materials may provide new insights into the design and development of multifunctional nanofibrous aerogels for various applications.

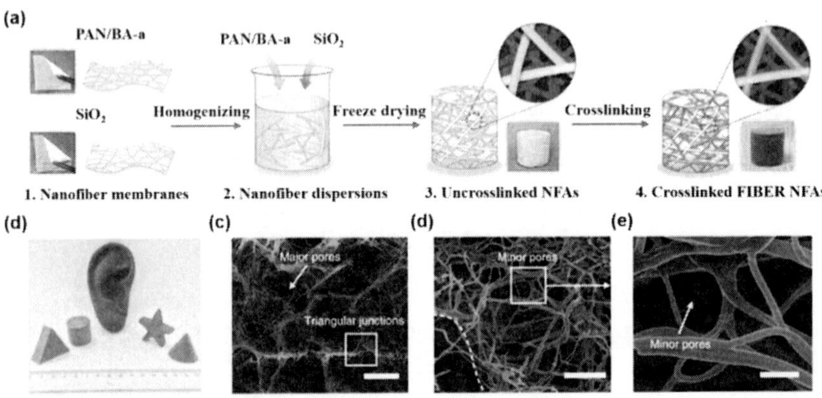

Figure 4. Design, processing and cellular architectures of FIBER nanofibrous aerogels. (**a**) Schematic showing the synthetic steps. (1) Flexible PAN/BA-a and SiO_2 nanofiber membranes are produced by electrospinning. (2) Homogeneous nanofiber dispersions are fabricated via high-speed homogenization. (3) Uncrosslinked nanofibrous aerogels (NFAs) are prepared by freeze drying nanofiber dispersions. (4) The resultant FIBER NFAs are prepared by the crosslinking treatment. (**b**) An optical photograph of FIBER NFAs with diverse shapes. (**c**–**e**) Microscopic architecture of FIBER NFAs at various magnifications, showing the hierarchical cellular fibrous structure. Reproduced with permission of Ref. [113] (Copyright Nature Publishing Group 2014).

3.1.2. Polysaccharide-Based Aerogels

Polysaccharides are regarded as key ingredients for the production of bio-based materials in life sciences (e.g., food, cosmetics, medicaldevices, pharmaceutics). The biodegradability and biocompatibility of these biopolymers, coupled to the large variety of chemical functionalities they encompass, make them promising carriers for drug delivery systems. In nature, polysaccharides are known to self-associate or self-order into particular structures, physical forms, or shapes (e.g., the starch granule, plant structures). Significantly for materials chemistry, polysaccharides can also self-associate upon inducement of an aqueous gel state. This provides the materials chemist the opportunity to exploit this ability for the preparation of new porous aerogel and xerogel materials. Polysaccharide-based aerogels are highly porous (ε = 90%–99%) and lightweight (ρ = 0.07–0.46 $g{\cdot}cm^{-3}$) with high surface area (S_{BET}= 70–680 $m^2{\cdot}g^{-1}$), able to provide enhanced drug bioavailability and drug loading capacity [114,115]. Alternatively, the resulting aerogel materials can also meet the performance criteria for other emerging niche markets, such as catalysts, catalyst supports or adsorbents, providing a new opportunity to obtain useful materials from one of the less energy-intensive sources of biomass.

In the late 1990s, Glenn *et al.* and Te Wierik *et al.* independently demonstrated that the low surface area of native starch could be expanded to generate xerogel materials with specific surface areas (S_{BET}) of between 25 and 145 $m^2{\cdot}g^{-1}$ depending on preparative route [116,117]. Clark and coworkers demonstrated the potential of cornstarch (≈73% amylopectin) in the production of low density, high surface area starch xerogels ($S_{BET} \approx 120$ $m^2{\cdot}g^{-1}$) for applications in normal-phase chromatography separations [118]. Recently, this work was extended to the microwave-assisted preparation of high surface area ($S_{BET} > 180$ $m^2{\cdot}g^{-1}$), highly mesoporous starch-derived materials (Vmeso > 0.6 $cm^3{\cdot}g^{-1}$; >95% mesoporosity) [119]. This research demonstrated that the key to the formation of the porous polysaccharide form in starch was the generation of polysaccharide gel.

Chitosan is a biopolymer obtained from the deacetylation of chitin, which is itself extracted from shells of crustacean and mollusks such as shrimps, crabs and squids. Chitin is the second-most-abundant polysaccharide next to cellulose and it is produced in approximately one hundred billion tons per year. The interest for chitosan stems from its applications in biomaterials, drug-delivery systems, food additives, water clarification, and as support for cells and enzymes [120]. Chitosan can form either chemical or physical hydrogel. Chemical hydrogels are formed by irreversible covalent links, as in covalently crosslinked chitosan hydrogels. Physical hydrogels are formed by various reversible links, as in ionically crosslinked hydrogels and polyelectrolyte complexes, or in entangled gels. The latter are formed by solubilization of chitosan in an acidic aqueous medium, and precipitation in an alkaline solution, which is the simplest way to prepare a chitosan hydrogel. The choice of the cross-linker and its concentration should be a compromise between aerogel stability and the required open porosity and homogeneity [121]. In general, chemical gels allow a processing with better control of the porous structure and swelling behavior than with physical gels, but at expense of higher raw materials and processing costs and more complex chemical characterization. After solvent exchange to an alcohol, chitin hydrogels can then be supercritically dried to obtain chitin aerogels of high porosity, high surface area and low density with changes in value depending on the chitin concentration and the alcohol solvent used in the original wet gel. For a given degree of deacetylation, the type of chitin affects the distribution of the residual acetylated glucosamines, the crystallinity of the polymer and its gelling properties. Generally, aerogels with higher surface area are obtained from hydrogels of chitosan obtained by deacetylation of α-chitin from crab shell than by deacetylation of β-chitin from squid pen [122].

The utilization of polysaccharides in the preparation of aerogel materials represents an interesting alternative to the materials prepared through conventional polymerization and cocondensation techniques. Furthermore, the use of polysaccharides derived from biomass potentially adds value to inexpensive and typically waste products from industry such as the food sector.

3.1.3. Resin-Based Aerogels

RF-based aerogels were first described by Pekala through the polycondensation of resorcinol (R) with formaldehyde (F) in an aqueous environment using Na_2CO_3 as a catalyst [15]. The polymerization reaction consists of two steps: (1) addition reaction between R and F to form hydroxymethyl resorcinol monomers; (2) $-CH_2-$ or $-CH_2OCH_2-$ bridging polymerization between monomers, producing water/formaldehyde or water. Continuous polymerization will produce RF clusters which further crosslink throughout the whole colloid system to gelate. The as-prepared RF aerogels consist of interconnected colloidal-like particles with diameters of approximately 10 nm and form the surface functionalized polymer "clusters" with a low density below 0.1 $g \cdot cm^{-3}$. The Na_2CO_3-catalyzed RF aerogels and corresponding sol-gel process are most widely adopted till now, because of the good formability, easy microstructure-controllability and adjustable properties [25,82,123,124].

However, the rate of these reactions is very slow at room temperature. Normally, a multiple stage heating process is used to accelerate the gelation process. Besides, catalysts were also applied to accelerate the polycondensation process [123,125,126,127,128,129,130,131]. The acid-catalyzed route, e.g., acetic acid, hydrochloric acid (HCl), oxalic acid or para-toluenesulfonic acid, was reported for the preparation of RF aerogels. Compared with the typical base-catalyzed RF aerogels, acid-catalyzed processes are more time-efficient. The usage of HCl as catalyst for the preparation of RF aerogels leads to organic aerogels in very short gelation times, varying from a few seconds to minutes [128]. The obtained RF aerogels have a 3D network that consists of approximately monodispersed spherical-like particles in the range of 700–1500 nm. On the other hand, the RF aerogels prepared by base-catalyzed route show a narrow pore size distribution (PSD) with the maximum position located between 10 and 20 nm, while the pores of RF synthesized by acid-catalyzed route are 10 times larger with a broader PSD [132]. Wu *et al.* [133] prepared phenol-furfural aerogels by dividing the sol-gel polymerization into two steps, *i.e.*, the formation of gel precursors by pre-polymerizing phenol and furfural with NaOH as the catalyst and subsequently the gelation of the precursors with HCl as the catalyst. It was found that the crosslinking density of the gel precursors, controlled by the catalytic time of NaOH, exerts a great influence on the yield, drying shrinkage, bulk density and texture of the aerogels obtained. An appropriate increase in the crosslinking density of the gel precursors helps to increase the yield and mesoporosity, leading to an increase of the bulk density. However, the organic aerogels obtained with NaOH-catalyzed pre-polymerization process experience a larger drying shrinkage than those without NaOH-catalysts. In addition, the application of ammonium carbonate as catalyst for the preparation of RF-aerogels leads to organic aerogels without metallic impurities in contrast to the conventional catalysts like sodium carbonate. The particle size and density of the synthesized RF aerogels are 0.15–4 μm and 0.37–0.42 $g \cdot cm^{-3}$, respectively [76].

A new class of phenolic resin, namely polybenzoxazine, is recently developed. Polybenzoxazine has excellent molecular design flexibility that allows the properties of the cured materials to be controlled for various applications. Polybenzoxazine has many advantageous characteristics compared with traditional phenolic resin, such as high mechanical strength, near-zero shrinkage upon polymerization of its monomer, and high char yield [134,135]. Interestingly, ring-opening polymerization of benzoxazine in solvents leads to gelation and formation of 3D polymer networks [136,137]. Lorjai *et al.* first reported the synthesis of polybenzoxazine gels by thermally-induced ring opening polymerization of benzoxazine in xylene at 130 °C [138]. The subsequent carbon aerogels prepared by the carbonization of polybenzoxazine aerogels exhibited a microporous structure with pore diameters less than 2 nm, the densitiy of 300 $kg \cdot m^{-3}$, and surface area of 384 $m^2 \cdot g^{-1}$, respectively. The most prominent characteristic of the polybenzoxazine-derived carbon aerogels is high proportion of micropore volume to micropore surface area, as compared with organic aerogels obtained from other precursors. Recently, Mahadik-Khanolkar *et al.* published a new method for preparation of polybenzoxazine gels at room temperature by using hydrochloric acid as the catalyst of cationic ring opening polymerization [139]. The scanning electron microscopy images revealed that the corresponding aerogels consisted of spherical polymer particles as the building block and that the diameter of spherical particles reduced with an increase in the concentration of benzoxazine in solution. Time-efficient gelation of benzoxazine using p-toluenesulfonic acid as the catalyst in several solvents and controllable pore structure formation in the resultant polybenzoxazine aerogels are reported by Gu and coworkers [140]. All of the aerogels synthesized in different solvents were found to be mesoporous materials, and the skeletal frameworks of polybenzoxazine aerogels can be tuned using appropriate solvents and suitable polymerization temperature.

3.1.4. Polyimide-Based Aerogels

Polyimides are widely used in industries for their high performance, such as ultralow dielectric constant, high mechanical strength and thermal stability at high temperatures [16,85,99,141,142,143,144,145]. Therefore, it would be ideal to design PI aerogels for use as highly porous, lightweight and high-temperature stable materials. PI aerogels prepared from aromatic dianhydrides and diamines, have been synthesized by gelation of chemically or thermally imidized PIs in dilute solutions, followed by drying. The PI aerogels exhibit a rich variety of nanoporous structures, such as crisp fragments [83], loose or close clusters of polymer fibers [99,143,144,146] in the range of tens of nanometers to several micrometers. It was found that the morphologies are highly dependent on the solvent interactions during polymerization, as well as the chain rigidity and chain length of monomers and crosslinkers. One merit of PI aerogels is their excellent mechanical properties, which meets the high standard for applications in aerospace. PI aerogels can be produced with very little shrinkage during the processing and have high compressive modulus ranging from 1 to 5 MPa, much higher modulus than typical polymer reinforced silica aerogels with similar density exhibiting compressive modulus ranging from 0.1 to 2 MPa. Additionally, PI aerogels are light yellow to orange yellow in appearance, which can be tailored into different shapes, as shown in Figure 5. Thin films (nominally 0.5 mm) of the aerogels from certain formulations are flexible to be rolled or folded backward and completely recover with no evidence of cracking or flaking. Thicker parts such as the 6.5 cm × 6.5 cm × 1.3 cm rectangular prism shown in Figure 5b are rigid and strong, even able to support the weight of a car, as seen in Figure 5c [99].

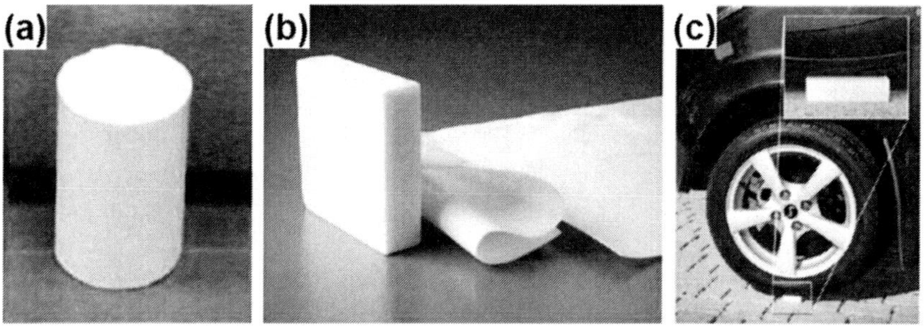

Figure 5. (**a**) Octa(aminophenyl)-silsesquioxane (OAPS) cross-linked polyimide aerogels; (**b**) polyimide aerogels cross-linked with 1,3,5-triaminophenoxybenzene (TAB) fabricated as flexible thin films or molded to a net shape; and (**c**) demonstrating strength of the aerogel by supporting the weight of a car. Reproduced with permission of Ref. [99] (Copyright American Chemical Society 2012).

Imidization is essential for the formation of PI aerogels during gelation, which can be induced either chemically or thermally. The aerogels prepared by thermal imidization exhibited very little shrinkage with a porosity of up to 90%. However, the gels tended to redissolve during thermal imidization and reform upon cooling, obviating the need to keep the gels in molds. This is not conducive to fabrication of freestanding films using roll-to-roll processing, and also suggests that part of the amic acid hydrolyzes during thermal imidization, which breaks up the network structure [83,99]. On the other hand, PI aerogels prepared by chemical imidization exhibit various structures and properties, which is strongly related to the choice of crosslinking agents. A series of octa(aminophenyl)-silsesquioxane (OAPS) crosslinked PI aerogels were prepared by Guo and coworkers [143]. The morphology of as-prepared PI aerogels resembles bundles of polymer fibers tangled together with fiber diameters in the range of 15–50 nm. The aerogels can be produced with very little shrinkage during the processing and have thermal conductivity comparable to silica aerogels with similar density (only 14 $mW\cdot m^{-1}\cdot K^{-1}$ for aerogels with density of 0.1 $g\cdot cm^{-3}$) but are much stronger with compressive modulus ranging from 1 to 5 MPa. Most notably, these aerogels can be fabricated as thin films that are flexible and foldable, being stable up to 400 °C for short-term exposure as well. PI aerogels using 1,3,5-triaminophenoxybenzene (TAB) as the crosslinker were also reported [16]. The as-prepared PI aerogels show a low dielectric constant value at 1.08 and high compressive moduli of 4–8 MPa (40 times higher than silica aerogels at the same density). In comparison to PI aerogels crosslinked using OAPS, the one crosslinked with TAB has a density about 26% higher, but modulus increases by a factor of 4 and surface areas are also significantly higher for these TAB crosslinked aerogels.

The property of PI aerogels could be affected by the backbone structures, thus aerogels with different combinations of diamine and dianhydride are examined [99]. A series of TAB crosslinked aerogels derived from three different diamines (*i.e.*, *p*-phenylene diamine (PPDA), 2,2'-dimethylbenzidine (DMBZ), and 4,4'-oxydianiline (ODA)) and two different dianhydrides (*i.e.*, benzophenone-3,3',4,4'-tetracarboxylic dianhydride (BTDA) and biphenyl-3,3',4,4'-tetracarboxylic dianydride (BPDA)) were prepared and studied. BPDA, PPDA, and DMBZ produces a rigid backbone in the main chain of PI, leading to higher glass transition temperatures, whereas ODA and BTDA provide flexible linking groups between phenyl rings, thus resulting in less rigid structures. Interestingly, the pore structures observed by SEM images shown in Figure 6 for the aerogels produced using different diamines and dianhydrides are quite dissimilar. Aerogels derived from BPDA (Figure 6a–c) all appear as collections of polymer strands ranging from 30–50 nm in thickness. The combination of BTDA and ODA (Figure 6d) produces aerogels that are the most similar to silica aerogels with clusters of 50–100 nm particles loosely connected together. The aerogels composed of BTDA and DMBZ (Figure 6e) also appears as clusters of nanoparticles with more uniform and smaller size (~50 nm). Physical and mechanical properties of TAB crosslinked aerogels polymerized from different dianhydrides and diamines also vary a lot. It was found that aerogel properties, such as density and mechanical properties, were highly dependent on the shrinking degree during gelation. Aerogels made using a combination of BPDA and DMBZ shrank the least and therefore had the lowest density, whereas using BTDA and ODA led to greater shrinkage and higher densities. Aerogels made using ODA also tended to be less brittle, whereas DMBZ-based aerogels tended to decompose at lower temperatures. The PI aerogels have a high onset decomposition temperature over 600 °C using PPDA as the backbone diamine with excellent tensile strengths ranging from 4 to 9 MPa which can support the weight of a car. Furthermore, it was found that using mixtures of diamines led to combinations of improved properties. For example, 100% ODA gives flexible aerogels that are hydrophilic while 100% DMBZ gives more hydrophobic aerogels that are brittle. A mixture of 50% ODA and 50% DMBZ used in conjunction with BPDA gives polyimide aerogels with better moisture resistance and flexibility.

Different from the methods of preparing PI aerogels via polycondensation of dianhydrides and diamines, Leventis *et al.* [146] prepared PI aerogels by ring-opening metathesis polymerization of a norbornene end-capped diimide. The as-prepared PI aerogels retain high porosity (50%–90%, *v*/*v*), high surface area (210–632 $m^2 \cdot g^{-1}$, of which up to 25% is traced to micropores), and uniform pore size distribution in the mesoporous range (20–33 nm). The skeletal framework consists of primary particles of 16–17 nm in diameter, assembling to form secondary aggregates of 60–85 nm in diameter. The aerogels show a thermal conductivity of 0.031 $W \cdot m^{-1} \cdot K^{-1}$ and high mechanical compressive strength of 168 MPa. Hence, they are reasonable multifunctional candidates for thermal/acoustic insulation materials.

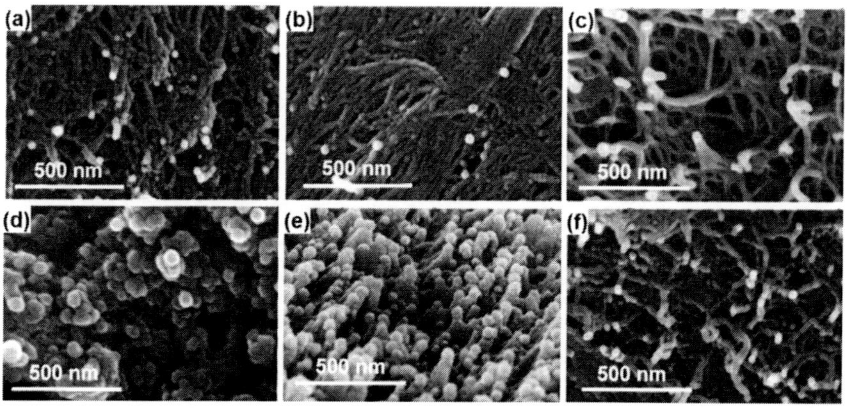

Figure 6. SEM images of PI aerogels synthesized from (**a**) ODA/BPDA; (**b**) PPDA/BPDA; (**c**) DMBZ/BPDA; (**d**) ODA/BTDA; (**e**) DMBZ/BTDA; and (**f**) PPDA/BTDA. Reproduced with permission of Ref. [99] (Copyright American Chemical Society 2012).

Overall, PI aerogels, one kind of the emerging aerogels, have attracted great interests for their unique thermal, mechanical and dielectric properties, which are ideal candidates for thermal or electrical insulation materials, especially in aerospace areas.

3.1.5. PVA-Based Aerogels

Poly(vinyl alcohol) is a highly hydrophilic and water-soluble polymer, synthesized by partial or complete hydrolysis of poly (vinyl acetate). The low toxicity, high water content, good mechanical properties and biocompatibility of PVA hydrogels make them widely used in various fields, such as drug delivery, contact lenses, wound bandages, tissue engineering and cell immobilization [17]. The preparation of PVA aerogel is usually simple and not diversified. Typically, an appropriate amount of PVA is dissolved in deionized water at 70–80 °C under vigorous stirring until the polymer is completely dissolved. Then, PVA hydrogels can be fabricated via different techniques, such as the freeze-thaw inducement of crystallization, acid-catalyzed dehydration, irradiation, radical production and chemical crosslinking. Subsequently, the hydrogels are usually freeze-dried to form aerogels. One advantage but also the limitation of PVA aerogels is that it can be fabricated in aqueous media followed by facile freeze-drying method, which is a green and time efficient approach to fabricate aerogel materials.

3.1.6. Polymer-Based Hybrid Aerogels

In order to strengthen the nanostructure and suppress the collapse of nanopores of polymer aerogels, nanofillers, such as SiO_2, clay, cellulose nanofibrils (CNFs), CNTs, and graphene or graphene oxide (GO) are usually incorporated into the polymer matrices to prepare nanofiller-polymer hybrid aerogels. The hybrid aerogels always exhibit enhanced mechanical or thermal properties as well as multifunctions as compared with pure polymer aerogels.

Clay-Polymer Hybrid Aerogels

In general, clay is introduced into aerogels to enhance the rigidity and flame retardancy properties of the aerogels. Sodium-montmorillonite (MMT) was incorporated into cellulose aerogels to obtain composite aerogels, which show a honeycomb-like porous structure and superior functionalities. The modulus of cellulose increases from 25.8 kPa and 2.13 MPa to 386 kPa and 3.86 MPa for the isotropic and anisotropic aerogels, respectively [34]. Additionally, the inclusion of MMT increased the heat endurance of the aerogels dramatically up to 800 °C while the mechanical properties were retained up to 300 °C with shape retention functions. A series of PVA/clay aerogel composites were investigated by Chen and coworkers, which were crosslinked either by radiation or chemical bonding to show a layered microstructure [17,36,37]. Furthermore, they indicate that crosslinking can increase the compressive modulus of aerogels and the strengthened PVA/clay aerogels possess very low flammability. PI/clay aerogel composites were also produced by freeze-drying of a PAA ammonium salt/clay precursor suspension, exhibiting layered structures and low densities in the range of 0.04–0.09 g·cm^{-3} [40].

CNT/CNF-Polymer Hybrid Aerogels

Introduction of CNTs into polymer aerogels can improve the electrical or thermal conductivity of the aerogels. Multi-walled carbon nanotubes (MWCNT)/RF composite aerogels were prepared by sol-gel polymerization of RF monomers in an aqueous suspension of CNTs, followed by supercritical CO_2 drying [23]. MWCNT can form a homogeneous uniform dispersion in aerogels at low concentrations, leading to a homogeneous material at the nanometer scale. The relative thermal conductivity increased slightly with the increasing MWCNT loading. The thermal conductivity of the composite gel exhibited 27% higher than that of the corresponding pure RF gel by incorporation of 1.04 vol % MWCNT. Due to the high surface area and aspect ratio, cellulose nanofibrils can easily form an entangled web-like structure to be used in a wide range of applications, such as antibacterial agents, thermal insulation and oil absorbents. Bringing CNFs into PVA aerogels could also lower the density (<15 $kg \cdot m^{-3}$) and increase the porosity (>98%) of the aerogels. More interestingly, the PVA/CNF aerogel became superhydrophobic and superoleophilic after being treated with methyltrichlorosilane [33]. The silane-treated, cross-linked PVA-CNF aerogels exhibit excellent absorption performance for various types of oils, organic solvents and heavy metal ions (e.g., Pb^{2+}, Hg^{2+}). Furthermore, these silane-treated PVA-CNF aerogels demonstrate excellent elasticity and mechanical durability. A ternary hybrid aerogel containing PVA, CNF and MWCNT was also prepared by the same group, which showed low density (<0.031 $g \cdot cm^{-3}$), high surface area (160–200 $m^2\ g^{-1}$), and very low thermal conductivity (<0.031 $W\ m^{-1}\ K^{-1}$) [39].

GO-Polymer Hybrid Aerogels

GO has drawn a great deal of attention as a reinforcing nanofiller in polymer composites due to its low cost, unique two-dimensional honeycomb layered structure, and excellent mechanical properties (fracture stress ~63 GPa) [147,148]. Compared with CNTs, nanoplatelets have a geometry that can offer isotropic reinforcement in more than one direction, and exhibit more effective reinforcement in polymer composites. Moreover, with numerous oxygen functional groups on the surfaces, GO can be readily dispersed in water to form a stable colloidal suspension and also form strong interfacial bondings with the polymer matrix [149]. As a consequence, GO can act as the "super gelator" in aqueous solution with a critical gelation concentration <0.5 wt %, thus obviously accelerating the gelation process to reduce both the drying shrinkage and aerogel density [150]. Guo *et al.* [24] prepared RF composite aerogels by a simple sol-gel polymerization method using graphene oxide as an anti-shrinkage additive. Morphological studies showed that GO can be well dispersed in the RF matrix owing to the good compatibility between the two elements, which effectively strengthen the RF skeleton. The linear shrinkage decreases progressively from 28.3% to 2.0% with GO loadings increasing from 0 wt % to 2 wt %, while the density decreases from 506 to 195 $kg \cdot m^{-3}$ simultaneously. This work offers a new method for the inhibition of aerogel shrinkage and broadens the application scope of GO as well. The gelation behavior of aqueous dispersions composed of GO and PVA was investigated through rheological study [151]. Strikingly, it was found that small amplitude oscillatory shear can induce the gelation of graphene dispersions and the rheological properties measured after the completion of SAOS-induced gelation was highly related with the density and the electrical conductivity of graphene aerogels. The as-prepared ultra-light aerogel exhibited low density of 4.0 $mg{\cdot}cm^{-3}$, low surface resistivity of 6.6 $U{\cdot}sq^{-1}$, high specific area of 1069 $m^2{\cdot}g^{-1}$, and high recoverable strain of 94% in compression. Moreover, the glass transition temperature of PVA in the aerogel was 49 K higher than that of pure PVA. Superhydrophobic and superoleophilic graphene/polyvinylidene fluoride (G/PVDF) aerogels were prepared by the solvothermal reduction of GO/PVDF mixed dispersions [32]. The as-prepared aerogel shows high specific surface area, eminent adsorption capacity for oils and organic solvents, superior water repelling ability, and excellent adsorption recyclability. Therefore, this kind of aerogel is a promising material for oil–water separation, oil spill cleanup and recovery of organic solvents.

In addition to those nanofillers mentioned above, various other inorganic components, such as metal or metal oxides, have been incorporated into polymer aerogels to meet the demand for multi-functions. For example, silica was incorporated into PVA aerogels to enhance the thermal stability and lower the flammability of the aerogels [36]. Flexible and porous zinc oxide bionanocomposite foams were prepared by controlling the hydrolysis and solvothermal crystallization in which cellulose was used as a template, showing excellent antibacterial activity [152].

3.2. Carbon Aerogels

Carbon aerogels are composed of nanostructured porous carbon materials in which the pores are filled with gas. Carbon aerogels have been extensively studied as they combine beneficial adsorption properties and structural stability with high thermal stability and, in principle, electronic conductivity, for potential applications in catalyst carriers, oils or organic solvents adsorption and energy storage [153,154]. One important feature of carbon aerogels is their ultralight nature, with the lowest density ever reported, 0.16 $mg{\cdot}cm^{-3}$ [155]. Besides being highly porous and lightweight, another advantage is the possibility to introduce electrical conductivity after being converted into carbon aerogels, thus allowing applications in electrical/electrochemical areas (e.g., batteries, supercapacitors, or conductive catalyst supports). Traditionally, carbon aerogels are fabricated from organic gels through controlled thermal annealing/carbonization under a non-oxidizing atmosphere (e.g., flowing N_2 or Ar). More recently, it has been reported that biomass-derived carbon aerogels can be synthesized through the aqueous gelation and self-assembly of suitable biomass-derived precursors (e.g., polysaccharides, proteins, sugars). Additionally, with the development of novel carbon materials, graphene or CNT based carbon aerogels with ultralow density and excellent conductivity have been produced as emerging novel aerogel materials, which further expand the application fields of aerogel materials.

3.2.1. Carbon Aerogels from Carbonization

The first carbon aerogel was born in 1989 by carbonization of RF aerogel. It is usually considered as a kind of highly-porous amorphous-graphite-based foam. The basic idea of preparing carbonized RF aerogel is to pyrolyze the high carbon-content precursor (RF aerogel) under high temperature (normally 800–1200 °C), ambient pressure and inert atmosphere. Like their parent organic aerogel systems, the carbonized gels have a network structure composed of interconnected nanosized primary particles. For conventional RF gel-derived carbon aerogels, micropores typically develop in the primary particles, whereas mesopores and macropores are generated as a result of the distances between the primary particles (originally occupied by solvent used in the synthesis of the parent organic gel). The porous dimensions of the parent organic gel do not automatically transfer directly into the carbon aerogel due to a variety of transformations (e.g., shrinkage, pore closure, aromatization/graphitization) may occur during the thermal annealing step. However, in principle, the amount of micropores and mesopores in these carbon aerogels can be dictated independently, a major advantage of carbon gels, through selection of the parent gel composition, drying method, curing time/temperature and, of course, the carbonization temperature.

More recently, biomass-based carbon aerogels were fabricated with the green chemistry concept, as shown in Figure 7. For example, Li *et al.* [156] fabricated carbon aerogels from the natural biomass, winter melon, via a hydrothermal and post-pyrolysis process (Figure 7a). The obtained carbon aerogel shows a low density of 0.048 $g{\cdot}cm^{-3}$, excellent hydrophobicity with a water contact angle of 135°, and selective adsorption for organic solvents and oils of 16–50 times its own weight. Similarly, watermelon was used as a renewable starting material to directly generate monolithic sponge-like carbon aerogel through a simple one-pot hydrothermal approach (Figure 7b) [157,158]. This approach does not need any additives as watermelons naturally contain water and carbohydrates. Directly cutting the native watermelon pieces could offer an easy shaping for hydrogels from 10 mL up to 1 L. After drying, low-density (*ca.* 0.58 $g{\cdot}cm^{-3}$), flexible, sponge-like aerogels made of nanospheres/nanofibers together with larger aggregated microparticles were obtained. In another case, highly porous carbon aerogels by using bagasse as a raw material were demonstrated inFigure 7c [159]. Macro and mesoporous carbon was first prepared by carbonizing the freeze-dried bagasse aerogel. Subsequently, microporous structure was created on the walls of the mesoporous carbon by chemical activation. Therefore, the as-prepared carbon aerogels with hierarchical porous structures exhibit high-energy storage ability as electrode materials in supercapacitors. Despite the fact that the direct approach based on fruit/vegetable is very simple and inexpensive, the control over the final morphology/porosity is quite limited and closely dependent on the native biomass. In addition, there might be some issues related to the competition with the food chain. Therefore, lightweight, hydrophobic and porous carbon microbelt aerogels and carbon fiber aerogels are produced via a facile route by using waste paper or cotton as the raw materials (Figure 7d) [160]. Importantly, the aerogel can adsorb a wide range of organic solvents and oils with a maximum sorption capacity up to 192 times of its own weight. Moreover, the aerogel also exhibits excellent recyclability, maintaining a high sorption capacity even after five cycles through distillation, burning or squeezing. The above synthesis strategies open a good pathway to employ abundant and renewable natural resources for manufacturing porous carbon materials, which may have great potentials for industrial applications and environmental protections.

The use of cellulose as a precursor for the preparation of aerogel and cryogels is well documented and has been discussed above. However, the transformation of cellulose aerogels into their carbonaceous or carbon equivalents has, until recently, not been significantly reported. Yu and coworkers have recently demonstrated the preparation of carbon aerogels using bacterial cellulose as the precursor polysaccharide for the preparation of bacterial cellulose aerogels and the thermal conversion to carbon aerogels [161,162]. Bacterial cellulose pellicles were first cut into the desired dimensions (e.g., $320 \times 240 \times 12$ mm^3) and purified, followed by freeze-drying to produce the bacterial cellulose aerogel. Thermal treatment under an inert (Ar) atmosphere between 700 and 1300 °C was employed to convert the polysaccharide aerogel into the carbon aerogel, denoted as the carbon nanofiber aerogel. After the pyrolysis step, the spatial volume of the obtained carbon nanofiber aerogel was only 15% of the original polysaccharide aerogel, and the density was found to decrease from 9–10 to 4–6 mg·cm^{-3} in the carbonized product. The resulting bacterial-cellulose-based carbon aerogels were composed of amorphous, turbostratic carbon nanofibers of 10–20 nm diameter, with an acceptable electrical conductivity of 20.6 S·m^{-1}. Interestingly, the carbon nanofiber aerogels prepared using this approach displayed high flexibility, which is not commonly displayed by conventional low-density, high-porosity materials (e.g., silica-based aerogels). It was possible to manually compress these bacterial cellulose-derived carbon nanofiber aerogels with more than 90% volume reduction, followed by near-total recovery of the original volume after release of the force, demonstrating the unusually compliant and elastic features. Although the presented carbon nanofiber aerogels derived from bacterial cellulose possess interesting and potentially unique mechanical properties, it is not clear exactly why morphology of the bacterial cellulose does not lose with such a low carbonization yield upon carbonization, given the general thermal decomposition chemistry of polysaccharides.

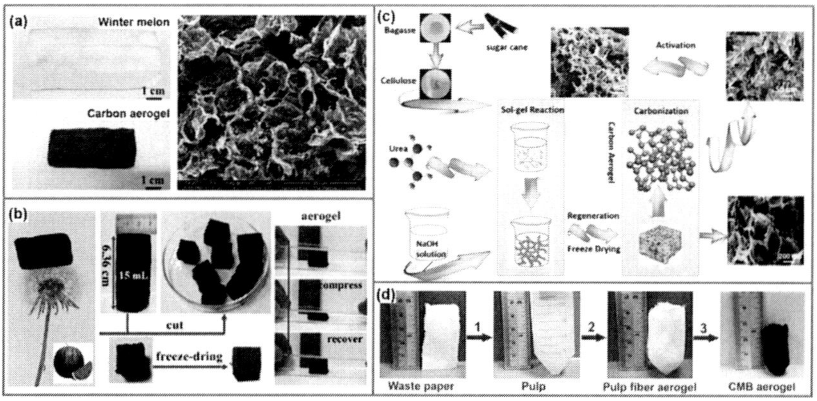

Figure 7. Carbon aerogels derived from different biomass: (**a**) Winter melon, reproduced with permission of Ref. [156] (Copyright American Chemical Society 2014). (**b**) Watermelon, reproduced with permission of Ref. [158] (Copyright American Chemical Society 2013). (**c**) Bagasse, reproduced with permission of Ref. [159] (Copyright The Royal Society of Chemistry 2014). (**d**) Waste paper, reproduced with permission of Ref. [160] (Copyright Wiley-VCH Verlag GmbH & Co. 2014).

To tune the electronic and surface chemical properties of carbon materials and improve their electrochemical performance, heteroatom-doped carbon aerogels were fabricated from bacterial cellulose. Specifically, P-doped, N,P-co-doped, and B,P-co-doped carbon nanofibers were successfully prepared by impregnating H_3PO_4, $NH_4H_2PO_4$, and H_3BO_3/H_3PO_4 into the bacterial cellulose pellicle, respectively, followed by carbonization in an inert atmosphere at 800 °C [163]. Nitrogen-doped carbon nanofiber aerogels were also reported by annealing them in NH_3 [164] or employing polyaniline coated bacterial cellulose as the precursor [165]. In the field of energy storage, Zhi *et al.* recently reported on the production of carbon aerogels from bacterial cellulose as an anode in Li-ion batteries in conjunction with deposited SnO_2 and Ge nanoparticles. Similarly, bacterial-cellulose-derived carbon nanofiber@ MnO_2 and nitrogen-doped carbon nanofiber aerogels were also employed as electrode materials in the development of asymmetric or all-solid state supercapacitors [166].

3.2.2. Graphene/CNT Based Carbon Aerogels

Novel carbon aerogels such as CNT and graphene aerogels are theoretically considered as the best candidates to make ultralight yet elastic and conductive aerogels, given their marvelous mechanical strength, low density, fine elasticity, good electrical conductivity, and extremely high aspect ratio [167,168,169]. CNT aerogel sheets are the sole component of new artificial muscles that provide giant elongations of 220% and elongation rates of (3.7×10^4)% per second, respectively, at operating temperatures from 80 to 1900 K [170]. Regardless of their ultralight density, the CNT or graphene aerogels are mechanically robust, which can support a counterpoise up to 1000 times of its own weight. More interestingly, some CNT or graphene aerogels exhibit superelasticity, which can be repeatedly compressed and recovers most of its volume after the release of compression. Besides, CNT aerogels have an electrical conductivity of 3.2×10^{-2} S·cm^{-1} that can be further increased to 0.67 S·cm^{-1} by a high-current pulse method without degrading their structures [171]. Therefore, this kind of novel carbon aerogels further expands the application fields of aerogel materials, including organic light-emitting displays, solar cells, sensors, novel electrodes and cold electron field emission.

Graphene-based aerogel is another kind of novel carbon aerogels, which was first prepared by Wang *et al.*, in 2009 [51]. GO solution was converted into the graphene aerogel by ultrasonic-induced gelation, drying and thermal reduction. Different from the physical crosslinking, Worsley *et al.* [172] reported the preparation of graphene aerogels crosslinked by chemical bonding through organic sol-gel chemistry. These graphene aerogels exhibit an improvement in bulk electrical conductivity (~1×10^2 S·m^{-1}), which is more than 2 orders of magnitude compared to that (~5×10^{-1} S·m^{-1}) of graphene assemblies with physical crosslinks alone. The graphene aerogels also possess large surface area (584 m^2·g^{-1}) and pore volume (2.96 cm^3·g^{-1}), being viable candidates for use in energy storage, catalysis, and sensing applications. Following this pioneering work, Shi and coworkers reported the formation of a stable GO/PVA hybrid hydrogel due to the strong hydrogen bonding interactions between hydroxyl-rich PVA chains and oxygen-containing groups on GO sheets. Subsequent studies in the field confirmed that DNA, protein, synthetic polymers with cationic charges and hydrogen bonding acceptors, small quaternary ammonium salts and metal ions are efficient crosslinkers for GO gelation, which can regulate the delicate balance between electrostatic repulsion, hydrophobic interactions and hydrogen bondings of GO-based colloid systems [173,174,175,176]. In another report, an aqueous GO suspension with appropriate concentration was directly submitted to hydrothermal treatment to form reduced graphene oxide (rGO) hydrogel without any other reagents [50]. After drying, the resulting rGO aerogel is conductive (5×10^3 S·cm^{-1}), thermally stable (25–100 °C) and mechanically strong (with storage modulus of 470 ± 20 kPa) (Figure 8a–d).

Besides 3D foam-like structures, graphene aerogels can be tailored into different macroscopic architectures, such as one-dimensional graphene aerogel fibers. Gao and coworkers first prepared continuous neat GO aerogel fibers with unique "porous core-dense shell" structure from flowing GO liquid crystals with lamellar ordering by the combination of spinning technology and ice-templating strategy [177]. After chemical reduction and annealing process, the aligned porous graphene aerogel fibers possess high specific surface area and fine electrical conductivity. Interestingly, graphene aerogel fibers have fine flexibility and can be folded, and the bent fiber can be stretched to its original straight shape without any fracture (Figure 8e,f). Graphene sheets stacked densely to form a wrinkled shell surface, and interconnected to form a porous part which is surrounded by the dense shell as shown by the SEM images in Figure 8g,h. This work solved the widely expressed conflicts between high porosity (meaning lightweight and high surface area) and high performance in both strength and electrical conductivity, which could be helpful in the future design and fabrication of advanced aerogel materials. An ultra-flyweight aerogel (UFA) was fabricated by the same group through freeze-drying aqueous solutions of CNTs and giant graphene oxide (GGO) sheets, followed by chemical reduction of GGO into graphene with hydrazine vapor [155]. SEM images show the UFA exhibits an interconnected, porous 3D framework of randomly oriented, crinkly sheets with continuous macropores ranged from hundreds of nanometers to tens of micrometers (Figure 8k,l). The density of the UFA achieves 0.16 $mg \cdot cm^{-3}$, which is the lowest density of ultralight materials ever reported. In addition, compression experiments on the UFAs showed a nearly complete recovery after 50%–82% compression, indicating superelasticity and complete fatigue resistance under large cyclic strains. The desirable multifunctional attributes of UFAs such as outstanding temperature-invariant elasticity, ultralow density, excellent thermal stability, and good electrical conductivity would further widen the applications of carbon aerogels.

3.2.3. Carbon-Based Hybrid Aerogels

The structures and properties of carbon aerogels can be easily tuned by incorporation of metal species into the carbon framework. Three main strategies have been used to introduce metal species into the carbon framework. The first consists of dissolving the metal precursor in the initial sol solution. The second involves the use of a monomer derivative containing an ion exchange moiety that can be polymerized using sol-gel techniques. The repeating unit of the organic polymer contains a binding site for metal ions, ensuring the uniform dispersion of the dopant. Finally, the third strategy is to deposit the metal precursor on the organic or carbon aerogel.

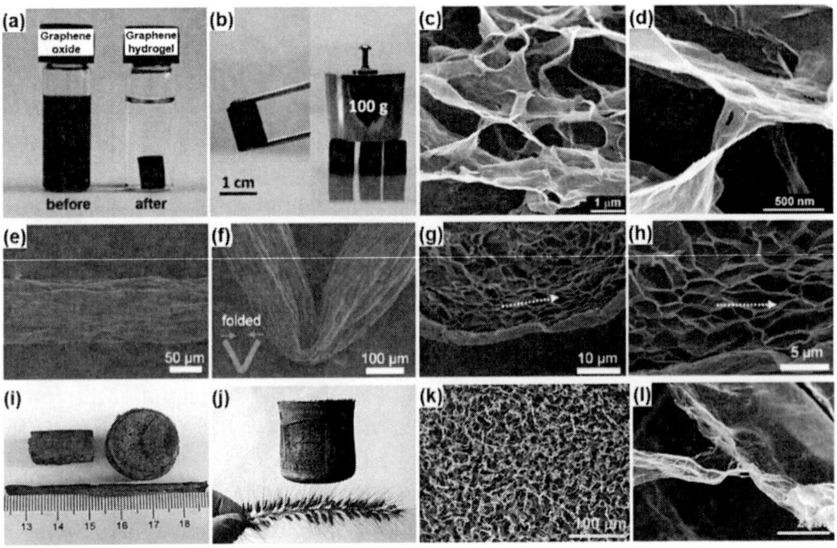

Figure 8. (**a**) Photograph of an aqueous GO dispersion (2 mg·mL^{-1}) before and after hydrothermal reduction at 180 °C for 12 h; (**b**) photographs of a strong rGO hydrogel allowing easy handling and supporting weight; and (**c**,**d**) SEM images of the rGO hydrogel showing interior microstructures. Reproduced with permission of Ref. [50]; (Copyright 2010 American Chemical Society.) (**e**) SEM images of the surface of graphene aerogel fibers and (**f**) folded/stretched graphene aerogel fibers; (**g**,**h**) SEM images of fracture morphology of graphene aerogel fibers. The arrows indicate the alignment direction of graphene sheets. Macroscopic and microscopic structures of UFAs. Reproduced with permission of Ref. [177] (Copyright American Chemical Society 2012). (**i**) Digital photograph of UFAs with diverse shapes; (**j**) A 100 cm^3 UFA cylinder standing on a flower like dog's tail; (**k**,**l**) Microscopically porous architecture of a UFA at different magnifications, showing CNT-coated graphene cell walls. Reproduced with permission of Ref. [155] (Copyright Wiley-VCH Verlag GmbH & Co. 2013).

Mo-doped carbon aerogels were obtained in the polycondensation reaction of aqueous resorcinol and formaldehyde by adding Mo-salt either into the initial mixture or the dried polymer aerogel by incipient wetting impregnation [178]. It was found that Mo-salt added during the polymerization process yielded a more compact gel structure with practically no mesoporosity. With post-impregnation, by contrast, mesopores with diameter of 3–15 nm were generated. Moreover, the content of Mo nanoparticle in the samples was also different. For the former one, Mo was lost during the solvent exchange process before supercritical drying (*i.e.*, Mo failed to bind chemically to the polymer matrix). The residual Mo congregated into bulk clusters of α-Mo_2C with size of 25–60 nm. For post-impregnation, finely dispersed α-Mo_2C and η-Mo_3C_2 crystals with the size of 8–20 nm formed on the surface of both carbon and Mo-based oxides. Ni-doped graphene/RF-based carbon cryogels were prepared using Ni^{2+} ions as catalysts for the gelation process to substitute the usually used alkaline carbonates [179]. The metal ions of Ni^{2+} effectively elevated the cross-linking between GO and RF skeletons, thus strengthening the whole cryogel. The as-formed 3D interconnected structures, which can be well-maintained after freeze-drying of the hydrogel precursor and subsequent carbonization under an inert atmosphere, exhibit good mechanical properties as shown in Figure 9. During the carbonization process, Ni^{2+} ions are converted into Ni nanoparticles and uniformly embedded in the interconnected structures. Therefore, the unique porosity within the interconnected structures endows the cryogels with good capability for extraction of oils and some organic solvents while the bulk form enables its recycling use.

In addition, the high surface area and porosity of carbon aerogels make them efficient scaffolds for immobilizing inorganic nanoparticles to meet the demand for multi-functions. Various inorganic components such as metal oxides (Fe_3O_4, Fe_2O_3, MnO_2, SnO_2, WO_3, $ZnMn_2O_4$, *etc.*), metal hydroxides ($Ni(OH)_2$), and metal sulfides (MoS_2, WS_2) have been incorporated into carbon aerogels, and applied in catalysis, energy storage or water treatment. For example, WO_3 decorated carbon aerogels were fabricated by carbonization of RF aerogel followed by a simple immersion-calcination process and used as electrode materials for supercapacitors [60]. WO_3 in the carbon aerogels show a form of single crystalline nanoparticles of 15–40 nm in size. One order of magnitude improvement in specific capacitance was achieved, from 54 $F{\cdot}g^{-1}$ for WO_3 nanoparticles to 700 $F{\cdot}g^{-1}$ for WO_3/carbon aerogel composites. 3D N-doped graphene aerogel-supported Fe_3O_4 nanoparticles are reported as efficient cathode catalysts for oxygen reduction reactions. As shown in Figure 10, the graphene hybrids exhibit an interconnected macroporous framework of graphene sheets with uniform dispersion of Fe_3O_4 nanoparticles, exhibiting improved oxygen reduction performance [62]. Yin *et al.* [61] fabricated spinel $ZnMn_2O_4$/carbon aerogel composites with a 3D porous structure through a facile solution immersion chemical route. The hybrids show enormous interfacial surface area, connected 3D framework, abundant porosity and much higher capacity than that of pure spinel $ZnMn_2O_4$. These works further demonstrate the importance of 3D macropores and high specific surface area of the carbon aerogel support for improving the overall performance of the hybrid materials.

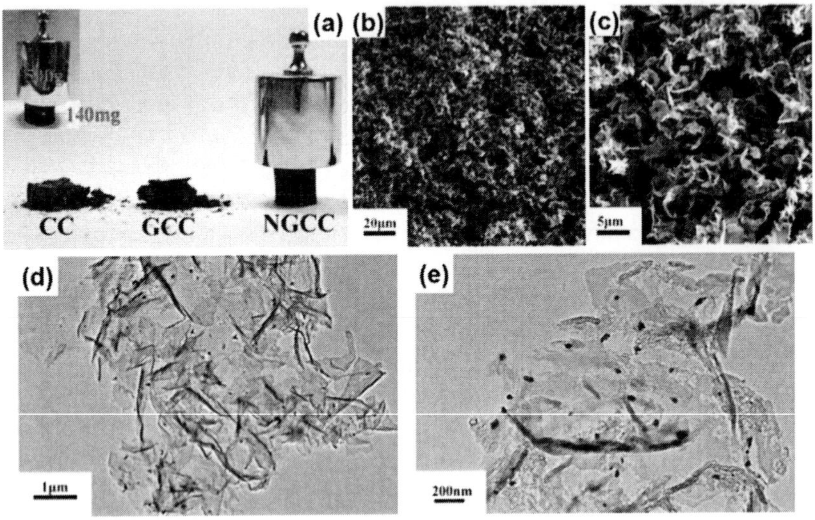

Figure 9. (**a**) Digital photographs of carbon cryogel (CC), graphene/carbon cryogel (GCC), and Ni-doped graphene/carbon cryogel (NGCC) with the same weight (90 mg) after loading with 200 g weight; (inset) 140 mg Ni-doped graphene/carbon cryogel monolith enduring a weight of 500 g. (**b**,**c**) SEM images and (**d**,**e**) TEM images of Ni-doped graphene/carbon cryogel. Reproduced with permission of Ref. [179] (Copyright American Chemical Society 2013).

4. APPLICATIONS OF POLYMER/CARBON-BASED AEROGELS

4.1. Energy Storage

The development of electrochemical systems with high power and energy densities for sustainable energy storage has been considered as one of the critical solutions to resolve global concerns of fossil fuel depletion and environmental deterioration. Carbon aerogels with 3D porous architectures are excellent candidates for high performance electrode materials due to their high specific surface area and developed porosity, which provide continuous channels to facilitate ion transport by shortening diffusion pathways and ensuring excellent electrical contact.

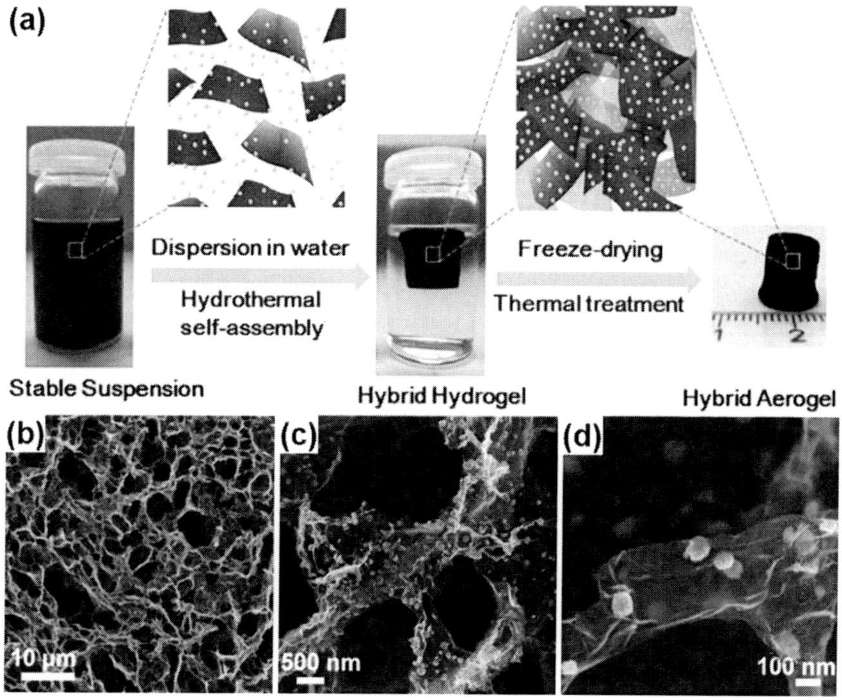

Figure 10. (**a**) Fabrication process for the 3D Fe_3O_4/N-doped graphene aerogel catalyst. (**b–d**) SEM images of Fe_3O_4/N-doped graphene aerogel revealing the 3D macroporous structure and uniform distribution of Fe_3O_4 nanoparticles in the graphene aerogels. Reproduced with permission of Ref. [62] (Copyright American Chemical Society 2012).

The pore size distribution of carbon aerogels strongly affects the specific surface area and subsequently the electrochemical behavior. It has been proved that carbon aerogels with a pore diameter in the range from 3 to 13 nm exhibit the best voltammetry characteristics and the highest capacitance values (70–150 $F \cdot g^{-1}$). The carbon aerogels obtained at temperatures over 900 °C show an aggravation of specific capacitance, whereas functionalization of the carbon surface by heating at 500 °C in air environment could cause an improvement of the capacitance through the pseudocapacitance effects [22]. Moreover, Hall *et al.* [180] studied the effect of pore size of activated carbon aerogels on the capacitance performance in ionic liquid electrolytes. The best capacitance performance was obtained for the activated carbon aerogel with average pore diameter of 3.5 nm, whereas the optimum rate performance was obtained for the activated carbon aerogel with average pore diameter of 6 nm. When combined in an electrochemical capacitor with ionic liquid electrolyte

$EMImBF_4$, a specific capacitance of 210 $F·g^{-1}$ was obtained for activated carbon aerogel with average pore diameter of 3.5 nm at an operating voltage of 3 V. On the other hand, the activated carbon aerogel with average pore diameter of 6 nm allowed for maximum capacitance retention of approximately 70% at 64 $mA·cm^{-2}$.

However, the specific capacitance of pure carbon aerogels is very limited due to its intrinsic electrical double layer behavior. By activating the carbon aerogels in different gases or chemical substance, the porous structure and surface area of the concerned carbon aerogels can be improved. It was reported that the porous structure and electrochemical performance of the carbon aerogels were significantly affected by chemical activation of KOH [159]. Macro and mesoporous carbon was first prepared by carbonizing the freeze-dried bagasse aerogel, and microporous structure was consequently created on the walls of the mesoporous carbon by chemical activation. The KOH-activated carbon aerogels have a typical hierarchical porous structure. The first level is large macropores, which are larger than 50 nm in diameter. The second level is the mesopores in the channel wall of the large pores, which is composed of several connections of carbon nanoparticles and is about 2–50 nm in diameter. And the third level is micropores in the carbon nanoparticles, which is composed of graphite-like structures and is smaller than 2 nm. As a result, the KOH-activated carbon aerogels show a high specific capacitance of 268.4 $F·g^{-1}$ at a scan rate of 2 $mV·s^{-1}$, which is about four times higher than that of the unactivated one. In addition, the capacitance of carbon aerogels could be improved by heteroatom doping. For example, N-doped carbon aerogels were synthesized by using watermelon as the starting materials and polypyrrole as the N-source via a hydrothermal method [157]. The obtained aerogels exhibit excellent electrochemical performance, in particular, an outstanding specific capacitance of 281.0 $F·g^{-1}$ can be achieved for the sample calcined at 600 °C. The improvement of electrochemical performance is attributed to the combinatorial effect of the appropriate N content and porous structure.

Another method is to incorporate electrically active materials such as metal oxides or conductive polymers into carbon aerogels to improve their electrochemical performance. MnO_2 loaded carbonaceous aerogel (MnO_2@CA) composites were also synthesized via the hydrothermal carbonaceous process [181]. The obtained MnO_2@CA displays the specific capacitance of 123.5 $F·g^{-1}$ at 5 $mV·s^{-1}$. The enhanced supercapacitance of MnO_2@CA nanocomposites could be ascribed to both electrochemical contributions of the loaded MnO_2 nanoparticles and the porous structure of 3D carbonaceous aerogels. $LiFePO_4$/carbon hybrid aerogels were prepared by mixing precursor solution with carbon aerogel via a simple solution impregnation method and used as cathode for lithium ion batteries [182]. $LiFePO_4$ nanowires forming on the ektexine of carbon aerogel intertwined with $LiFePO_4$/CA particles and formed a special web structure, which can improve the initial discharge capacity up to 139.3 mAh·g^{-1}. The added carbon aerogel induces the growth $LiFePO_4$ nanowire and increases the electronic conductivity while the formation of 3D nano-network increases the contact area between active materials and electrolyte. Thus, both the specific capacity and the cycle stability of $LiFePO_4$ at high discharge rate can be significantly improved.

Recently, graphene aerogels draw great attention owing to their unique properties and excellent performance in energy-related fields. As an electrode material for supercapacitors, the obtained graphene foam has electrical conductivity of 5 × 10^{-3} S·cm^{-1}, and exhibit a specific capacitance of 175 F·g^{-1} in an aqueous electrolyte [50]. Subsequently, in order to obtain higher conductivity of graphene foam, the authors used a reducing agent (hydrazine or hydroiodic acid) for further reduction of the foam [183]. In this case, the chemically reduced graphene foam possessed high conductivity of 1.3–3.2 S·m^{-1}, high specific capacitance of 220 F·g^{-1} and higher rate capability compared to their previous work. Following this pioneering work, graphene incorporated or crosslinked carbon aerogels were extensively studied. Worsley *et al.* [172] demonstrated that GO aqueous suspension can be cross-linked with RF using sodium carbonate as catalyst by sol-gel chemistry, followed by supercritical CO_2 drying and pyrolysis at 1050 °C under nitrogen atmosphere. The as-prepared macroscopic 3D graphene foam exhibited excellent electrical conductivity (100 S·m^{-1}), high surface area (584 m^2·g^{-1}) and large pore volume (2.96 cm^3·g^{-1}). Subsequently, the porous structure of graphene foam was adjusted by increasing the RF content, and as a result, the obtained foam had a high specific surface area of 1200 m^2·g^{-1} and a large pore volume of 6 cm^3·g^{-1}, and maintained its high conductivity [184]. Graphene-carbon aerogel composites with tailored pore structure and morphology were synthesized by a facile yet effective method with ionic liquids as templates [92]. They indicated that the incorporation of graphene can increase the specific capacity of carbon aerogels from 140 F·g^{-1} to 230 F·g^{-1} at a current density of 0.1 F·g^{-1} without using other activation agents. Furthermore, the electrode materials prepared from graphene-carbon aerogels show high cyclic stability even after 5000 cycles. Graphene/carbon aerogels with multimodal pores were reported via carbonization of graphene crosslinked PI aerogels [185]. Compared to most carbon aerogels prepare previously, this preparation process is facilitated by the exclusion of harmful formaldehyde. Moreover, as a powerful crosslinking agent, graphene is beneficial for acceleration of the gelation process, improvement of the porous structures inside carbon aerogels and enhancement of the specific surface area and electrical conductivity of carbon aerogels. SEM observation (Figure 11) shows the multimodal pores and 3D nano-network of carbon aerogels, which provide high electroactive regions and short diffusion lengths for both charge and ion transport. Owing to the incorporation of graphene, the as-prepared carbon aerogels possess high specific surface area up to 998.7 $m^2 \cdot g^{-1}$ and specific capacitance up to 178.1 $F \cdot g^{-1}$ at a current density of 1 $A \cdot g^{-1}$, which is much higher than that of pure carbon aerogels (193.6 $m^2 \cdot g^{-1}$ and 104.2 $F \cdot g^{-1}$). This work provides a new and facile method for fabricating high performance carbon aerogels with hierarchical structures and extends the potential applications of polyimide.

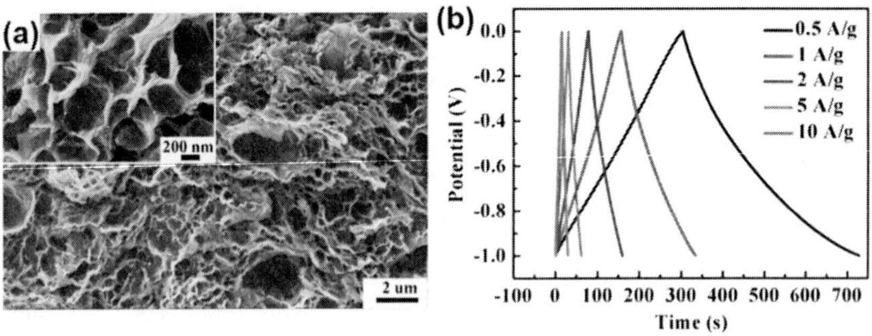

Figure 11. (**a**) SEM images and (**b**) galvanostatic charge–discharge curves of graphene/carbon aerogels prepared from carbonization of graphene crosslinked PI aerogels. Reproduced with permission of Ref. [185] (Copyright The Royal Society of Chemistry 2015).

Overall, electrodes based on carbon aerogels possess good electrochemical performance, high reversibility and high specific capacitance. Moreover, the utilization efficiency of carbon aerogels is promoted due to their excellent electronic conductivity and extensive mesoporous network of carbon aerogels. Therefore, carbon aerogels with high surface area and electrical conductivity are considered as promising electrode materials for applications in high-performance supercapacitors.

4.2. Absorption

It is urgent to develop an economic and feasible strategy to solve the environmental and ecological problems arising from the increasing pollution from crude oils, petroleum products and toxic organic solvents. Due to their high porosity and high hydrophobicity, carbon aerogels have been extensively investigated as candidates for removal of pollutants and separation of oil from water [156,160,186,187]. As listed in Table 1, carbon aerogels usually exhibit much higher sorption capacity (up to 900 $g \cdot g^{-1}$) compared with other solid materials such as polymers, graphite or activated carbon, which only have sorption capacities less than 100 g·g^{-1}.

Table 1. Comparison of various aerogel materials for adsorption.

Sorbent Materials	Absorbed Substances	Sorption Capacity (g·g^{-1})	Ref.
Graphene/PVDF aerogels	Oils and organic solvents	20–70	[32]
PVA/CNF aerogels	Oils and organic solvents	44–96	[33]
CNF aerogels	Water, oils and organic solvents	139–375	[111]
Ultra-flyweight aerogels	Oils and organic solvents	215–913	[155]
Winter melon carbon aerogels	Oils and organic solvents	16–50	[156]
Carbon microbelt aerogels	Oils and organic solvents	50–188	[160]
Ni-doped graphene/carbon cryogels	Oils and organic solvents	22.2–23.2	[179]
Twisted carbon fiber aerogels	Oils and organic solvents	50–192	[186]
Vegetable fiber	Crude oil	1–100	[188]
Nanowire membrane	Oils and some organic solvents	4–20	[189]
Wool-based nonwoven	Diesel, crude oil, SN 150	9–15	[190]
Polymers	Oils and organic solvents	5–25	[191]
Exfoliated graphite	Heavy oil	60–90	[192]
Activated carbons	Benzene, toluene	<1	[193]
Carbon nanotube sponges	Oils and organic solvents	80–180	[194]
Graphene/CNT foam	Compressor oil, organic solvents	80–140	[195]
CNT sponge doped with boron	Oils and organic solvents	25–125	[196]
Graphene-based sponges	Oils and organic solvents	60–160	[197]
Reduced graphite oxide foam	Cyclohexane, chlorobenzene, toluene, petroleum, motor oil	5–40	[198]
Nitrogen doped graphene foam	Oils and organic solvents	200–600	[199]
Carbonaceous nanofiber aerogels	Oils and organic solvents	40–115	[200]

The hydrophobicity of carbon aerogels makes it an ideal candidate for removal of pollutants and separation of oil and water, because its hydrophobic surface can selectively and effectively adsorb oily target compounds mixed with water. Lightweight, hydrophobic and porous aerogels made from twisted carbon fibers (TCF) are produced by Bi and coworkers via a facile route by using cotton fibers as raw materials [186]. Notably, the TCF aerogel can adsorb a wide range of organic solvents and oils with a maximum sorption capacity up to 192 times of the weight of the pristine TCF aerogel (Figure 12). Moreover, the TCF aerogel also exhibits the excellent recyclability, and maintains a high sorption capacity even after five cycles through distillation, burning or squeezing. Thereafter, the same group produced aerogels made of carbon microbelts (CMBs) by using waste paper as the raw material [160]. The CMB aerogel can adsorb a wide range of organic solvents and oils with a maximum sorption capacity up to 188 times of the weight of the pristine CMB aerogel. Most importantly, the abundant source and simple preparation method make the CMB aerogel cost-effective for possible industrial applications, such as barrier separation and water purification.

Figure 12. (**a**) Photographs showing the sorption process of heptane by using a TCF aerogel taken at intervals of 10 s. Heptane stained with Sudan red 5B floating on water was completely absorbed within 40 s. (**b**) Photographs showing the sorption process of chloroform by using a TCF aerogel. Chloroform stained with Sudan red 5B at the bottom of water was completely absorbed within 5 s. Reproduced with permission of Ref. [186] (Copyright Wiley-VCH Verlag GmbH & Co. 2014).

Nitrogen-rich carbon (NRC) aerogels with highly hydrophobic surfaces were reported by Yang and coworkers [201]. The NRC aerogels can be used as oil-sorbent materials to adsorb oils or organic solvents in oil-spilled accidents by an adsorption/distillation or adsorption/combustion ways. The NRC aerogels can maintain 100% of their initial adsorption capacities after 100 adsorption/distillation cycles and 61.2% after 100 adsorption/combustion cycles, which thus has the highest recyclability of the reported carbon aerogels. Gao *et al.* [179] prepared Ni-doped graphene/carbon cryogels and further used as versatile sorbents. The cryogel monoliths exhibit ultrafast absorption rate and high uptake capacity for oils and organic solvents, such as trichloromethane (close to 50 $g{\cdot}g^{-1}$), with a good recyclability. MWNT-carbon hybrid aerogels with 3D interconnected frameworks and low density of about 0.02 $g{\cdot}cm^{-3}$ were via a one-pot hydrothermal carbonization process [202]. The uptake capacity of a thermally treated MWNT-carbon hybrid aerogel is up to about 37 times of its weight for various lubricating oils, and up to about 23 times for vegetable oil (as shown in Figure 13). It is worth to mention that the thermally treated MWNT-carbon hybrid aerogel shows a good recyclability to these solvents, as about 98% of oil removal efficiency can still be retained even after five cycles of adsorption–desorption process by using hexane as an extractant.

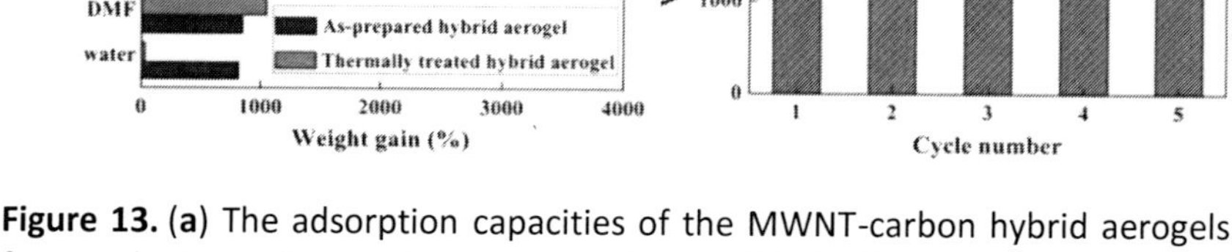

Figure 13. (**a**) The adsorption capacities of the MWNT-carbon hybrid aerogels for a selection of several common solvents. (**b**) Cyclic performance of the thermally treated MWNT-carbon hybrid aerogels for oil adsorption in five cycles. Reproduced with permission of Ref. [202] (Copyright The Royal Society of Chemistry 2013).

Except oils and organic solvents, carbon aerogels can also be applied to remove heavy metal ions from water. It was found that the removal of heavy metal ions such as Cd(II), Pb(II), Hg(II), Cu(II), Ni(II), Mn(II) and Zn(II) by carbon aerogel was influenced by concentration, pH, contact time, adsorbent dose and temperature [203]. Superhydrophobic and superoleophilic PVA/CNF hybrid aerogels were reported after being treated with methyltrichlorosilane via a simple thermal chemical vapor deposition process [33]. The silane-treated PVA/CNF aerogels not only exhibited excellent absorption performance for various types of oils (e.g., crude oil) or organic solvents, but also showed a remarkable scavenging capability for several types of heavy metal ions. The scavenging capacities of the PVA/CNF aerogels for Hg^{2+}, Pb^{2+}, Cu^{2+}, and Ag^{+} were 157.5 $mg{\cdot}g^{-1}$, 110.6 $mg{\cdot}g^{-1}$, 151.3 $mg{\cdot}g^{-1}$, and 114.3 $mg{\cdot}g^{-1}$, respectively, which are greatly improved compared to pure PVA aerogels. It is believed that metal ion scavenging in aqueous solutions by porous materials such as aerogels is mainly driven by electrostatic interaction and complexation between the metal ions and the carboxyl groups present in the porous materials. The remarkable scavenging heavy metal ions capability exhibited by PVA/CNF aerogels makes them versatile absorbents for various potential applications including water purification.

4.3. Adsorption

Gas purification represents a major challenge in a time of concerns for growing air pollution based on emission of gases (especially toxic and greenhouse gases) from various industrial sources. In particular, the subject of carbon dioxide (CO_2) capture, utilization, and storage has received widespread attention because of the interest in reducing the amount of released CO_2 as a greenhouse gas. The amount of CO_2 present in the atmosphere contributes to 60% of global warming effects [204]. Hence, it is particularly urgent to prevent the release of carbon dioxide and lower their concentration in the atmosphere. Many CO_2 capture technologies, such as absorption, cryogenic, adsorption, and membranes, have been investigated. Among them, adsorption on solid adsorbents is considered to be one of the most promising approaches to deal with environment related problems. There is intense focus on developing simple and cheap synthesis methodologies that deliver highly porous materials with appropriate properties for use as CO_2 capture. In this regard, aerogels are attractive due to their high porosity, versatility, low bulk density, and high internal surface area [205]. Therefore, recent investigations have focused on using aerogels as new CO_2 capture materials. Masika developed a carbon aerogel with a high surface area of 1100 $m^2{\cdot}g^{-1}$, and the CO_2 adsorption capacity obtained in pure CO_2 at 25 °C and 1 bar was only 2.2 $mmol{\cdot}g^{-1}$ [206]. Chemical activation of the carbon aerogel with KOH generates activated carbon aerogels with surface area between 915 and 1980 $m^2{\cdot}g^{-1}$ and pore volume of up to 2.03 $cm^3{\cdot}g^{-1}$ with the extent of the increase in textural properties. The activated carbon aerogels show a micropore size distribution centred at ca. 8 and 13 Å, with a large increase in pore volume and proportion of microporosity.

As a result, the activated carbon aerogels exhibit attractive CO_2 uptake of between 2.7 and 3.0 mmol·g^{-1} at 25 °C and 1 bar under flowing pure CO_2 conditions [207]. 3D porous carbon through steam activation of graphene aerogel was also reported for CO_2 uptake [208]. The steam activated graphene aerogel exhibits a 3D network structure with interconnected pores and possesses a high BET specific surface area (830–1230 m^2·g^{-1}) and a large pore volume (2.2–3.6 cm^3·g^{-1}). CO_2 adsorption capacity (10.8 wt %) for steam activated graphene aerogel is much higher than that (4.1 wt %) of graphene aerogel prepared without steam activation, indicating that steam activation is of great importance to the enhancement of CO_2 adsorption. A hybrid aerogel material composed of chitosan and graphene oxide was developed by Alhwaige and coworkers [209]. It is demonstrated that dispersing GO into chitosan matrix, in the form of aerogel, leads to increased surface area with subsequent increase in CO_2 sorption. The amount of CO_2 adsorbed at 25 °C increases from 1.92 to 4.15 mol·kg^{-1} with the addition of 20 wt % GO. Adsorption–desorption cycles exhibit the stability of the hybrid aerogels during prolonged cyclic operations, suggesting excellent potential for CO_2 capture technology.

4.4. Thermal Insulations

The energy needed to maintain a pleasant interior atmosphere accounts for more than 10% of the gross global energy consumption. Thermal insulation plays a major role in controlling the energy efficiency of buildings or aerospace and it is needed to substantially reduce thermal conductivity (λ) values below those of currently used insulation materials such as expanded polystyrene (EPS; λ = 30–40 mW·m^{-1}·K^{-1}), polyurethane (λ = 20–30 mW·m^{-1}·K^{-1}), glass fiber (λ = 33–44 mW·m^{-1}·K^{-1}) and mineral wool (λ = 30–40 mW·m^{-1}·K^{-1}). In fact, significantly reducing λbelow the value for air (25 mW·m^{-1}·K^{-1}) is desirable to minimize the required space and materials without severely compromising the architectural design. Approaches to obtain super-insulating materials include the replacement of air with another gas or vacuum or by reducing the pore size below the mean free path of air [210,211]. However, it is challenging to maintain a special gas or vacuum inside insulating panels for an extended time, which can result in a loss of the initially low thermal conductivity with time. Therefore, nanoporous aerogels are ideal candidates for both ambient and high temperature insulation applications owing to their unique nanostructure, high porosity and lightweight. Generally, heat is transferred by radiation, conduction or convection. Due to the small pore sizes of aerogels (1–100 nm), the convective and gas conduction pathways are virtually eliminated. Hence, what remains is primarily radiative heat transfer and conduction through the aerogel network.

Silica aerogels can be used under ambient conditions and still maintain a low λ of ~17–21 $mW\cdot m^{-1}\cdot K^{-1}$, but they are brittle and difficult to prepare in large sizes. Biopolymer-based materials such as wood chips, recycled paper and cork were extensively used for thermal insulation before the introduction of fossil-fuel-based foams, but their insulating performance is relatively poor with thermal conductivities of 40–50 $mW\cdot m^{-1}\cdot K^{-1}$. However, with the emergence of nanocellulose extracted from wood and other sources, there are now possibilities for the nanoscale engineering of renewable materials to generate more efficient thermal insulators. Cellulose aerogels were fabricated from nanocellulose fibers with different morphologies and surface properties derived from biomass resources [212]. Though the structures of aerogels are different, all cellulose aerogels show a thermal conductivity lower than 0.016 $W\cdot m^{-1}\cdot K^{-1}$, which can be used as thermal insulation materials. Pectin aerogels were prepared via dissolution-gelation-coagulation and subsequent drying with supercritical CO_2 [87]. The resulting pectin aerogels have a thermal conductivity ranging from 0.016 to 0.022 $W\cdot m^{-1}\cdot K^{-1}$, which is below that of air in ambient conditions, making them new thermal super-insulating fully biomass-based materials. However, it remains a big challenge for biopolymer-based insulation materials due to their poor resistance to fire and their moisture sensitivity. Recent work has shown that nanomaterials such as clays and graphenoids can provide organic polymer-based nanocomposites with good fire retardancy and excellent mechanical properties. Thus, graphene oxide and sepiolite nanorods were incorporated into the suspensions of cellulose nanofibers to prepare cellulose-based aerogels [35]. The aerogels are super-insulating, showing a thermal conductivity of 15 $mW\cdot m^{-1}\cdot K^{-1}$ and excellent combustion resistance, which is about half of that of expanded polystyrene. At 30 °C and 85% relative humidity, the foams retained more than half of their initial strength. The results show that nanoscale engineering is a promising strategy for producing foams with excellent properties by using cellulose and other renewable nanosized fibrous materials. Similarly, nanoporous hybrid aerogels based on cellulose nanofibers and nanozeolite particles were fabricated to achieve lower thermal conductivity of these materials. Thermal conductivity value as low as 18 $mW\cdot m^{-1}\cdot K^{-1}$ was obtained that confirms the superinsulation ability of these new fibrous aerogels. Synergism on the thermal conductivity properties was shown by adjunction of nanozeolites to cellulose microfibrils by reaching pore size lower than 100 nm that significantly reduces the thermal conductivity of the hybrid aerogels [213].

PI aerogels are ideal candidate for thermal insulation materials, especially in aerospace applications when much higher use temperatures are needed, such as insulation for launch vehicles or for planetary entry, descent, and landing systems. Aramid aerogels prepared by efficient reaction of carboxylic acids and isocyanates show a low thermal conductivity of 0.028 $W\cdot m^{-1}\cdot K^{-1}$ [214]. Based on this result, the authors further synthesized PI aerogels by ring-opening metathesis polymerization of a norbornene end-capped diimide [146]. The new materials combine facile one-step synthesis with heat resistance up to 200 °C, high mechanical compressive strength and specific energy absorption (168 MPa and 50 $J{\cdot}g^{-1}$, respectively, at 0.39 $g{\cdot}cm^{-3}$ and 88% ultimate strain), low speed of sound (351 $m{\cdot}s^{-1}$ at 0.39 $g{\cdot}cm^{-3}$) and styrofoam-like thermal conductivity (0.031 $W{\cdot}m^{-1}{\cdot}K^{-1}$ at 25 °C). Therefore, they are reasonable multifunctional candidate materials for further exploration as thermal/acoustic insulation at elevated temperatures. Apart from thermal conductivity, decomposition temperature of aerogels as well as its dependency on external conditions is usually discussed for different classes of thermal insulation aerogels. It is reported that the onset decomposition temperature of crosslinked aromatic PI aerogels can reach up to 525 °C to 625 °C [143,144,146]. Consequently, PI aerogels can be used for aircraft engine applications because of their high temperature stability, low density and low thermal conductivity.

In summary, polymer aerogels have a similar thermal conductivity with that of silica aerogels, which is about 0.016 $W{\cdot}m^{-1}{\cdot}K^{-1}$, but much stiffer and stronger than silica aerogels. Therefore, polymer aerogels have potential applications as thermal insulation materials with improved thermal stability and mechanical properties, which can meet the requirement in aerospace.

4.5. Flame Retardant Materials

Even though polymer aerogels are excellent thermal insulation materials, their applications in flame retardant materials are limited by their poor resistance. Therefore, modifications are required to decrease their flammability through the addition of flame retardant compounds. Environmental regulation has restricted the use of some halogenated flame retardant additives, initiating a search for alternative flame retardant additives. Recent work has shown that nanoparticle fillers, such as clays, graphene and CNTs, are highly attractive for this purpose, since they can simultaneously improve both the mechanical properties and flammability of the polymer aerogels.

Preparation of polymer/clay aerogel materials through a freeze drying process was reported and intensively studied by Schiraldi and coworkers [17,36,37,215,216]. They fabricated low flammable aerogels based on renewable ammonium alginate and MMT clay through a simple, environmentally-friendly freeze-drying process [215]. Alginate is a polyelectrolyte derived from seaweed that is considered to be biocompatible, non-toxic, non-immunogenic and biodegradable, and it has been proved to be an inherently flame-retardant material with limited oxygen index value of 48.0, compared with 20.0 for viscose fiber. During cone calorimetry tests, the peak heat release rate and effective heat of combustion of alginate fiber were 4.99

$kW \cdot m^{-2}$ and 0.46 $MJ \cdot kg^{-1}$, respectively, compared with 168.75 $kW \cdot m^{-2}$ and 12.06 $MJ \cdot kg^{-1}$ for viscose fiber. With addition of clay, the flammability of alginate/clay aerogels was further reduced. No open fire was observed during testing of these aerogels, and 53 wt % of sample weight remained with the original sample shape remaining intact. With a high concentration of clay (50 wt % of the solid), the silicate layers with high aspect ratio served as heat and mass transport barriers, which would largely decrease the burning rate. Furthermore, ceramic transformation of high concentration of clay particles helps to form stable structure of sample, and inhibit the flame spread. Thereafter, a series of PVA hybrid aerogels were prepared by using different nanofillers, e.g., nanoscale silica, halloysite, MMT and laponite through the above mentioned freeze-drying method [36]. The aerogels with layered structures have lower flammability than those non-layered aerogels and the heat release values of aerogels are much lower than those for commercial expanded polystyrene foam. Based on these results, they further prepared PVA/MMT aerogels followed by radiation crosslinking to increase the mechanical properties of aerogels [37]. The cross-linked PVA/MMT aerogels possess very low flammability and high compressive modulus, especially at the absorbed dose of 30 kGy. For the mechanism of flammability, they indicated that with increasing clay content, the heat and mass transport barriers are developed, simultaneously low levels of thermal conductivity are maintained during burning. Furthermore, a facile preparation of PVA/MMT hybrid aerogels was introduced by using water as solvent and divinylsulfone as crosslinking agent [17]. The compressive modulus of PVA/MMT aerogels can increase 29-fold upon crosslinking, which show a low flammability and a great potential for applications in flame retarded materials.

High-performance flame retardant materials from renewable resources are desirable for improving the energy efficiency of buildings. Traditional fossil-fuel-derived flame retardant materials such as expanded polystyrene and polyurethane have thermal conductivities that are too high for retrofitting or for building new, surface-efficient passive houses. Therefore, lightweight, highly porous cellulose nanofibers (CNF)-GO-sepiolite nanorods (SEP) foams were produced by freeze-casting and exhibit excellent combustion resistance [35]. Freeze-casting offers a versatile approach to producing highly anisotropic porous materials, in which ice crystals grow with the temperature gradient and eventually produce a frozen material consisting of anisotropic ice crystals surrounded by the walls formed by the dispersed nanoparticles. As a consequence, the anisotropic nanocomposite foams are mechanically stiff in the freezing direction and able to sustain a considerable load. Vertical burning tests (UL 94) show that nanocellulose-based composite foams with an optimized addition of GO (10 wt %), SEP (10 wt %) and boric acid (BA) (3 wt %) display excellent fire retardancy, where the flame does not self-propagate (as shown in Figure 14). For the optimal foam composition, the limiting oxygen index, which gives the oxygen concentration (in %) needed to keep a material burning, is as high as 34, that is 60% higher than the O_2 level in air (21%). In addition, it was found that CNF-GO-BA-SEP nanocomposite foams lie on the border between only smoldering and ignition when exposed to a defined heat flux of 35 $kW{\cdot}m^{-2}$, in contrast even to halogenated polyurethane foams or layer-by-layer modified polymer foams, which always ignite under similar conditions. The results provide substantial motivation to continue the development of high-performance thermal insulating materials based on renewable or widely abundant resources for the improvement of energy efficiency and reduction of the environmental impact of buildings.

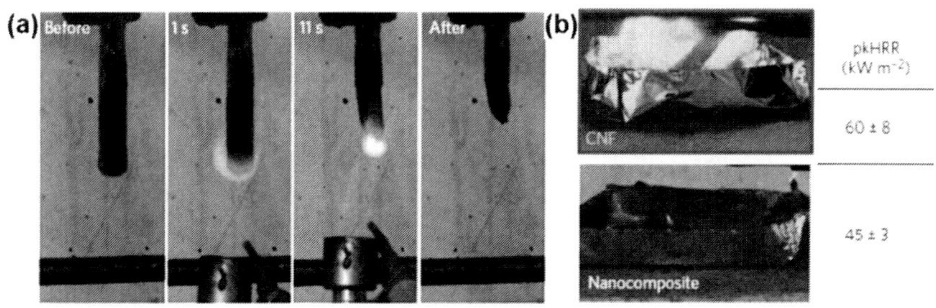

Figure 14. (**a**) Vertical burning test of a nanocomposite foam containing 77% CNF, 10% GO, 10% SEP and 3% BA (in wt %). The panel shows the foam before the test, after 11 s of application of a methane flame, and the foam after the test, showing high fire retardancy. (**b**) Photographs of CNF and CNF-GO-BA-SEP nanocomposite foams during the cone calorimetry test together with the corresponding peak of heat release rates (pkHRR). The test reveals high combustion resistance for the nanocomposite foam at the limit of non-ignitability, while CNF foams are entirely combusted. Reproduced with permission of Ref. [35] (Copyright Nature Publishing Group 2014).

PI aerogels not only have the typical characteristics of ordinary polymer aerogels but also show good heat resistance. As a result, PI could be used to strengthen clay-based aerogels while retaining the good thermal stability of these materials. However, almost all PIs cannot be dissolved in water, which restricts the direct introduction of PI into clay aerogels. In order to solve this problem, Wu *et al.*[40] prepared PAA ammonium salt to be used as a water-soluble precursor to produce PI/MMT hybrid aerogels. The resulted PI/MMT aerogels show a "layer-strut" bracing structure, which can improve the mechanical performance and heat resistance of aerogels. Besides, the onset decomposition temperature of the aerogels reached up to 410 °C. The combination of PI and MMT raises the possibility that these modified hybrid aerogels could be used as high-temperature insulation and flame-retarded materials.

In addition to polymer aerogels, carbon aerogels with hydrophobic surfaces and graphite skeletons also exhibit excellent fire resistant properties. Nitrogen-rich carbon (NRC) aerogels from alkaline lignin exhibit highly hydrophobic surfaces and excellent fire resistant property, making the NRC aerogels can be readily cleaned for reuse by direct combustion in air [201]. The excellent fire resistant properties are attributed to their graphite skeletons generated during the pyrolysis process. In addition, heteroatom doping can also attribute to the enhancement of fire resistant performance. Sugar-based molecules and polysaccharide biomass can be turned into porous functional carbonaceous products at comparably low temperatures of 400 °C under a nitrogen atmosphere in the presence of an ionic liquid (IL) or a poly(ionic liquid) (PIL) [217]. The IL and PIL act as “activation agents” with own structural contribution, and effectively promote the conversion and pore generation in the biomaterials even at a rather low doping ratio (7 wt %). Moreover, the nitrogen atoms incorporated in the final products from the IL/PIL precursors further improve the oxidation stability in the fire-retardant tests. The fire resistance of carbon aerogels from cotton with or without PIL were tested by wetting the foam with 500 wt % of ethanol and burning it in air (Figure 15a). As observed, the carbon aerogels derived from pure cotton already collapsed after the 1st cycle, while the PIL/cotton carbon remained intact. The mass *vs.* firing cycle plot (Figure 15b) indicates that the 1st firing cycle burnt away 80.3 wt % of the cotton-based carbon product while only 4.1 wt % of mass loss occurred for the cotton/PIL carbon aerogel. Within three cycles, only the cotton/PIL carbon aerogel preserved its shape and 85 wt % of its mass. Under even harsh conditions, for examples, the aerogel maintained its original shape and size without quenching fire when repeatedly exposed to the flame of a butane/propane gas burner (Figure 15c). This illustrates that the material is even robust against co-firing, a really exceptional property for high surface area carbons.

4.6. Catalyst Carriers

In order to improve the performance (general physicochemical characteristics as well as reactivity for catalysis) of catalysts, the most important characteristics are high surface area and phase stability of catalyst nanoparticles. Therefore, carbon aerogels with high conductivity, high mesoporosity and environmental stability are promising materials for application in catalysis [94].

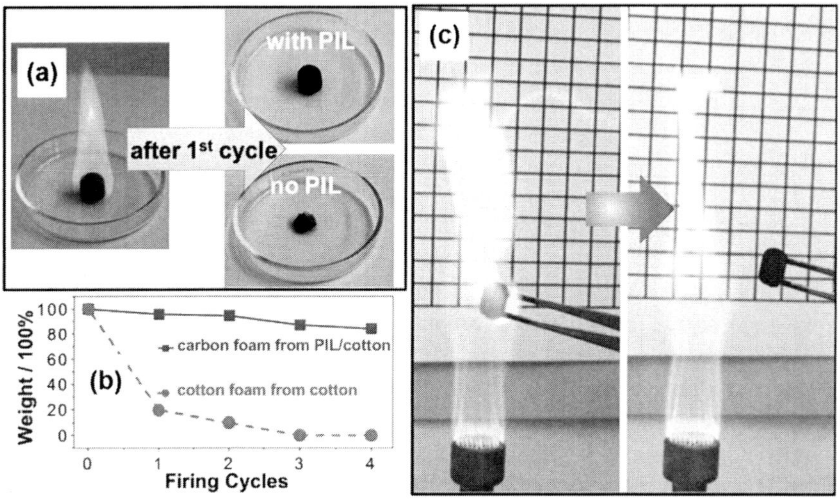

Figure 15. (**a**) Photographs of the 1st firing cycle test of the carbon foams (absorb 500 wt % of ethanol and burn in air): (**Left**) a burning ethanol-wetted foam and (**Right**) foams after the 1st firing cycle test. (**b**) Plot of the carbon foam mass *vs.* the firing cycle. (**c**) Photographs illustrating the fire-retardancy of a carbonaceous foam by repeatedly firing the sample using a butane/propane gas burner. Reproduced with permission of Ref. [217] (Copyright the Royal Society of Chemistry 2015).

The incorporation of metal or metal oxide species into the carbon framework can modify the structure, conductivity and catalytic activity of the carbon aerogel. Different metal-containing carbon aerogels have already been prepared and characterized, including Ce- and Zr-doped carbon aerogels and Cr-, Mo-, W-, Fe-, Ru-, Co-, Ni-, Pd-, Pt-, Cu- and Ag-containing carbon aerogels. The catalytic behavior of these metal-doped carbon aerogels with good activity and selectivity have been tested in a few reactions, including the isomerization of 1-butene, toluene combustion, synthesis of methyl-tert-butyl ether, the formation of nanofilaments from CO decomposition and cathode catalysts in proton exchange membrane fuel cell. For example, Mo-doped carbon aerogels were prepared in the polycondensation reaction of aqueous resorcinol and formaldehyde by adding Mo-salt at two different stages of synthesis [178]. The results show that the catalytic activity of carbon aerogels toward acetic acid hydroconversion reaction is enhanced by molybdenum. The more open pore structure, higher concentration and finer Mo distribution as well as its chemical form may all contribute to the greater conversion and higher value products obtained with the post-impregnated sample. γ-Fe_2O_3/α-Fe_2O_3/carbon aerogel is a magnetic photocatalyst, which shows strong Rhodamine B dye adsorption ability [218]. The removal capacity of RhB dyes of the as-prepared mesoporous aerogels is increased from 83.5% to 91% under visible light irradiation. The work shows that the as-prepared trifunctional mesoporous hybrid structures have great potential applications in wastewater treatments.

4.7. Hydrogen Storage

With the aggravation of the energy crisis, hydrogen energy has recently attracted increasing attentions. However, the main worldwide issue is to explore economic, safe, practical and convenient methods for hydrogen storage. The porous adsorbents for hydrogen storage are basically divided into four categories, *i.e.*, carbon-based porous materials, non-carbon nanotube materials, minerals and metal-organic porous materials. The basic requirements of hydrogen storage materials for porous solid adsorption are large surface area and high porosity. Therefore, carbon aerogels with large surface area, high porosity and structure stability have been studied for their potential applications in hydrogen storage [219]. However, the research of carbon aerogels as a hydrogen storage material is still in its initial stages. RF-based carbon aerogels are often used as the scaffold to store hydrogen. It is reported that carbon aerogels or aerogels catalyzed by acetic acid and activated by CO_2 show a higher hydrogen uptake of 4.65 wt % at 77 K and 3.9 MPa than that of uncatalyzed or unactivited aerogels.

Solid-state hydrogen storage materials such as MgH_2, $LiBH_4$ and $NaAlH_4$ offer high hydrogen storage capacities, which are required for a compact and efficient hydrogen storage system. However, until now, solid-state hydrogen storage has suffered from difficulties such as combination of high storage capacities in light element hydrides with hydrogen release. One approach is confinement of metal hydrides in inert mesoporous scaffolds with pore sizes in the range of 2–50 nm [220,221,222,223,224]. Therefore, carbon aerogels with nano-pore sizes are the right choice. Recently, Gosalawit-Utke and coworkers illustrated that confinement of MgH_2, $LiBH_4$ and $NaAlH_4$ in carbon based scaffolds can facilitate improved hydrogen release, uptake kinetics, reversible hydrogen storage capacity and more favorable thermodynamic properties [220,221,222]. To increase their surface area, carbon aerogels were first activated by CO_2 before the confinement of $NaAlH_4$ [224]. It was found that the hydrogen desorption kinetics decreased with increasing surface area and the hydrogen storage capacity is more stable and decreases less during continuous hydrogen release and uptake cycles. In fact, the available amount of hydrogen (2.7 wt % H_2) was more than doubled compared to the nanoconfinement in the non-activated carbon aerogel (1.3 wt % H_2). Hydrogen desorption kinetics and reversible hydrogen storage properties of $0.55LiBH_4$-$0.45Mg(BH_4)_2$ melt-infiltrated nanoporous carbon aerogels were also studied [223].

Nanoconfinement of $0.55LiBH_4$-$0.45Mg(BH_4)_2$ in high surface area carbon aerogel appears to facilitate hydrogen release illustrated by release of 13.3 wt % H_2 (93%) and only 8.4 wt % H_2 (58%) from bulk hydride in the first cycle under the same physical condition. Notably, nanoconfinement also appear to have a beneficial effect on hydrogen uptake, since 8.3 wt % H_2 (58%) is released from the high surface area scaffold and only 3.1 wt % H_2(22%) from the bulk sample during the fourth hydrogen release.

5. CONCLUSIONS AND OUTLOOK

5.1. Conclusions

Polymer aerogels and their derivative aerogels have been used to bridge the nanoscale porous structures of aerogel materials to practical macro-scale applications, due to their unique structures and properties such as 3D porous networks, high specific surface area, low density, low dielectric constant and high mechanical properties. In this review, the most recent progress in preparation, structures and properties of polymer- or carbon-based aerogels as well as their potential applications in various fields including energy storage, adsorption, thermal insulation and flame retardancy are introduced in details. Furthermore, with successful implementation, these highly porous structures of aerogels offer a larger range of applicability and, often, enhanced performance than their primitive organic precursors.

Regarding the preparation, generally, the most elementary step is the sol-gel process, *i.e.*, the transition of a system of colloidal particles in a solution into a disordered, branched, continuous network. Another important process is drying, which has been successfully developed, demonstrating the feasibility in preparing high quality polymer- and carbon-based aerogels. Among various drying methods, supercritical drying technique is the most effective method to obtain well-defined structures, while freeze-drying is a much easier, more economic and environmentally friendly way. Ambient pressure drying can be applied in large-scale industrial production, whereas it may result in environment pollution for the evaporation of solvents. For microwave drying and vacuum drying, they usually produce large pores or even induce the collapse of materials. Therefore, most of the drying methods reviewed in this paper require further improvement to produce aerogels with high porosity, low density and better mechanical properties through a low cost and environmentally friendly way. In order to achieve this goal, a better understanding of the preparation and drying mechanism of aerogels is necessary.

In the aspect of potential applications, the unique structures and properties of polymer/carbon aerogels and their hybrid aerogels make them promising candidates in many practical areas. Due to their lightweight, low thermal conductivity and low speed of sound, aerogels are hopeful thermal and acoustic insulation materials, which can be applied in building construction, vehicles, automobiles or space suit. The high surface area, high porosity and low density derived from the 3D porous network of aerogels ensure their practical applications in the fields of catalysis, sensors, adsorption and fuel storage. More importantly, unlike traditional silica aerogels, polymer-based aerogels with much enhanced mechanical properties can meet the requirements of industry and even severe demand in aerospace. Meanwhile, carbon aerogels are electrically conductive, which can be further applied in the fields of energy storage such as supercapacitors and lithium ion batteries. Finally, the feasibility of fabricating polymer-based hybrid aerogels by incorporating nanofillers can endow the aerogels with multi-functions. Therefore, polymer- or carbon-based aerogels and their derived hybrid aerogels with diversified structures and properties significantly extend the practical applications of aerogel-based materials.

5.2. Challenges and Outlook

Although new kinds of aerogel are continuously added to the aerogel family, the species of aerogels are still limited. Therefore, creating novel aerogels, either single-component or hybrid, is urgently needed. The studies on the single-component aerogel are fundamental, scientific and potentially applicable, while the studies on the hybrid aerogels are practical, technical and direct applicable. To create a single-component aerogel with the novel composition is relatively difficult but fundamentally valuable. A series of new aerogels (e.g., chalcogenide aerogel, CNT aerogel, graphene aerogel, diamond aerogel, silicon imidonitride aerogel, *etc.*) were successively invented in the past few decades. The next booming class may be carbide aerogel or single-element (mainly metal) aerogels. However, single-component aerogels are restricted in applications by their unicity. Hence, from a more practical view, hybrid aerogels with practical and smart functions should be designed and prepared. Combining polymer and carbon aerogels with nanofillers (e.g., graphene, carbon nanotube, nanofiber and nanoclays) is a promising strategy for producing aerogels with excellent properties and multi-functions.

Compared with the other applications, industrial applications can directly benefit human beings. However, due to the bad formability or mechanical properties, most of the aerogels are difficult to use in industry at the present stage. Besides, low yield and high cost also limit the commercialization of the aerogels. Thus, aerogel research is still developing and has a long way to go. The preparation of novel single-component aerogels, material design of hybrid aerogels and industrial application need to be further studied and investigated so as to give this state of matter a bright future.

ACKNOWLEDGMENTS

The authors are grateful for financial support from the National Natural Science Foundation of China (51125011, 51433001) as well as from Saint-Gobain Research (Shanghai) Co., Ltd.

AUTHOR CONTRIBUTIONS

Tianxi Liu and Wei Fan conceived the idea for this paper. Lizeng Zuo and Youfang Zhang were instrumental in collecting and putting all the literature together in the form of this review. Lizeng Zuo and Wei Fan contributed by guiding and structuring the complete document into a comprehensive review paper and checking for the language coherency. Longsheng Zhang and Yue-E Miao contributed by checking the language and helping in editing the final document.

CONFLICTS OF INTEREST

The authors declare no conflict of interest.

REFERENCES

1. Du, A.; Zhou, B.; Zhang, Z.H.; Shen, J. A special material or a new state of matter: A review and reconsideration of the aerogel. Materials 2013, 6, 941–968.

2. Burger, T.; Fricke, J. Aerogels: Production, modification and applications. Phys. Chem. Chem. Phys. 1998, 102, 1523–1528.

3. Fricke, J.; Emmerling, A. Aerogels—Preparation, properties, applications. Struct. Bond. 1992, 77, 37–87.

4. Pierre, A.C.; Pajonk, G.M. Chemistry of aerogels and their applications. Chem. Rev. 2002, 102, 4243–4265.

5. Cuce, E.; Cuce, P.M.; Wood, C.J.; Riffat, S.B. Toward aerogel based thermal superinsulation in buildings: A comprehensive review. Renew. Sust. Energy Rev. 2014, 34, 273–299.

6. Bheekhun, N.; Abu Talib, A.; Hassan, M.R. Aerogels in aerospace: An overview. Adv. Mater. Sci. Eng. 2013, 466, 577–585.

7. Koebel, M.; Rigacci, A.; Achard, P. Aerogel-based thermal superinsulation: An overview. J. Sol-Gel Sci. Technol. 2012, 63, 315–339.

8. Baetens, R.; Jelle, B.P.; Gustavsen, A. Aerogel insulation for building applications: A state-of-the-art review. Energy Build.2011, 43, 761–769.

9. Hrubesh, L.W. Aerogel applications. J. Non-Cryst. Solids 1998, 225, 335–342.

10. Fricke, J.; Emmerling, A. Aerogels—Recent progress in production techniques and novel applications. J. Sol-Gel Sci. Technol.1998, 13, 299–303.

11. Kistler, S.S. Coherent expanded aerogels and jellies. Nature 1931, 127, 741–741.

12. Nicolaon, G.A.; Teichner, S.J. On a new process of preparation of silica xerogels and aerogels and their textural properties.Bull. Soc. Chim. Fr. 1968, 5, 1900–1906.

13. Russo, R.E.; Hunt, A.J. Comparison of ethyl versus methyl sol gels for silica aerogels using polar nephelometry. J. Non-Cryst. Solids 1986, 86, 219–230.

14. Tewari, P.H.; Hunt, A.J.; Lofftus, K.D. Ambient-temperature supercritical drying of transparent silica aerogels. Mater. Lett.1985, 3, 363–367.

15. Pekala, R.W. Organic aerogels from the polycondensation of resorcinol with formaldehyde. J. Mater. Sci. 1989, 24, 3221–3227.

16. Meador, M.A.B.; Wright, S.; Sandberg, A.; Nguyen, B.N.; Van Keuls, F.W.; Mueller, C.H.; Rodriguez-Solis, R.; Miranda, F.A.; Aal, N.A. Dielectric and other properties of polyimide aerogels containing fluorinated blocks. ACS Appl. Mater. Interfaces2014, 6, 6062–6068.

17. Chen, H.B.; Hollinger, E.; Wang, Y.Z.; Schiraldi, D.A. Facile fabrication of poly(vinyl alcohol) gels and derivative aerogels.Polymer 2014, 55, 380–384.

18. Yamashita, J.; Ojima, T.; Shioya, M.; Hatori, H.; Yamada, Y. Organic and carbon aerogels derived from poly(vinyl chloride).Carbon 2003, 41, 285–294.

19. Banerjee, S.; Das, R.K.; Terech, P.; de Geyer, A.; Aymonier, C.; Loppinet-Serani, A.; Raffy, G.; Maitra, U.; del Guerzo, A.; Desvergne, J. Hybrid organogels and aerogels from co-assembly of structurally different low molecular weight gelators. J. Mater. Chem. C 2013, 1, 3305–3316.

20. Terech, P.; Aymonier, C.; Loppinet-Serani, A.; Bhat, S.; Banerjee, S.; Das, R.; Maitra, U.; del Guerzo, A.; Desvergne, J. Structural relationships in 2,3-bis-n-decyloxyanthracene and 12-hydroxystearic acid molecular gels and aerogels processed in supercritical CO2. J. Phys. Chem. 2010, 114, 11409–11419.

21. Sangeetha, N.M.; Maitra, U. Supramolecular gels: Functions and uses. Chem. Soc. Rev. 2005, 34, 821–836.

22. Frackowiak, E.; Beguin, F. Carbon materials for the electrochemical storage of energy in capacitors. Carbon 2001, 39, 937–950.

23. Haghgoo, M.; Yousefi, A.A.; Mehr, M.J.Z.; Celzard, A.; Fierro, V.; Leonard, A.; Job, N. Characterization of multi-walled carbon nanotube dispersion in resorcinol-formaldehyde aerogels. Microporous Mesoporous Mater. 2014, 184, 97–104.

24. Guo, K.; Song, H.H.; Chen, X.H.; Du, X.; Zhong, L. Graphene oxide as an anti-shrinkage additive for resorcinol-formaldehyde composite aerogels. Phys. Chem. Chem. Phys. 2014, 16, 11603–11608.

25. Attia, S.M.; Sharshar, T.; Abd-Elwahed, A.R.; Tawfik, A. Study of transport properties and conduction mechanism of pure and composite resorcinol formaldehyde aerogel doped with Co-ferrite. Mater. Sci. Eng. B 2013, 178, 897–910.

26. Irin, F.; Das, S.; Atore, F.O.; Green, M.J. Ultralow percolation threshold in aerogel and cryogel templated composites.Langmuir 2013, 29, 11449–11456.

27. Oh, Y.; Bag, S.; Malliakas, C.D.; Kanatzidis, M.G. Selective surfaces: High-surface-area zinc tin sulfide chalcogels. Chem. Mater. 2011, 23, 2447–2456.

28. Shafaei-Fallah, M.; He, J.; Rothenberger, A.; Kanatzidis, M.G. Ion-exchangeable cobalt polysulfide chalcogel. J. Am. Chem. Soc. 2011, 133, 1200–1202.

29. Gill, S.K.; Brown, P.; Hope-Weeks, L.J. Gold modified cadmium sulfide aerogels. J. Sol-Gel Sci. Technol. 2011, 57, 68–75.

30. Bag, S.; Kanatzidis, M.G. Chalcogels: Porous metal-chalcogenide networks from main-group metal ions. Effect of surface polarizability on selectivity in gas separation. J. Am. Chem. Soc. 2010, 132, 14951–14959.

31. Yu, H.T.; Liu, Y.; Brock, S.L. Tuning the optical band gap of quantum dot assemblies by varying network density. ACS Nano2009, 3, 2000–2006.

32. Li, R.; Chen, C.B.; Li, J.; Xu, L.M.; Xiao, G.Y.; Yan, D.Y. A facile approach to superhydrophobic and superoleophilic graphene/polymer aerogels. J. Mater. Chem. A 2014, 2, 3057–3064.

33. Zheng, Q.F.; Cai, Z.Y.; Gong, S.Q. Green synthesis of polyvinyl alcohol (PVA)-cellulose nanofibril (CNF) hybrid aerogels and their use as superabsorbents. J. Mater. Chem. A 2014, 2, 3110–3118.

34. Donius, A.E.; Liu, A.; Berglund, L.A.; Wegst, U.G.K. Superior mechanical performance of highly porous, anisotropic nanocellulose-montmorillonite aerogels prepared by freeze casting. J. Mech. Behav. Biomed. Mater. 2014, 37, 88–99.

35. Wicklein, B.; Kocjan, A.; Salazar-Alvarez, G.; Carosio, F.; Camino, G.; Antonietti, M.; Bergström, L. Thermally insulating and fire-retardant lightweight anisotropic foams based on nanocellulose and graphene oxide. Nat. Nanotechnol. 2014, 10, 277–283.

36. Chen, H.B.; Wang, Y.Z.; Schiraldi, D.A. Preparation and flammability of poly(vinyl alcohol) composite aerogels. ACS Appl. Mater. Interfaces 2014, 6, 6790–6796.

37. Chen, H.B.; Liu, B.; Huang, W.; Wang, J.S.; Zeng, G.; Wu, W.H.; Schiraldi, D.A. Fabrication and properties of irradiation-cross-linked poly(vinyl alcohol)/clay aerogel composites. ACS Appl. Mater. Interfaces 2014, 6, 16227–16236.

38. Chen, H.B.; Chiou, B.S.; Wang, Y.Z.; Schiraldi, D.A. Biodegradable pectin/clay aerogels. ACS Appl. Mater. Interfaces 2013, 5, 1715–1721.

39. Zheng, Q.F.; Javadi, A.; Sabo, R.; Cai, Z.Y.; Gong, S.Q. Polyvinyl alcohol (PVA)-cellulose nanofibril (CNF)-multiwalled carbon nanotube (MWCNT) hybrid organic aerogels with superior mechanical properties. RSC Adv. 2013, 3, 20816–20823.

40. Wu, W.; Wang, K.; Zhan, M.S. Preparation and performance of polyimide-reinforced clay aerogel composites. Ind. Eng. Chem. Res. 2012, 51, 12821–12826.

41. Kabiri, S.; Tran, D.N.H.; Altalhi, T.; Losic, D. Outstanding adsorption performance of graphene-carbon nanotube aerogels for continuous oil removal. Carbon 2014, 80, 523–533.

42. Kolla, P.; Lai, C.L.; Mishra, S.; Fong, H.; Rhine, W.; Smirnova, A. CVD grown CNTs within iron modified and graphitized carbon aerogel as durable oxygen reduction catalysts in acidic medium. Carbon 2014, 79, 518–528.

43. Aliev, A.E.; Mayo, N.K.; Baughman, R.H.; Avirovik, D.; Priya, S.; Zarnetske, M.R.; Blottman, J.B. Thermal management of thermoacoustic sound projectors using a free-standing carbon nanotube aerogel sheet as a heat source. Nanotechnology 2014,25, 293–295.

44. Wang, C.H.; He, X.D.; Shang, Y.Y.; Peng, Q.Y.; Qin, Y.Y.; Shi, E.Z.; Yang, Y.B.; Wu, S.T.; Xu, W.J.; Du, S.Y.; et al. Multifunctional graphene sheet-nanoribbon hybrid aerogels. J. Mater. Chem. A 2014, 2, 14994–15000.

45. Kang, T.J.; Kim, T.; Jang, E.Y.; Im, H.; Lepro-Chavez, X.; Ovalle-Robles, R.; Oh, J.; Kozlov, M.E.; Baughman, R.H.; Lee, H.H.; et al. Nanotube aerogel sheet flutter for actuation, power generation, and infrasound detection. Sci. Rep. 2014, 4, 6105–6109.

46. Fan, Z.; Tng, D.Z.Y.; Lim, C.X.T.; Liu, P.; Nguyen, S.T.; Xiao, P.F.; Marconnet, A.; Lim, C.Y.H.; Duong, H.M. Thermal and electrical properties of graphene/carbon nanotube aerogels. Colloids Surf. A 2014, 445, 48–53.

47. Guo, K.; Hu, Z.J.; Song, H.H.; Du, X.; Zhong, L.; Chen, X.H. Low-density graphene/carbon composite aerogels prepared at ambient pressure with high mechanical strength and low thermal conductivity. RSC Adv. 2015, 5, 5197–5204.

48. Liang, J.F.; Cai, Z.; Li, L.D.; Guo, L.; Geng, J.X. Scalable and facile preparation of graphene aerogel for air purification. RSC Adv. 2014, 4, 4843–4847.

49. Chen, L.; Feng, M.; Zhan, H.B. Fundamental electrochemistry of three-dimensional graphene aerogels. RSC Adv. 2014, 4, 30689–30696.

50. Xu, Y.X.; Sheng, K.X.; Li, C.; Shi, G.Q. Self-assembled graphene hydrogel via a one-step hydrothermal process. ACS Nano2010, 4, 4324–4330.

51. Wang, J.; Ellsworth, M.W. Graphene aerogels. ECS Trans. 2009, 5, 241–247.

52. Chen, K.; Bao, Z.H.; Liu, D.; Zhu, X.R.; Zhang, Z.H.; Zhou, B. Confined synthesis and properties of porous silicon from silica aerogel templates by magnesiothermic reduction. Acta Phys. Chim. Sin. 2011, 27, 2719–2725.

53. Leventis, N.; Sadekar, A.; Chandrasekaran, N.; Sotiriou-Leventis, C. Click synthesis of monolithic silicon carbide aerogels from polyacrylonitrile-coated 3D silica networks. Chem. Mater. 2010, 22, 2790–2803.

54. Worsley, M.A.; Kuntz, J.D.; Cervantes, O.; Han, T.; Gash, A.E.; Satcher, J.H.; Baumann, T.F. Route to high surface area TiO2/C and TiCN/C composites. J. Mater. Chem. 2009, 19, 7146–7150.

55. Worsley, M.A.; Kuntz, J.D.; Pauzauskie, P.J.; Cervantes, O.; Zaug, J.M.; Gash, A.E.; Satcher, J.H.; Baumann, T.F. High surface area carbon nanotube-supported titanium carbonitride aerogels. J. Mater. Chem. 2009, 19, 5503–5506.

56. Davis, M.; Zhang, K.; Wang, S.R.; Hope-Weeks, L.J. Enhanced electrical conductivity in mesoporous 3D indium-tin oxide materials. J. Mater. Chem. 2012, 22, 20163–20165.

57. Chervin, C.N.; Clapsaddle, B.J.; Chiu, H.W.; Gash, A.E., Jr.; Satcher, J.H.; Kauzlarich, S.M. Role of cyclic ether and solvent in a non-alkoxide sol-gel synthesis of yttria-stabilized zirconia nanoparticles. Chem. Mater. 2006, 18, 4865–4874.

58. Chervin, C.N.; Clapsaddle, B.J.; Chiu, H.W.; Gash, A.E.; Satcher, J.H.; Kauzlarich, S.M. Aerogel synthesis of yttria-stabilized zirconia by a non-alkoxide sol-gel route. Chem. Mater. 2005, 17, 3345–3351.

59. Colvin, J.D.; Fournier, K.B.; May, M.J.; Scott, H.A. A computational study of X-ray emission from laser-irradiated Ge-doped foams. Phys. Plasmas 2010, 17, 131–144.

60. Wang, Y.H.; Wang, C.C.; Cheng, W.Y.; Lu, S.Y. Dispersing WO3 in carbon aerogel makes an outstanding supercapacitor electrode material. Carbon 2014, 69, 287–293.

61. Yin, L.W.; Zhang, Z.W.; Li, Z.Q.; Hao, F.B.; Li, Q.; Wang, C.X.; Fan, R.H.; Qi, Y.X. Spinel ZnMn2O4 nanocrystal-anchored 3D hierarchical carbon aerogel

hybrids as anode materials for lithium ion batteries. Adv. Funct. Mater. 2014, 24, 4176–4185.

62. Wu, Z.S.; Yang, S.B.; Sun, Y.; Parvez, K.; Feng, X.L.; Muellen, K. 3D nitrogen-doped graphene aerogel-supported Fe3O4nanoparticles as efficient eletrocatalysts for the oxygen reduction reaction. J. Am. Chem. Soc. 2012, 134, 9082–9085.

63. Rhine, W.; Wang, J.; Begag, R. Polyimide Aerogels, Carbon Aerogels, and Metal Carbide Aerogels and Methods of Making Same. US Pantent 20040132845, 8 July 2004.

64. Hench, L.L.; West, J.K. The sol-gel process. Chem. Rev. 1990, 90, 33–72.

65. Leventis, N.; Koebel, M.M. Aerogels Handbook, 1st ed.; Springer: New York, NY, USA, 2011.

66. Einarsrud, M.A.; Nilsen, E.; Rigacci, A.; Pajonk, G.M.; Buathier, S.; Valette, D.; Durant, M.; Chevalier, B.; Nitz, P.; Ehrburger-Dolle, F. Strengthening of silica gels and aerogels by washing and aging processes. J. Non-Cryst. Solids 2001, 285, 1–7.

67. Carrott, P.J.M.; Marques, L.M.; Carrott, M. Core-shell polymer aerogels prepared by co-polymerisation of 2,4-dihydroxybenzoic acid, resorcinol and formaldehyde. Microporous Mesoporous Mater. 2012, 158, 170–174.

68. Wu, D.C.; Fu, R.W.; Sun, Z.Q.; Yu, Z.Q. Low-density organic and carbon aerogels from the sol-gel polymerization of phenol with formaldehyde. J. Non-Cryst. Solids 2005, 351, 915–921.

69. Qin, G.T.; Wei, W.; Guo, S.C. Semi-continuous drying of RF gels with supercritical acetone. Carbon 2003, 41, 851–853.

70. Li, W.C.; Lu, A.H.; Guo, S.C. Characterization of the microstructures of organic and carbon aerogels based upon mixed cresol-formaldehyde. Carbon 2001, 39, 1989–1994.

71. Liang, C.H.; Sha, G.Y.; Guo, S.C. Resorcinol-formaldehyde aerogels prepared by supercritical acetone drying. J. Non-Cryst. Solids 2000, 271, 167–170.

72. Tamon, H.; Ishizaka, H.; Araki, T.; Okazaki, M. Control of mesoporous structure of organic and carbon aerogels. Carbon 1998,36, 1257–1262.

73. Bock, V.; Emmerling, A.; Fricke, J. Influence of monomer and catalyst concentration on RF and carbon aerogel structure. J. Non-Cryst. Solids 1998, 225, 69–73.

74. Bock, V.; Emmerling, A.; Saliger, R.; Fricke, J. Structural investigation of resorcinol formaldehyde and carbon aerogels using SAXS and BET. J. Porous Mater. 1997, 4, 287–294.

75. Liu, L.; Yang, J.; Meng, Q.H. The preparation and characterization graphene-cross-linked phenol-formaldehyde hybrid carbon xerogels. J. Sol-Gel Sci. Technol. 2013, 67, 304–311.

76. Reuss, M.; Ratke, L. RF-aerogels catalysed by ammonium carbonate. J. Sol-Gel Sci. Technol. 2010, 53, 85–92.

77. Zhu, Y.D.; Hu, H.Q.; Li, W.C.; Zhang, X.Y. Cresol-formaldehyde based carbon aerogel as electrode material for electrochemical capacitor. J. Power Sources 2006, 162, 738–742.

78. Bruck, S.; Ratke, L. RF-aerogels: A new binding material for foundry application. J. Sol-Gel Sci. Technol. 2003, 26, 663–666.

79. Brandt, R.; Petricevic, R.; Probstle, H.; Fricke, J. Acetic acid catalyzed carbon aerogels. J. Porous Mater. 2003, 10, 171–178.

80. Tonanon, N.; Wareenin, Y.; Siyasukh, A.; Tanthapanichakoon, W.; Nishihara, H.; Mukai, S.R.; Tamon, H. Preparation of resorcinol formaldehyde (RF) carbon gels: Use of ultrasonic irradiation followed by microwave drying. J. Non-Cryst. Solids2006, 352, 5683–5686.

81. Yang, J.; Li, S.K.; Yan, L.L.; Liu, J.X.; Wang, F.C. Compressive behaviors and morphological changes of resorcinol-formaldehyde aerogel at high strain rates. Microporous Mesoporous Mater. 2010, 133, 134–140.

82. Schwan, M.; Ratke, L. Flexibilisation of resorcinol-formaldehyde aerogels. J. Mater. Chem. A 2013, 1, 13462–13468.

83. Kawagishi, K.; Saito, H.; Furukawa, H. Superior nanoporous polyimides via supercritical CO2 drying of jungle-gym-type polyimide gels. Macromol. Papid Commun. 2007, 28, 96–100.

84. Matson, D.W.; Smith, R.D. Supercritical fluid technologies for ceramic-processing applications. J. Am. Ceram. Soc. 1989, 72, 871–881.

85. Shen, D.X.; Liu, F.G.; Yang, H.X.; Yang, S.Y. Intrinsically highly hydrophobic semi-alicyclic fluorinated polyimide aerogel with ultra low dielectric constants. Chem. Lett. 2013, 42, 1230–1232.

86. Kobayashi, Y.; Saito, T.; Isogai, A. Aerogels with 3D ordered nanofiber skeletons of liquid-crystalline nanocellulose derivatives as tough and transparent insulators. Angew. Chem. Int. Ed. 2014, 53, 10394–10397.

87. Rudaz, C.; Courson, R.; Bonnet, L.; Calas-Etienne, S.; Sallee, H.; Budtova, T. Aeropectin: Fully biomass-based mechanically strong and thermal superinsulating aerogel. Biomacromolecules 2014, 15, 2188–2195.

88. Pajonk, G.M.; Repellinlacroix, M.; Abouarnadasse, S.; Chaouki, J.; Klvana, D. From sol-gel to aerogels and cryogels. J. Non-Cryst. Solids 1990, 121, 66–67.

89. Kalinin, S.V.; Kheifets, L.I.; Mamchik, A.I.; Knot'Ko, A.G.; Vertegel, A.A. Influence of the drying technique on the structure of silica gels. J. Sol-Gel Sci. Technol. 1999, 15, 31–35.

90. Jin, H.; Nishiyama, Y.; Wada, M.; Kuga, S. Nanofibrillar cellulose aerogels. Colloids Surf. A 2004, 240, 63–67.

91. Li, G.Y.; Wang, Z.G. Microporous polyimides with uniform pores for adsorption and separation of CO2 gas and organic vapors.Macromolecules 2013, 46, 3058–3066.

92. Ling, Z.; Wang, G.; Dong, Q.; Qian, B.Q.; Zhang, M.D.; Li, C.P.; Qiu, J.S. An ionic liquid template approach to graphene-carbon xerogel composites for supercapacitors with enhanced performance. J. Mater. Chem. A 2014, 2, 14329–14333.

93. Al-Muhtaseb, S.A.; Ritter, J.A. Preparation and properties of resorcinol-formaldehyde organic and carbon gels. Adv. Mater.2003, 15, 101–114.

94. Moreno-Castilla, C.; Maldonado-Hodar, F.J. Carbon aerogels for catalysis applications: An overview. Carbon 2005, 43, 455–465.

95. Zhang, S.Q.; Wang, J.; Shen, J.; Deng, Z.S.; Lai, Z.Q.; Zhou, B.; Attia, S.M.; Chen, L.Y. The investigation of the adsorption character of carbon aerogels. Nanostruct. Mater. 1999, 11, 375–381.

96. Pekala, R.W.; Farmer, J.C.; Alviso, C.T.; Tran, T.D.; Mayer, S.T.; Miller, J.M.; Dunn, B. Carbon aerogels for electrochemical applications. J. Non-Cryst. Solids 1998, 225, 74–80.

97. Lin, C.; Ritter, J.A. Carbonization and activation of sol-gel derived carbon xerogels. Carbon 2000, 38, 849–861.

98. Lin, C.; Ritter, J.A. Effect of synthesis pH on the structure of carbon xerogels. Carbon 1997, 35, 1271–1278.

99. Meador, M.A.B.; Malow, E.J.; Silva, R.; Wright, S.; Quade, D.; Vivod, S.L.; Guo, H.Q.; Guo, J.; Cakmak, M. Mechanically strong, flexible polyimide aerogels cross-linked with aromatic triamine. ACS Appl. Mater. Interfaces 2012, 4, 536–544.

100. Kistler, S.S. Coherent expanded aerogels. J. Phys. Chem. 1932, 36, 52–64.

101. Cai, J.; Kimura, S.; Wada, M.; Kuga, S.; Zhang, L. Cellulose aerogels from aqueous alkali hydroxide-urea solution.ChemSusChem 2008, 1, 149–154.

102. Zhou, Z.H.; Zhang, X.X.; Lu, C.H.; Lan, L.D.; Yuan, G.P. Polyaniline-decorated cellulose aerogel nanocomposite with strong interfacial adhesion and enhanced photocatalytic activity. RSC Adv. 2014, 4, 8966–8972.

103. Deng, M.L.; Zhou, Q.; Du, A.K.; van Kasteren, J.; Wang, Y.Z. Preparation of nanoporous cellulose foams from cellulose-ionic liquid solutions. Mater. Lett. 2009, 63, 1851–1854.

104. Hoepfner, S.; Ratke, L.; Milow, B. Synthesis and characterisation of nanofibrillar cellulose aerogels. Cellulose 2008, 15, 121–129.

105. Li, J.; Lu, Y.; Yang, D.; Sun, Q.; Liu, Y.; Zhao, H. Lignocellulose aerogel from wood-ionic liquid solution (1-allyl-3-methylimidazolium chloride) under freezing and thawing conditions. Biomacromolecules 2011, 12, 1860–1867.

106. Sescousse, R.; Gavillon, R.; Budtova, T. Aerocellulose from cellulose-ionic liquid solutions: Preparation, properties and comparison with cellulose-NaOH and cellulose-NMMO routes. Carbohydr. Polym. 2011, 83, 1766–1774.

107. Gavillon, R.; Budtova, T. Aerocellulose: New highly porous cellulose prepared from cellulose-NaOH aqueous solutions.Biomacromolecules 2008, 9, 269–277.

108. Cai, H.L.; Sharma, S.; Liu, W.Y.; Mu, W.; Liu, W.; Zhang, X.D.; Deng, Y.L. Aerogel microspheres from natural cellulose nanofibrils and their application as cell culture scaffold. Biomacromolecules 2014, 15, 2540–2547.

109. Nguyen, S.T.; Feng, J.D.; Ng, S.K.; Wong, J.P.W.; Tan, V.B.C.; Duong, H.M. Advanced thermal insulation and absorption properties of recycled cellulose aerogels. Colloids Surf. A 2014, 445, 128–134.

110. Kow, K.W.; Yusoff, R.; Aziz, A.R.A.; Abdullah, E.C. From bamboo leaf to aerogel: Preparation of water glass as a precursor. J. Non-Cryst. Solids 2014, 386, 76–84.

111. Jiang, F.; Hsieh, Y.L. Amphiphilic superabsorbent cellulose nanofibril aerogels. J. Mater. Chem. A 2014, 2, 6337–6342.

112. Lin, J.Y.; Yu, L.B.; Tian, F.; Zhao, N.; Li, X.H.; Bian, F.G.; Wang, J. Cellulose nanofibrils aerogels generated from jute fibers.Carbohydr. Polym. 2014, 109, 35–43.

113. Si, Y.; Yu, J.Y.; Tang, X.M.; Ge, J.L.; Ding, B. Ultralight nanofibre-assembled cellular aerogels with superelasticity and multifunctionality. Nat. Commun. 2014, 5.

114. Garcia-Gonzalez, C.A.; Alnaief, M.; Smirnova, I. Polysaccharide-based aerogels—Promising biodegradable carriers for drug delivery systems. Carbohydr. Polym. 2011, 86, 1425–1438.

115. Quignard, F.; Valentin, R.; di Renzo, F. Aerogel materials from marine polysaccharides. New J. Chem. 2008, 32, 1300–1310.

116. Wierik, G.; Bergsma, J.; ArendsScholte, A.W.; Boersma, T.; Eissens, A.C.; Lerk, C.F. A new generation of starch products as excipient in pharmaceutical tablets. 1. Preparation and binding properties of high surface area potato starch products. Int. J. Pharm. 1996, 134, 27–36.

117. Glenn, G.M.; Irving, D.W. Sstarch-based microcellular foams. Cereal Chem. 1995, 72, 155–161.

118. Budarin, V.; Clark, J.H.; Deswarte, F.; Hardy, J.; Hunt, A.J.; Kerton, F.M. Delicious not siliceous: Expanded carbohydrates as renewable separation media for column chromatography. Chem. Commun. 2005, 23, 2903–2905.

119. White, R.J.; Budarin, V.L.; Clark, J.H. Tuneable mesoporous materials from alpha-D-polysaccharides. ChemSusChem 2008, 1, 408–411.

120. El Kadib, A.; Bousmina, M. Chitosan bio-based organic-inorganic hybrid aerogel microspheres. Chem. Eur. J. 2012, 18, 8264–8277.

121. Ricci, A.; Bernardi, L.; Gioia, C.; Vierucci, S.; Robitzer, M.; Quignard, F. Chitosan aerogel: A recyclable, heterogeneous organocatalyst for the asymmetric direct aldol reaction in water. Chem. Commun. 2010, 46, 2903–2905.

122. Lamarque, G.; Viton, C.; Domard, A. Comparative study of the first heterogeneous deacetylation of alpha- and beta-chitins in a multistep process. Biomacromolecules 2004, 5, 992–1001.

123. Chen, K.; Bao, Z.H.; Du, A.; Zhu, X.R.; Shen, J.; Wu, G.M.; Zhang, Z.H.; Zhou, B. One-pot synthesis, characterization and properties of acid-catalyzed resorcinol/formaldehyde cross-linked silica aerogels and their conversion to hierarchical porous carbon monoliths. J. Sol-Gel Sci. Technol. 2012, 62, 294–303.

124. Tian, H.Y.; Buckley, C.E.; Paskevicius, M.; Wang, S.B. Carbon aerogels from acetic acid catalysed resorcinol-furfural using supercritical drying for hydrogen storage. J. Supercrit. Fluids 2011, 55, 1115–1117.

125. Tian, H.Y.; Buckley, C.E.; Paskeuicius, M.; Sheppard, D.A. Acetic acid catalysed carbon xerogels derived from resorcinol-furfural for hydrogen storage. Int. J. Hydrog. Energy 2011, 36, 671–679.

126. Mayes, R.T.; Tsouris, C.; Kiggans, J.O.; Mahurin, S.M.; DePaoli, D.W.; Dai, S. Hierarchical ordered mesoporous carbon from phloroglucinol-glyoxal and its application in capacitive deionization of brackish water. J. Mater. Chem. 2010, 20, 8674–8678.

127. Conceicao, F.L.; Carrott, P.J.M.; Carrott, M. New carbon materials with high porosity in the 1–7 nm range obtained by chemical activation with phosphoric acid of resorcinol-formaldehyde aerogels. Carbon 2009, 47, 1874–1877.

128. Reuss, M.; Ratke, L. Subcritically dried RF-aerogels catalysed by hydrochloric acid. J. Sol-Gel Sci. Technol. 2008, 47, 74–80.

129. Mulik, S.; Sotiriou-Leventis, C.; Leventis, N. Time-efficient acid-catalyzed synthesis of resorcinol-formaldehyde aerogels.Chem. Mater. 2007, 19, 6138–6144.

130. Fairen-Jimenez, D.; Carrasco-Marin, F.; Moreno-Castilla, C. Porosity and surface area of monolithic carbon aerogels prepared using alkaline carbonates and organic acids as polymerization catalysts. Carbon 2006, 44, 2301–2307.

131. Wu, D.C.; Fu, R.W. Fabrication and physical properties of organic and carbon aerogel derived from phenol and furfural. J. Porous Mat. 2005, 12, 311–316.

132. Barbieri, O.; Ehrburger-Dolle, F.; Rieker, T.P.; Pajonk, G.M.; Pinto, N.; Rao, A.V. Small-angle X-ray scattering of a new series of organic aerogels. J. Non-Cryst. Solids 2001, 285, 109–115.

133. Wu, D.C.; Fu, R.M. Synthesis of organic and carbon aerogels from phenol-furfural by two-step polymerization. Microporous Mesoporous Mater. 2006, 96, 115–120.

134. Lorjai, P.; Wongkasemjit, S.; Chaisuwan, T.; Jamieson, A.M. Significant enhancement of thermal stability in the non-oxidative thermal degradation of bisphenol-A/aniline based polybenzoxazine aerogel. Polym. Degrad. Stabil. 2011, 96, 708–718.

135. Alhwaige, A.A.; Agag, T.; Ishida, H.; Qutubuddin, S. Biobased chitosan/polybenzoxazine cross-linked films: Preparation in aqueous media and synergistic improvements in thermal and mechanical properties. Biomacromolecules 2013, 14, 1806–1815.

136. Chaisuwan, T.; Komalwanich, T.; Luangsukrerk, S.; Wongkasemjit, S. Removal of heavy metals from model wastewater by using polybenzoxazine aerogel. Desalination 2010, 256, 108–114.

137. Thubsuang, U.; Ishida, H.; Wongkasemjit, S.; Chaisuwan, T. Self-formation of 3D interconnected macroporous carbon xerogels derived from polybenzoxazine by selective solvent during the sol-gel process. J. Mater. Sci. 2014, 49, 4946–4961.

138. Lorjai, P.; Chaisuwan, T.; Wongkasemjit, S. Porous structure of polybenzoxazine-based organic aerogel prepared by sol-gel process and their carbon aerogels. J. Sol-Gel Sci. Technol. 2009, 52, 56–64.

139. Mahadik-Khanolkar, S.; Donthula, S.; Sotiriou-Leventis, C.; Leventis, N. Polybenzoxazine aerogels. 1. High-yield room-temperature acid-catalyzed synthesis of robust monoliths, oxidative aromatization, and conversion to microporous carbons.Chem. Mater. 2014, 26, 1303–1317.

140. Gu, S.; Li, Z.; Miyoshi, T.; Jana, S.C. Polybenzoxazine aerogels with controllable pore structures. RSC Adv. 2015, 5, 26801–26805.

141. Kim, S.Y.; Noh, Y.J.; Lim, J.; You, N. Silica aerogel/polyimide composites with preserved aerogel pores using multi-step curing.Macromol. Res. 2014, 22, 108–111.

142. Meador, M.A.B.; Wright, S.; Sandberg, A.; Nguyen, B.N.; van Keuls, F.W.; Mueller, C.H.; Rodriguez-Solis, R.; Miranda, F.A. Low dielectric polyimide

aerogels as substrates for lightweight patch antennas. ACS Appl. Mater. Interfaces 2012, 4, 6346–6353.

143. Guo, H.Q.; Meador, M.A.B.; McCorkle, L.; Quade, D.J.; Guo, J.A.; Hamilton, B.; Cakmak, M.; Sprowl, G. Polyimide aerogels cross-linked through amine functionalized polyoligomeric silsesquioxane. ACS Appl. Mater. Interfaces 2011, 3, 546–552.

144. Chidambareswarapattar, C.; Larimore, Z.; Sotiriou-Leventis, C.; Mang, J.T.; Leventis, N. One-step room-temperature synthesis of fibrous polyimide aerogels from anhydrides and isocyanates and conversion to isomorphic carbons. J. Mater. Chem. 2010, 20, 9666–9678.

145. Simpson, J.O.; St Clair, A.K. Fundamental insight on developing low dielectric constant polyimides. Thin Solid Films 1997, 308, 480–485.

146. Leventis, N.; Sotiriou-Leventis, C.; Mohite, D.P.; Larimore, Z.J.; Mang, J.T.; Churu, G.; Lu, H.B. Polyimide aerogels by ring-opening metathesis polymerization (ROMP). Chem. Mater. 2011, 23, 2250–2261.

147. Geim, A.K.; Novoselov, K.S. The rise of graphene. Nat. Mater. 2007, 6, 183–191.

148. Novoselov, K.S.; Geim, A.K.; Morozov, S.V.; Jiang, D.; Zhang, Y.; Dubonos, S.V.; Grigorieva, I.V.; Firsov, A.A. Electric field effect in atomically thin carbon films. Science 2004, 306, 666–669.

149. Stankovich, S.; Dikin, D.A.; Piner, R.D.; Kohlhaas, K.A.; Kleinhammes, A.; Jia, Y.; Wu, Y.; Nguyen, S.T.; Ruoff, R.S. Synthesis of graphene-based nanosheets via chemical reduction of exfoliated graphite oxide. Carbon 2007, 45, 1558–1565.

150. Bai, H.; Li, C.; Wang, X.L.; Shi, G.Q. A pH-sensitive graphene oxide composite hydrogel. Chem. Commun. 2010, 46, 2376–2378.

151. Kim, J.E.; Lee, H.S. Oscillatory shear induced gelation of graphene-poly(vinyl alcohol) composite hydrogels and rheological premonitor of ultra-light aerogels. Polymer 2014, 55, 287–294.

152. Wang, P.P.; Zhao, J.; Xuan, R.F.; Wang, Y.; Zou, C.; Zhang, Z.Q.; Wan, Y.Z.; Xu, Y. Flexible and monolithic zinc oxide bionanocomposite foams by a bacterial cellulose mediated approach for antibacterial applications. Dalton Trans. 2014, 43, 6762–6768.

153. Biener, J.; Stadermann, M.; Suss, M.; Worsley, M.A.; Biener, M.M.; Rose, K.A.; Baumann, T.F. Advanced carbon aerogels for energy applications. Energy Environ. Sci. 2011, 4, 656–667.

154. White, R.J.; Brun, N.; Budarin, V.L.; Clark, J.H.; Titirici, M. Always look on the "light" side of life: Sustainable carbon aerogels. ChemSusChem 2014, 7, 670–689.

155. Sun, H.Y.; Xu, Z.; Gao, C. Multifunctional, ultra-flyweight, synergistically assembled carbon aerogels. Adv. Mater. 2013, 25, 2554–2560.

156. Li, Y.Q.; Samad, Y.A.; Polychronopoulou, K.; Alhassan, S.M.; Liao, K. Carbon aerogel from winter melon for highly efficient and recyclable oils and organic solvents absorption. ACS Sustain. Chem. Eng. 2014, 2, 1492–1497.

157. Ren, Y.M.; Zhang, J.M.; Xu, Q.; Chen, Z.M.; Yang, D.Y.; Wang, B.; Jiang, Z. Biomass-derived three-dimensional porous N-doped carbonaceous aerogel for efficient supercapacitor electrodes. RSC Adv. 2014, 4, 23412–23419.

158. Wu, X.L.; Wen, T.; Guo, H.L.; Yang, S.B.; Wang, X.K.; Xu, A.W. Biomass-derived sponge-like carbonaceous hydrogels and aerogels for supercapacitors. ACS Nano 2013, 7, 3589–3597.

159. Hao, P.; Zhao, Z.H.; Tian, J.; Li, H.D.; Sang, Y.H.; Yu, G.W.; Cai, H.Q.; Liu, H.; Wong, C.P.; Umar, A. Hierarchical porous carbon aerogel derived from bagasse for high performance supercapacitor electrode. Nanoscale 2014, 6, 12120–12129.

160. Bi, H.C.; Huang, X.; Wu, X.; Cao, X.H.; Tan, C.L.; Yin, Z.Y.; Lu, X.H.; Sun, L.T.; Zhang, H. Carbon microbelt aerogel prepared by waste paper: An efficient and recyclable sorbent for oils and organic solvents. Small 2014, 10, 3544–3550.

161. Wu, Z.Y.; Li, C.; Liang, H.W.; Chen, J.F.; Yu, S.H. Ultralight, flexible, and fire-resistant carbon nanofiber aerogels from bacterial cellulose. Angew. Chem. Int. Ed. 2013, 52, 2925–2929.

162. Chen, L.F.; Huang, Z.H.; Liang, H.W.; Yao, W.T.; Yu, Z.Y.; Yu, S.H. Flexible all-solid-state high-power supercapacitor fabricated with nitrogen-doped carbon nanofiber electrode material derived from bacterial cellulose. Energy Environ. Sci. 2013,6, 3331–3338.

163. Chen, L.F.; Huang, Z.H.; Liang, H.W.; Gao, H.L.; Yu, S.H. Three-dimensional heteroatom-doped carbon nanofiber networks derived from bacterial cellulose for supercapacitors. Adv. Funct. Mater. 2014, 24, 5104–5111.

164. Liang, H.W.; Wu, Z.Y.; Chen, L.F.; Li, C.; Yu, S.H. Bacterial cellulose derived nitrogen-doped carbon nanofiber aerogel: An efficient metal-free oxygen reduction electrocatalyst for zinc-air battery. Nano Energy 2015, 11, 366–376.

165. Long, C.L.; Qi, D.P.; Wei, T.; Yan, J.; Jiang, L.L.; Fan, Z.J. Nitrogen-doped carbon networks for high energy density supercapacitors derived from polyaniline coated bacterial cellulose. Adv. Funct. Mater. 2014, 24, 3953–3961.

166. Chen, L.F.; Huang, Z.H.; Liang, H.W.; Guan, Q.F.; Yu, S.H. Bacterial-cellulose-derived carbon nanofiber@MnO2 and nitrogen-doped carbon nanofiber electrode materials: An asymmetric supercapacitor with high energy and power density. Adv. Mater.2013, 25, 4746–4752.

167. Wei, H.; Xu, M.W.; Bao, S.J.; Yang, F.; Chai, H. Design and synthesis of carbonized polypyrrole-coated graphene aerogel acting as an efficient metal-free catalyst for oxygen reduction. RSC Adv. 2014, 4, 16979–16984.

168. Zhang, Z.H.; Wu, P.Y. A facile one-pot route towards three-dimensional graphene-based microporous N-doped carbon composites. RSC Adv. 2014, 4, 45619–45624.

169. Pham, V.H.; Nguyen-Phan, T.D.; Jang, J.; Vu, T.; Lee, Y.J.; Song, I.K.; Shin, E.W.; Chung, J.S. Nitrogen-doped mesoporous reduced graphene oxide for high-performance supercapacitors. RSC Adv. 2014, 4, 22455–22462.

170. Aliev, A.E.; Oh, J.Y.; Kozlov, M.E.; Kuznetsov, A.A.; Fang, S.L.; Fonseca, A.F.; Ovalle, R.; Lima, M.D.; Haque, M.H.; Gartstein, Y.N.; et al. Giant-stroke, superelastic carbon nanotube aerogel muscles. Science 2009, 323, 1575–1578.

171. Zou, J.H.; Liu, J.H.; Karakoti, A.S.; Kumar, A.; Joung, D.; Li, Q.A.; Khondaker, S.I.; Seal, S.; Zhai, L. Ultralight multiwalled carbon nanotube aerogel. ACS Nano 2010, 4, 7293–7302.

172. Worsley, M.A.; Pauzauskie, P.J.; Olson, T.Y.; Biener, J.; Satcher, J.H.; Baumann, T.F. Synthesis of graphene aerogel with high electrical conductivity. J. Am. Chem. Soc. 2010, 132, 14067–14069.

173. Cong, H.P.; Ren, X.C.; Wang, P.; Yu, S.H. Macroscopic multifunctional graphene-based hydrogels and aerogels by a metal ion induced self-assembly process. ACS Nano 2012, 6, 2693–2703.

174. Xu, Y.X.; Wu, Q.; Sun, Y.Q.; Bai, H.; Shi, G.Q. Three-dimensional self-assembly of graphene oxide and DNA into multifunctional hydrogels. ACS Nano 2010, 4, 7358–7362.

175. Huang, C.C.; Bai, H.; Li, C.; Shi, G.Q. A graphene oxide/hemoglobin composite hydrogel for enzymatic catalysis in organic solvents. Chem. Commun. 2011, 47, 4962–4964.

176. Bai, H.; Li, C.; Wang, X.L.; Shi, G.Q. On the gelation of graphene oxide. J. Phys. Chem. C 2011, 115, 5545–5551.

177. Xu, Z.; Zhang, Y.; Li, P.G.; Gao, C. Strong, conductive, lightweight, neat graphene aerogel fibers with aligned pores. ACS Nano2012, 6, 7103–7113.

178. Nagy, B.; Abraham, D.; Dobos, G.; Madarasz, J.; Onyestyak, G.; Safran, G.; Geissler, E.; Laszlo, K. Molybdenum doped carbon aerogels with catalytic potential. Carbon 2014, 66, 210–218.

179. Gao, W.; Miao, Y.E.; Zhang, C.; Yang, Z.; Liu, Z.Y.; Tjiu, W.W.; Liu, T.X. Ni-doped graphene/carbon cryogels and their applications as versatile sorbents for water purification. ACS Appl. Mater. Interfaces 2013, 5, 7584–7591.

180. Sillars, F.B.; Fletcher, S.I.; Mirzaeian, M.; Hall, P.J. Effect of activated carbon xerogel pore size on the capacitance performance of ionic liquid electrolytes. Energy Environ. Sci. 2011, 4, 695–706.

181. Ren, Y.M.; Xu, Q.; Zhang, J.M.; Yang, H.X.; Wang, B.; Yang, D.Y.; Hu, J.H.; Liu, Z.M. Functionalization of biomass carbonaceous aerogels: Selective preparation of MnO2@CA composites for supercapacitors. ACS Appl. Mater. Interfaces 2014,6, 9689–9697.

182. Zhang, H.; Liu, D.; Qian, X.Z.; Zhao, C.J.; Xu, Y.L. A novel nano structured LiFePO4/C composite as cathode for Li-ion batteries. J. Power Sources 2014, 249, 431–434.

183. Zhang, L.; Shi, G.Q. Preparation of highly conductive graphene hydrogels for fabricating supercapacitors with high rate capability. J. Phys. Chem. C 2011, 115, 17206–17212.

184. Worsley, M.A.; Olson, T.Y.; Lee, J.R.I.; Willey, T.M.; Nielsen, M.H.; Roberts, S.K.; Pauzauskie, P.J.; Biener, J., Jr.; Satcher, J.H.; Baumann, T.F. High surface area, sp2-cross-linked three-dimensional graphene monoliths. J. Phys. Chem. Lett. 2011, 2, 921–925.

185. Zhang, Y.F.; Fan, W.; Huang, Y.P.; Zhang, C.; Liu, T.X. Graphene/carbon aerogels derived from graphene crosslinked polyimide as electrode materials for supercapacitors. RSC Adv. 2015, 5, 1301–1308.

186. Bi, H.C.; Yin, Z.Y.; Cao, X.H.; Xie, X.; Tan, C.L.; Huang, X.; Chen, B.; Chen, F.T.; Yang, Q.L.; Bu, X.Y.; et al. Carbon fiber aerogel made from raw cotton: A novel, efficient and recyclable sorbent for oils and organic solvents. Adv. Mater. 2013, 25, 5916–5921.

187. Ling, S.K.; Tian, H.Y.; Wang, S.B.; Rufford, T.; Zhu, Z.H.; Buckley, C.E. KOH catalysed preparation of activated carbon aerogels for dye adsorption. J. Colloid Interface Sci. 2011, 357, 157–162.

188. Annunciado, T.R.; Sydenstricker, T.; Amico, S.C. Experimental investigation of various vegetable fibers as sorbent materials for oil spills. Mar. Pollut. Bull. 2005, 50, 1340–1346.

189. Yuan, J.K.; Liu, X.G.; Akbulut, O.; Hu, J.Q.; Suib, S.L.; Kong, J.; Stellacci, F. Superwetting nanowire membranes for selective absorption. Nat. Nanotechnol. 2008, 3, 332–336.

190. Radetic, M.M.; Jocic, D.M.; Jovancic, P.M.; Petrovic, Z.L.; Thomas, H.F. Recycled wool-based nonwoven material as an oil sorbent. Environ. Sci. Technol. 2003, 37, 1008–1012.

191. Li, A.; Sun, H.X.; Tan, D.Z.; Fan, W.J.; Wen, S.H.; Qing, X.J.; Li, G.X.; Li, S.Y.; Deng, W.Q. Superhydrophobic conjugated microporous polymers for separation and adsorption. Energy Environ. Sci. 2011, 4, 2062–2065.

192. Toyoda, M.; Inagaki, M. Heavy oil sorption using exfoliated graphite—New application of exfoliated graphite to protect heavy oil pollution. Carbon 2000, 38, 199–210.

193. Lillo-Rodenas, M.A.; Cazorla-Amoros, D.; Linares-Solano, A.; Beguin, F.; Clinard, C.; Rouzaud, J.N. HRTEM study of activated carbons prepared by alkali hydroxide activation of anthracite. Carbon 2004, 42, 1305–1310.

194. Gui, X.C.; Wei, J.Q.; Wang, K.L.; Cao, A.Y.; Zhu, H.W.; Jia, Y.; Shu, Q.K.; Wu, D.H. Carbon nanotube sponges. Adv. Mater.2010, 22, 617–621.

195. Dong, X.C.; Chen, J.; Ma, Y.W.; Wang, J.; Chan-Park, M.B.; Liu, X.M.; Wang, L.H.; Huang, W.; Chen, P. Superhydrophobic and superoleophilic hybrid foam of graphene and carbon nanotube for selective removal of oils or organic solvents from the surface of water. Chem. Commun. 2012, 48, 10660–10662.

196.Hashim, D.P.; Narayanan, N.T.; Romo-Herrera, J.M.; Cullen, D.A.; Hahm, M.G.; Lezzi, P.; Suttle, J.R.; Kelkhoff, D.; Munoz-Sandoval, E.; Ganguli, S.; et al. Covalently bonded three-dimensional carbon nanotube solids via boron induced nanojunctions.Sci. Rep. 2012, 2, 172–179.

197.Duc Dung, N.; Tai, N.H.; Lee, S.B.; Kuo, W.S. Superhydrophobic and superoleophilic properties of graphene-based sponges fabricated using a facile dip coating method. Energy Environ. Sci. 2012, 5, 7908–7912.

198.Niu, Z.Q.; Chen, J.; Hng, H.H.; Ma, J.; Chen, X.D. A leavening strategy to prepare reduced graphene oxide foams. Adv. Mater.2012, 24, 4144–4150.

199.Zhao, Y.; Hu, C.G.; Hu, Y.; Cheng, H.H.; Shi, G.Q.; Qu, L.T. A versatile, ultralight, nitrogen-doped graphene framework.Angew. Chem. Int. Ed. 2012, 51, 11371–11375.

200.Liang, H.W.; Guan, Q.F.; Chen, L.F.; Zhu, Z.; Zhang, W.J.; Yu, S.H. Macroscopic-scale template synthesis of robust carbonaceous nanofiber hydrogels and aerogels and their applications. Angew. Chem. Int. Ed. 2012, 51, 5101–5105.

201.Yang, Y.; Tong, Z.; Ngai, T.; Wang, C.Y. Nitrogen-rich and fire-resistant carbon aerogels for the removal of oil contaminants from water. ACS Appl. Mater. Interfaces 2014, 6, 6351–6360.

202.Zhang, C.; Tjiu, W.W.; Liu, T.X. One-pot hydrothermal synthesis and reusable oil-adsorbing properties of porous carbonaceous monoliths using multi-walled carbon nanotubes as templates. RSC Adv. 2013, 3, 14938–14941.

203.Meena, A.K.; Mishra, G.K.; Rai, P.K.; Rajagopal, C.; Nagar, P.N. Removal of heavy metal ions from aqueous solutions using carbon aerogel as an adsorbent. J. Hazard. Mater. 2005, 122, 161–170.

204.Sui, Z.Y.; Cui, Y.; Zhu, J.H.; Han, B.H. Preparation of three-dimensional graphene oxide-polyethylenimine porous materials as dye and gas adsorbents. ACS Appl. Mater. Interfaces 2013, 5, 9172–9179.

205.Kong, Y.; Jiang, G.; Fan, M.; Shen, X.; Cui, S.; Russell, A.G. A new aerogel based CO2 adsorbent developed using a simple sol-gel method along with supercritical drying. Chem. Commun. 2014, 50, 12158–12161.

206.Masika, E.; Mokaya, R. High surface area metal salt templated carbon aerogels via a simple subcritical drying route: Preparation and CO2 uptake properties. RSC Adv. 2013, 3, 17677–17681.

207. Robertson, C.; Mokaya, R. Microporous activated carbon aerogels via a simple subcritical drying route for CO2 capture and hydrogen storage. Microporous Mesoporous Mat. 2013, 179, 151–156.

208. Sui, Z.; Meng, Q.; Li, J.; Zhu, J.; Cui, Y.; Han, B. High surface area porous carbons produced by steam activation of graphene aerogels. J. Mater. Chem. A 2014, 2, 9891–9898.

209. Alhwaige, A.A.; Agag, T.; Ishida, H.; Qutubuddin, S. Biobased chitosan hybrid aerogels with superior adsorption: Role of graphene oxide in CO2 capture. RSC Adv. 2013, 3, 16011–16020.

210. Jelle, B.P. Traditional, state-of-the-art and future thermal building insulation materials and solutions—Properties, requirements and possibilities. Energy Build. 2011, 43, 2549–2563.

211. Fernandez, J.E. Materials for aesthetic, energy-efficient, and self-diagnostic buildings. Science 2007, 315, 1807–1810.

212. Chen, W.S.; Li, Q.; Wang, Y.C.; Yi, X.; Zeng, J.; Yu, H.P.; Liu, Y.X.; Li, J. Comparative study of aerogels obtained from differently prepared nanocellulose fibers. ChemSusChem 2014, 7, 154–161.

213. Bendahou, D.; Bendahou, A.; Seantier, B.; Grohens, Y.; Kaddami, H. Nano-fibrillated cellulose-zeolites based new hybrid composites aerogels with super thermal insulating properties. Ind. Crop. Prod. 2015, 65, 374–382.

214. Leventis, N.; Chidambareswarapattar, C.; Mohite, D.P.; Larimore, Z.J.; Lu, H.B.; Sotiriou-Leventis, C. Multifunctional porous aramids (aerogels) by efficient reaction of carboxylic acids and isocyanates. J. Mater. Chem. 2011, 21, 11981–11986.

215. Chen, H.B.; Wang, Y.Z.; Sanchez-Soto, M.; Schiraldi, D.A. Low flammability, foam-like materials based on ammonium alginate and sodium montmorillonite clay. Polymer 2012, 53, 5825–5831.

216. Wang, Y.T.; Liao, S.F.; Shang, K.; Chen, M.J.; Huang, J.Q.; Wang, Y.Z.; Schiraldi, D.A. Efficient approach to improving the flame retardancy of poly(vinyl alcohol)/clay aerogels: Incorporating piperazine-modified ammonium polyphosphate. ACS Appl. Mater. Interfaces 2015, 7, 1780–1786.

217. Men, Y.J.; Siebenburger, M.; Qiu, X.L.; Antonietti, M.; Yuan, J.Y. Low fractions of ionic liquid or poly(ionic liquid) can activate polysaccharide

biomass into shaped, flexible and fire-retardant porous carbons. J. Mater. Chem. A 2013, 1, 11887–11893.

218. Lin, Y.F.; Chang, C.Y. Magnetic mesoporous iron oxide/carbon aerogel photocatalysts with adsorption ability for organic dye removal. RSC Adv. 2014, 4, 28628–28631.

219. Thomas, K.M. Hydrogen adsorption and storage on porous materials. Catal. Today 2007, 120, 389–398.

220. Gosalawit-Utke, R.; Niesen, T.K.; Pranzas, K.; Saldan, I.; Pistidda, C.; Karimi, F.; Laipple, D.; Skibsted, J.; Jensen, T.R.; Klassen, T.; et al. 2LiBH4-MgH2 in a resorcinol-furfural carbon aerogel scaffold for reversible hydrogen storage. J. Phys. Chem. C 2012, 116, 1526–1534.

221. Gosalawit-Utke, R.; Milanese, C.; Javadian, P.; Girella, A.; Laipple, D.; Puszkiel, J.; Cattaneo, A.S.; Ferrara, C.; Wittayakhun, J.; Skibsted, J.; et al. 2LiBH4–MgH2–0.13TiCl4 confined in nanoporous structure of carbon aerogel scaffold for reversible hydrogen storage. J. Alloy. Compd. 2014, 599, 78–86.

222. Gosalawit-Utke, R.; Meethom, S.; Pistidda, C.; Milanese, C.; Laipple, D.; Saisopa, T.; Marini, A.; Klassen, T.; Dornheim, M. Destabilization of LiBH4 by nanoconfinement in PMMA-co-BM polymer matrix for reversible hydrogen storage. Int. J. Hydrog. Energy 2014, 39, 5019–5029.

223. Javadian, P.; Jensen, T.R. Enhanced hydrogen reversibility of nanoconfined LiBH4-Mg(BH4)2. Int. J. Hydrog. Energy 2014, 39, 9871–9876.

224. Nielsen, T.K.; Javadian, P.; Polanski, M.; Besenbacher, F.; Bystrzycki, J.; Skibsted, J.; Jensen, T.R. Nanoconfined NaAlH4: Prolific effects from increased surface area and pore volume. Nanoscale 2014, 6, 599–607.

Chapter 2

Polymer/Carbon Nanotube Nanocomposites

Veena Choudhary[1] and Anju Gupta

[1] Centre for Polymer Science and EngineeringIndian Institute of Technology Delhi, India

1. INTRODUCTION

Polymer is a versatile material having many unique properties like low density, reasonable strength, flexibility, easy processibilty, etc. However, the mechanical properties of these materials are inadequate for many engineering applications. Hence, there is a continuous search towards new polymeric materials with improved properties. Initially, blending of different class of polymer was used to fabricate new materials with unique properties. However, blending lead to only marginal improvement in physical properties which were still inadequate for engineering applications. So to improve the strength and stiffness of polymer materials different kinds of organic and inorganic fillers were used. It was observed that strength and stiffness of long fibers reinforced thermosetting polymer is comparable to metals at a fraction of their weight. As a result of which these material were used in aircraft and in sport equipment. However, processing of these materials is very difficult; therefore small fiber or particle reinforced composites were developed. The common particle fillers used were silica, carbon black, metal particles etc. But significantly high filler loading was required to achieve desired mechanical property, which thus increased cost and made processibility difficult. So to achieve high mechanical properties at lower filler loading, nanofillers were used. The nanofiller reinforced polymer matrix is known as polymer nanocomposite.

Polymer nanocomposites are a new class of composite materials, which is receiving significant attention both in academia and industry. As nano fillers are only a few nanometers (~10,000 times finer than a human hair) in dimension, it offers ultra-large interfacial area per volume between the nano-element and polymer matrix. As a result, the nanofiller reinforced composites exhibit enhanced toughness without sacrificing stiffness or optical clarity. It also possesses greater thermal and oxidative stability, better barrier, mechanical properties as well as unique properties like self-extinguishing behavior. Compared to different range of nanofillers, carbon nanotubes (CNTs) have emerged as the most promising nanofiller for polymer composites due to their remarkable mechanical and electrical properties (Ishikawa, 2001; Kracke & Damaschke, 2000). Currently, one of the most intriguing applications of CNTs is the CNT/polymer nanocomposites (Cai, 2000; Fiege, 1999; Gomes, 1999;Hersam, 1998; Ruiz, 1998). For the last two decades, a lot of research work has been done on evaluating the potential of CNTs as filler for polymer nanocomposites. In the present chapter, we will briefly discuss on CNTs and their properties, different fabrication methods of polymer nanocomposites and their mechanical, electrical and thermal properties.

2. CARBON NANOTUBES (CNTS)

CNTs are long cylinders of covalently bonded carbon atoms which possess extraordinary electronic and mechanical properties. There are two basic types of CNTs: single-wall carbon nanotubes (SWCNTs) which are the fundamental cylindrical structure and multi-wall carbon nanotubes (MWCNTs) which are made of coaxial cylinders (Fig. 1), having interlayer spacing close to that of the interlayer distance in graphite (0.34 nm). These cylindrical structures are only few nanometre in diameter, but the cylinder can be tens of microns long, with most end capped with half of a fullerene molecule. It was first discovered by M. Endo in 1978, as part of his Ph.D. work at the University of Orleans in France, but real interest in CNTs started when Iijima (1991) first reported it in 1991. The field thrives after that and the first polymer composites using CNT as filler was reported by Ajayan et al (1994).

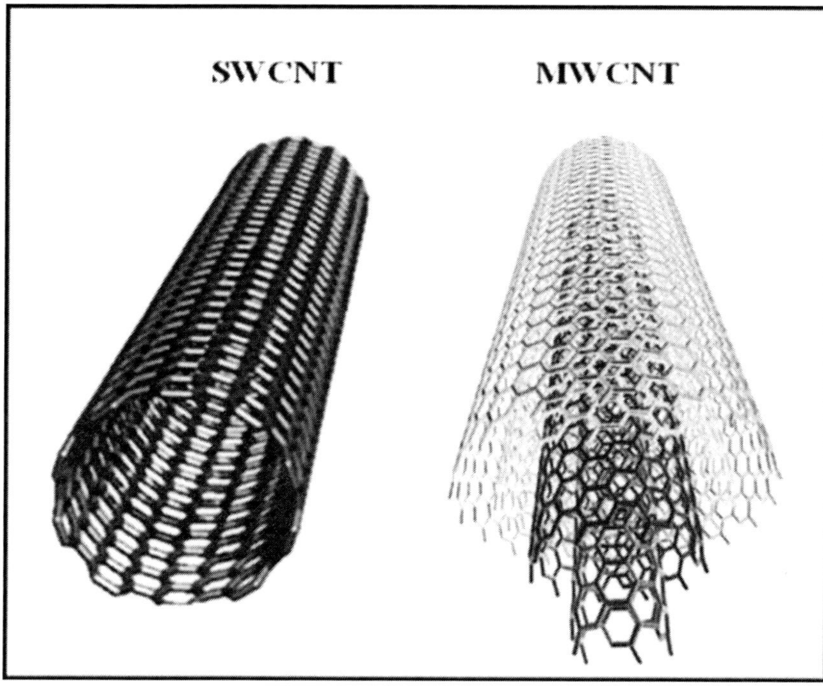

Figure 1. Schematic diagrams of single-wall carbon nanotube (SWCNT) and multi-wall carbon nanotube (MWCNT)

2.1. Synthesis Of Carbon Nanotubes

CNTs can be prepared using three methods which includes arc discharge, laser ablation and chemical vapor deposition (CVD). Most of these processes take place in vacuum or with process gases. CVD growth of CNTs can occur in vacuum or at atmospheric pressure. High quality and large quantities of nanotubes can be synthesized by these methods.

i. Arc discharge

The carbon arc discharge method, initially used for producing C_{60} fullerenes, is the most common and perhaps easiest way to produce CNTs. But this technique produces graphitic impurities such as carbon soot containing amorphous carbon, anions and fullerens. In this method an inert gas atmosphere is created in the reaction vessel by passing an inert gas at controlled pressure. Two graphitic rods constitute the electrodes, between which a potential difference is applied. As the rods are brought closer, a discharge occurs, resulting in formation of plasma (Fig. 2). The deposit, which contains CNTs, forms on the large negative electrode (cathode) while the smaller positive electrode (anode) is consumed. When a metal catalyst is used along with graphite, a hole is drilled in the carbon anode and it is filled with a mixture of metal and graphite powder. In this case, most nanotubes are found in soot deposited on the arc chamber wall.

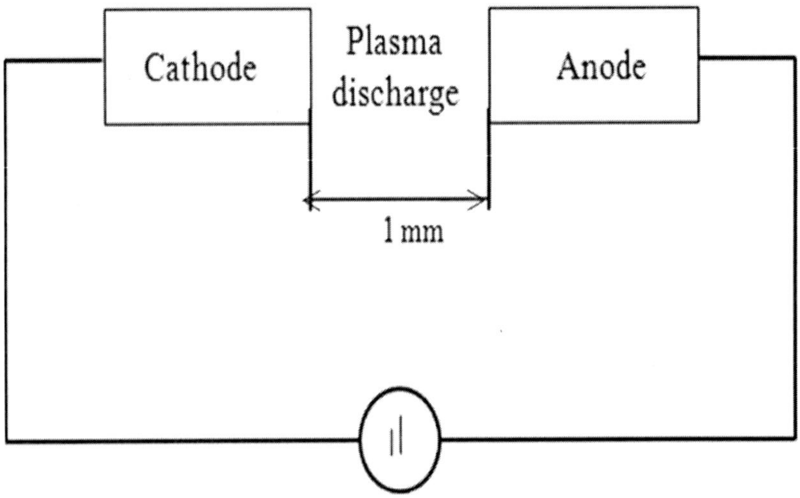

Figure 2. Plasma arc discharge setup

ii. Laser ablation

Laser ablation uses an intense laser pulse to vaporize a carbon target, which also contains small amount of metals such as nickel and cobalt and is placed in a tube furnace at 1200º C. As the target is ablated, inert gas is passed through the chamber carrying the grown nanotubes on a cold finger for collection (Fig. 3). This method mainly produces SWCNT in the form of ropes.

iii. Chemical vapor deposition

In this process a mixture of hydrocarbon, metal catalyst along with inert gas is introduced into the reaction chamber (Fig. 4a). During the reaction, nanotubes form on the substrate by the decomposition of hydrocarbon at temperatures 700–900º C and atmospheric pressure. The diameter of nanotubes that are to be grown are related to the size of the metal particles. This mechanism of CNT growth is still being studied. In Figure 4b two growth modes can be seen. The first 'tip growth mode' where, the catalyst particles can stay at the tips of the growing nanotube during the growth process and second 'base growth mode' where, catalyst particles remain at the nanotube base, depending on the adhesion between the catalyst particle and the substrate (Fig. 4b). This technique offers more control over the length and structure of the produced nanotubes compared to arc and laser methods. This process can also be scaled up to produce industrial quantities of CNTs. A number of reviews (Awasthi, 2005;Monthioux, 2002; Thostenson, 2001) are available which briefly discusses on these production techniques.

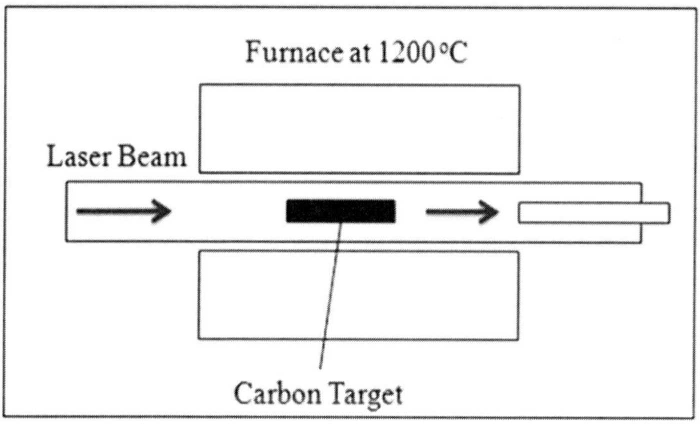

Figure 3. Laser ablation setup

2.2. Properties of carbon nanotubes

The determination of physical properties of CNTs is relatively more difficult compared to other fillers due to very small size of CNTs. However a number of experimental studies have been carried out on the direct determination of mechanical properties of individual CNTs (Yu et al., 2000a). The stiffness of CNTs was first determined by observing the amplitude of thermal vibrations in transmission electron microscopy (TEM) and the average stiffness values of 1.8 TPa and 1.25 TPa (Yu et al., 2000b) were reported for MWCNTs and SWCNTs, respectively. The in-situ tensile tests on individual MWCNTs and ropes of SWCNTs was performed by carrying out a stress-strain measurement using a "nanostressing stage" operating inside a scanning electron microscope (SEM). Experimental results showed that strength of outer shell of MWCNT ranged from 11 to 63 GPa at fracture strains of up to 12% and modulus values ranged from 270 to 950 GPa (Yu et al., 2000b). It was observed that strength of nanotubes depends on the number of defects, as well as interlayer interactions in MWCNTs and bundles of SWCNTs. Structural defects as well as bends or twists significantly affect mechanical strength of CNTs (Kane & Mele, 1997). Theoretical studies of the electronic properties of SWCNTs, suggest that nanotube shells can be either metallic or semiconducting, depend critically on helicity (Fig. 5) (Saito, 1992; Hamada, 1992; Mintmire, 1992).Tans et al. *(*1997) first, experimentally showed that there are indeed metallic and semiconducting SWCNTs, which verified the theoretical predictions. It was observed that due to poor control on synthesis, on average, approximately 1/3 of SWCNTs formed are metallic and 2/3 semiconductors (Odom et al., 1998). The room temperature conductivity of metallic SWCNT was found to be 10^5 to 10^6 S/m and for semiconducting tubes about 10 S/m. The conductivity of SWCNT is close to the in-plane conductivity of graphite [10^6 S/m (Charlier & Issi 1996)]. Conductivities of individual MWCNTs have been reported in the range of 10^7 to 10^8 S/m (Ebbesen et al., 1996), depending on the helicities of the outermost shells or the presence of defects (Dai et al., 1996). The axial thermal conductivity of individual, perfect CNTs is reported to be as high as 3300 W/m/K (Kim et al., 2001).

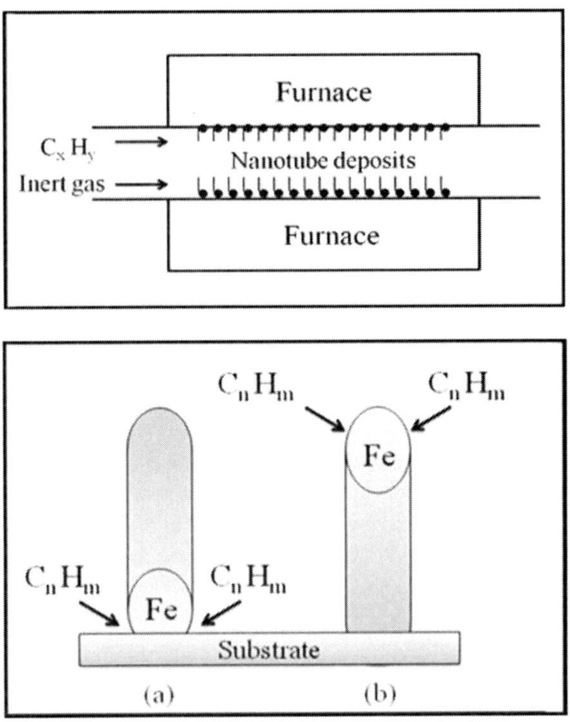

Figure 4. a. Chemical vapor deposition setup; b. Growth modes for CVD (a) base growth mode and (b) tip growth mode

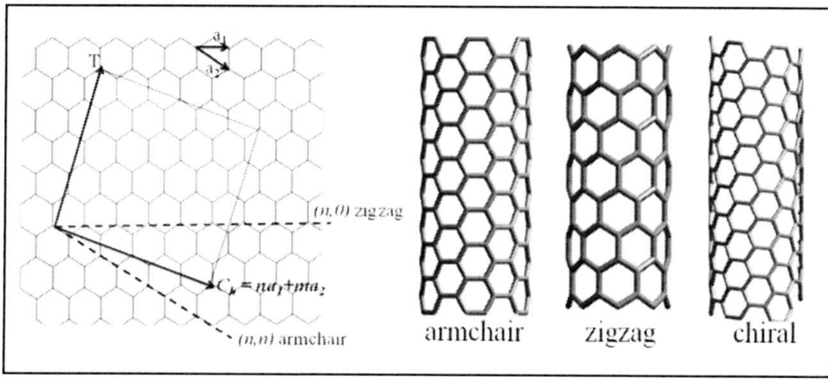

Figure 5. A sheet of graphene rolled to show formation of different types of single walled carbon nanotube

3. FUNCTIONALIZATION OF CARBON NANOTUBES

For a nanocomposite, a good dispersion of the filler within the host matrix is very important. At the same time it is also important to stabilize the dispersion to prevent re-aggregation of the filler. These tasks are particularly very challenging in case of nanofillers since the extremely large surface area lead to strong tendency to form agglomerates. CNTs are very well known to form aggregates during compounding and hence various techniques have been used to overcome this problem like use of sonication or mechanical mixing during the fabrication of nanocomposite which generally help in dispersing CNTs. But the most effective way to resolve this problem is surface functionalization of CNTs. Surface functionalization helps in stabilizing the dispersion, since it can prevent re-aggregation of nanotubes and also leads to coupling of CNT with polymeric matrix. Coupling between CNT and polymer matrix is also very important for efficient transfer of external stress to nanotube. In recent years, various methods have been developed for surface functionalization of CNT which includes functionalization of defect groups, covalent functionalization of sidewalls, non-covalent functionalization, e.g., formation of supramolecular adducts with surfactants or polymers (Fig. 6). Although surface functionalization leads to significant improvement in CNT dispersion and stress transfer but this method also causes deterioration of intrinsic properties of CNTs. Covalent functionalization often lead to tube fragmentation and the non-covalent functionalization results in poor exfoliation. Alteration of CNT properties lead to poor reinforcement and conductivity. Hence, it becomes obvious that dispersion and stabilization are not simple issues and compromises have to be made depending on the applications (Hirsch, 2002).

3.1. Covalent Functionalization Of Sidewall

Local strain in CNTs, which arises from either pyramidalization or misalignment of π-orbitals of the sp^2-hybridized carbon atoms, makes nanotubes more reactive than a flat graphene sheet, thereby paving the path to covalently attach chemical species to nanotubes (Banerjee et al., 2005). Covalent functionalization of CNTs can be achieved by introducing some functional groups on defect sites of CNTs (Fig. 7) by using oxidizing agents such as strong acids, which results in the formation of carboxylic or hydroxyl groups (-COOH, -OH) on the surface of nanotubes (Coleman, 2000, 2006, Singh, 2009). This type of functionalization is known as defect group functionalization. Such functionalization improves nanotube dispersion in solvents and polymers and imparts high stability in polar solvents. For example, Feng et al. (2008) reported that oxidation of MWCNTs with HNO_3/H_2O_2 and $HNO_3/$ H_2SO_4 introduces some carboxylic groups on CNTs, which enhanced their stability in water at room temperature for more than 100 days. As a result, the water-stable nanotubes can easily be embedded in water soluble polymers such as poly(vinyl alcohol) (PVA), giving CNT/polymer nanocomposites with homogeneous dispersion of CNTs (Zhao et al., 2008). Oxidized nanotubes also show excellent stability in other solvents such as caprolactam, which is used in the production of polyamide (PA6) (Gao et al., 2005). Studies on acid functionalization of CNTs have shown significant improvement in interfacial bonding between CNTs and polymer matrices, which have been shown to give a stronger nanotube–polymer interaction, leading to improved Young's modulus and mechanical strength (Gao, 2006; Sui, 2008; Yuen, 2008a, 2008b, 2008c; Luo, 2008;Rasheed, 2006a, 2006b; Sahoo, 2006; Wong, 2007).

Polymer molecules can also be grafted on the surface of CNTs in presence of these active functional groups (–COOH, $-NH_2$, -OH). Grafting of polymer chain on CNTs can be carried out either by 'grafting from' or 'grafting to' technique (Coleman, 2000, 2006; Liu, 2005). The grafting-from approach is based on the initial immobilization of initiators onto the nanotube surface, followed by in-situ polymerization with the formation of polymer molecules attached to CNTs. The advantage of this approach is that polymer-functionalized nanotubes with high grafting density can be prepared. However, this process needs a strict control of the amounts of initiator and substrate. The grafting from technique is widely used for the preparation of poly(methyl methacrylate) (PMMA) and related polymer grafted nanotubes. For example, Qin et al. (2004) reported the preparation of poly(n-butyl methacrylate) grafted SWCNTs by attaching n-butyl methacrylate (nBMA) to the ends and sidewalls of SWCNT via atom transfer radical polymerization (ATRP) using methyl 2-bromopropionate as the free radical initiator. A similar approach was reported by Hwang et al. (2004) for the synthesis of PMMA grafted MWCNTs by potassium persulfate initiated emulsion polymerization reactions and use of the grafted nanotubes as a reinforcement for commercial PMMA by solution casting.

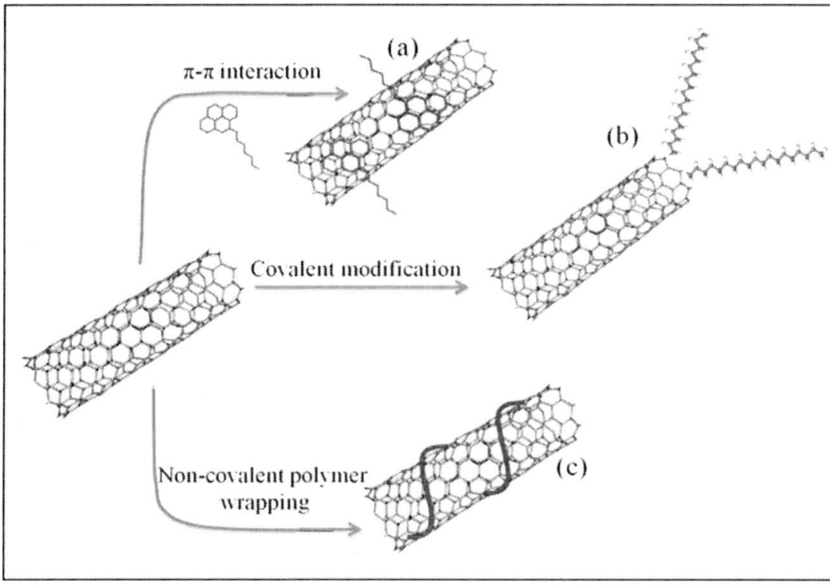

Figure 6. Possibilities for the functionalization of SWCNTs a) л-л interaction; b) defect group functionalization ; c) non-covalent functionalization with polymers

On the other hand, in case of grafting-to approach, attachment of already functionalized polymer molecules to functionalized nanotube surface takes place via appropriate chemical reactions. One of the first examples of 'grafting to' approach was published by Fu et al (2001). In this work carboxylic acid groups on the nanotube surface were converted to acyl chlorides by refluxing the samples with thionyl chloride. Then the acyl chloride functionalized CNTs were reacted with hydroxyl groups of dendritic poly(ethylene glycol) (PEG) polymers via the esterification reactions. The grafting-to method was applied for the preparation of epoxy-polyamidoamine–SWCNT composites (Sun et al., 2008). An advantage of this method is that, commercially available polymers containing reactive functional groups can be utilized.

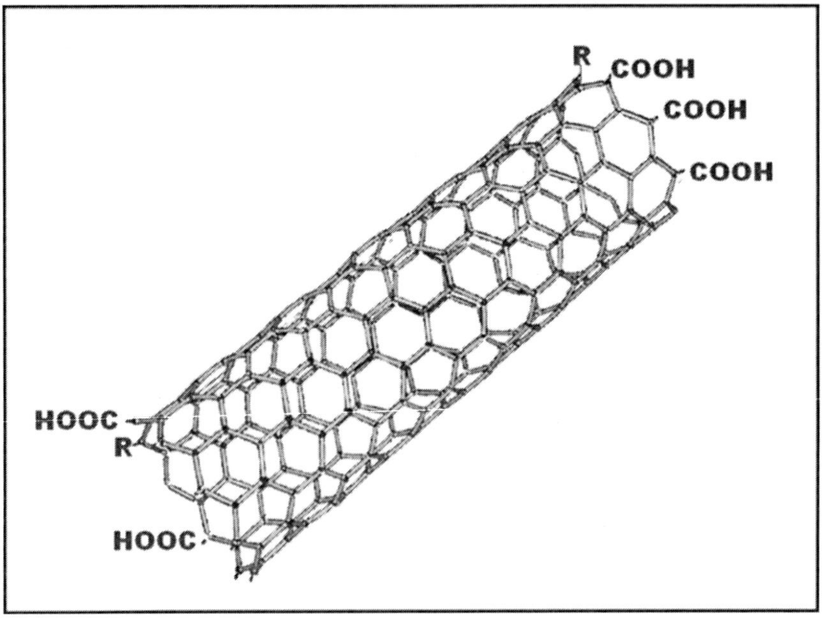

Figure 7. Covalent functionalization of carbon nanotubes on defects sites

3.2. Non-Covalent Functionalization With Surfactant Or Polymer

The non-covalent functionalization has unique ability of preserving the intrinsic properties of nanotube which is very important for its electrical and thermal conductivity. Various studies have shown that surfactant or wrapping with polymer can lead to individualization of nanotube in aqueous or organic solvent. (Moore, 2003; Matarredona, 2003; Vigolo, 2000; Regev, 2004; Grossiord, 2005; Curran, 1998; Coleman, 2000; O'Connell, 2001). Moore *et al.* studied various anionic, cationic and non-ionic surfactants and polymers to determine their ability to disperse nanotube in aqueous media. They reported that size of the hydrophilic group of surfactant or polymer play a key role in nanotube dispersion. It was also observed that surfactant alone is not capable of suspending nanotubes effectively and vigorous sonication is required (Matarredona et al., 2003). Polymers such as poly(*m*-phenylene-*co*- 2,5-dioctoxy-*p*-phenylenevinylene) (PmPV) can be used to wrap around nanotube in organic solvents such as $CHCl_3$ (Coleman et al., 2000). Polar side chain containing polymer, such as poly(vinyl pyrrolidone) [PVP] or poly(styrene sulfonate) [PSS] gave stable solutions of SWCNT/polymer complexes in water (O'Connell et al., 2001). The thermodynamic driving force for wrapping of polar polymer on nanotube is the need to avoid unfavorable interactions between the non-polar tube walls and the polar solvent (water). To disperse the nanotube in non-polar polymer matrices, such as polyolefins, polymerization-filling technique (PFT) was used. In this process nanotube is first dispersed with catalyst and co-catalyst and followed by polymerization (Dubois & Alexandre, 2006). This technique leads to individualization of nanotube and allows the homogeneous dispersion of nanotubes upon melt blending.

4. CNT/POLYMER NANOCOMPOSITES

In order to utilize CNTs and their extra ordinary properties in real-world applications, CNT/polymer nanocomposites were developed. Currently, polymer composite is the biggest application area for CNTs. These nanocomposites are being utilized in different fields including transportation, automotive, aerospace, defense, sporting goods, energy and infrastructure sectors. Such wide range applications of such materials are due to their high durability, high strength, light weight, design and process flexibility. CNT/polymer nanocomposites are also used as electrostatic discharge (ESD) and electromagnetic interference (EMI) shielding material because of high electrical conductivity of this material. However, the effective utilization of CNTs for fabricating nanocomposites strongly depends on the homogeneous dispersion of CNTs throughout the matrix without destroying their integrity. Furthermore, good interfacial bonding is also required to achieve significant load transfer across the CNT–matrix interface, a necessary condition for improving the mechanical properties of composites. So it is very important to achieve high degree of CNT dispersion during processing without affecting its property. In the following section we will discuss about the different processing techniques and properties of CNT/polymer nanocomposite.

4.1. Processing Of Cnt/Polymer Nanocomposites

From the above discussion, it is very clear that CNTs have strong tendency to form aggregates due to their large surface area. These aggregates persist unless high shear forces are applied e.g., vigorous mixing of the polymer. But such mixing often damages nanotube structures, compromising their properties. Therefore, the biggest challenge is to fully disperse individual nanotubes in the matrices to realize full potential of CNTs. Surface modification of CNTs have somewhat helped in dispersing CNT but long term stability is still a real challenge. Nevertheless, several approaches have been successfully adopted to obtain intimate mixing of nanotubes with polymer matrices, including dry powder mixing, solution blending, melt mixing, in-situ polymerization and surfactant-assisted mixing.

i. Solution processing

The most common method for preparing CNT/polymer nanocomposites involves mixing of CNT and polymer in a suitable solvent. The benefit of solution blending is rigorous mixing of CNTs with polymer in a solvent which facilitates nanotube de-aggregation and dispersion. This method consists of three steps: dispersion of nanotubes in a suitable solvent, mixing with the polymer (at room temperature or elevated temperature) and recovery of the nanocomposite by precipitating or casting a film. Both organic and aqueous medium have been used to produce CNT/polymer nanocomposites [Bandyopadhyaya, 2002; Pei, 2008; Wu, 2007; Cheng, 2008]. In this method dispersion of nanotube can be achieved by magnetic stirring, shear mixing, reflux or most commonly, ultrasonication. Sonication can be provided in two forms, mild sonication in a bath or high-power sonication. The use of high-power ultrasonication for a long period of time can shorten the nanotube length, i.e. reduces the aspect ratio, which is detrimental to the composite properties (Badaire et al., 2004).

To minimize this problem, surfactants have been used to disperse higher loadings of nanotubes (Islam, 2003; Barrau, 2003; Bryning, 2005a). Islam et al. (2003) reported that SWCNT (20 mg/mL) can be dispersed in water by using 1% sodium dodecylbenzene sulfonate as surfactant and low power, high-frequency (12 W, 55 kHz) sonication for 16-24 h. However, it has a major drawback of retaining surfactant in the nanocomposites which deteriorate the transport properties of nanocomposites. Bryning et al. (2005a) prepared SWCNT/epoxy nanocomposites and showed that the thermal conductivity of composite is much lower if surfactant is used for SWCNT dispersion.

In solvent blending, slow evaporation step often lead to CNT aggregation. To overcome this problem, CNT/polymer suspension can be kept on a rotating substrate [spin-casting (Lamy de la Chapelle et al., 1999)] or can be dropped on a hot substrate [drop-casting (Benoit, et al., 2001)] to expedite the evaporation step. Coagulation [developed by Du et al. (Du et al., 2003)] is another method, which involves pouring of a CNT/polymer suspension into an excess of non-solvent. This lead to entrapment of SWCNT by precipitating polymer chains which inturn prevents the SWCNT from bundling. The method is very successful in case of PMMA and polyethylene [PE] nanocomposites (Haggenmueller et al., 2006).

ii. Melt blending

While solution processing is a valuable technique for both nanotube dispersion and nanocomposite formation, it is less suitable for industrial scale processes. For industrial applications, melt processing is a preferred choice because of its low cost and simplicity to facilitate large scale production for commercial applications. In general, melt processing involves the melting of polymer pellets to form a viscous liquid and application of high shear force to disperse the nanofillers such as CNTs. Successful examples of melt blending include MWCNT/polycarbonate (Poetschke et al., 2003) MWCNT/ nylon-6, (Liu, 2004; Zhang, 2004) SWCNT/polypropylene, (Bhattacharya et al., 2003) and SWCNT/polyimide (Siochi et al., 2004) nanocomposites. Although melt blending is very simple, but high shear force and high temperature can deteriorate nanocomposite property, as high shear force which is required to achieve CNT dispersion can also lead to CNT fragmentation. So an optimum shear stress is required to achieve desired dispersion at lowest possible damage of CNTs.

The use of high temperature is also very critical, as high temperature enhances CNT dispersion by lowering the viscosity but too high temperature can degrade the polymer intrinsic properties. So optimization of temperature is also very important. To overcome these challenges many modification in melt compounding has been made like : Haggenmueller et al. (2000) combined the solution and melt blending by subjecting a solvent cast SWCNT/polymer film to several cycles of melt pressing. An approach developed by Jin et al (2002) introduces polymer-coated MWCNT (rather than pristine MWCNT) into the polymer melt to promote compatibilization. However optimization of processing conditions is an important issue, not just for different nanotube types, but for the whole range of polymer–nanotube combinations (Dubois & Alexandre, 2006).

iii. In-situ polymerization

In recent years, in-situ polymerization has been extensively explored for the preparation of polymer grafted nanotubes and processing of corresponding polymer composite materials. The main advantage of this method is that it enables grafting of polymer macromolecules onto the walls of CNTs. In addition, it is a very convenient processing technique, which allows the preparation of nanocomposites with high nanotube loading and very good miscibility with almost each polymer matrix. This technique is particularly important for the preparation of insoluble and thermally unstable polymers, which cannot be processed by solution or melt processing. Depending on required molecular weight and molecular weight distribution of polymers, chain transfer, radical, anionic, and ring-opening metathesis polymerizations can be used for in-situ polymerization processing. Initially, in-situ radical polymerization was applied for the synthesis of PMMA/MWCNT nanocomposites (Jia, 1999; Velasco-Santos, 2003; Putz, 2004). More recently (Wu, 2009) studied the mechanical and thermal properties of hydroxyl functionalized MWCNTs/acrylic acid grafted PTT nanocomposites and showed a significant enhancement in thermal and mechanical properties of PTT matrix due to the formation of ester bonds between –COOH groups of acrylic acid grafted PTT and –OH groups of MWCNTs.

In-situ polymerization was also very useful for the preparation of polyamide/CNT polymer nanocomposites. Park et al. (2002) also reported the synthesis of SWCNT reinforced polyimide nanocomposites by in-situ polymerization of diamine and dianhydride under sonication. Epoxy nanocomposites comprise the majority of reports using in-situ polymerization methods, (Schadler, 1998; Zhu, 2003, 2004; Gong, 2000; Ajayan, 2000; Moniruzzaman, 2006a) where the nanotubes are first dispersed in the resin followed by curing the resin with the hardener. Zhu et al. (2003) prepared epoxy nanocomposites by this technique using carboxylated end-cap SWCNT and an esterification reaction to produce a composite with improved tensile modulus. It is important to note that as polymerization progresses and the viscosity of the reaction medium increases, the extent of in-situ polymerization reactions might be limited.

In general, in -situ polymerization can be applied for the preparation of almost any polymer nanocomposites containing CNTs which can be non-covalently or covalently bound to polymer matrix. Non-covalent binding between polymer and nanotube involves physical adsorption and wrapping of polymer molecules through van der Waals and л–л interactions. The role of covalently functionalized and polymer grafted nanotubes will be considered in more detail below.

5. ALIGNMENT OF CARBON NANOTUBES IN NANOCOMPOSITES

The superior properties of CNTs offer exciting opportunities for new nanocomposites, but the important limitation for some potential applications of CNTs come from the fact that randomly oriented nanotubes embedded in polymer matrices have exhibited substantially lower electrical and thermal conductivities than expected (Fischer, 1997; Hone, 1999). Nanotube alignment can be obtained prior to composite fabrication or during composite fabrication or after composite fabrication by in-situ polymerization (Raravikar, 2005; Feng, 2003), mechanical stretching (Jin et al., 1998), melt fiber spinning (Haggenmueller, 2000, 2003), electrospinning (Gao, 2004; Hou, 2005; Ko, 2003) and application of magnetic or electric field (Ma, 2008; Camponeschi,

2007). Haggenmueller et al. (2000) have tried a combination of solvent casting and melt mixing methods to disperse single-walled CNTs in PMMA films and subsequently melt spun into fibers. However, only the melt mixing method was found to be successful in forming continuous fibers. Ma et al. (2008) studied alignment and dispersion of functionalized nanotube composites of PMMA induced by electric field and obtained significant enhancement in dispersion quality and alignment stability for oxidized MWCNTs as compared to pristine MWCNTs. Camponeschi et al. (2007) found that orientation and alignment of CNTs embedded in the epoxy under a magnetic field increased and showed improvement in mechanical properties of the resulting nanocomposites. Gao et al. (2004) prepared SWCNT/poly(vinyl pyrrolidone) fibres by electrospinning using electrostatic forces and found SWCNT exhibit good alignment and dispersion.Xie et al. (2005) showed that enhanced dispersion and alignment of CNTs in polymer matrices greatly improve mechanical, electrical, thermal, electrochemical, optical and super-hydrophobic properties of CNT/polymer nanocomposites. Safadi et al. (2002) prepared PS/MWCNT nanocomposite films by spin casting at high speed (2200 rpm) and found that MWCNTs were aligned in specific angles relative to the radial direction: 45º and 135º on average. The presence of ~2.5 vol.% MWCNTs doubles the tensile modulus and transforms the film from insulating to conducting. Thostenson & Chou (2002) found that tensile strength and modulus of melt drawn PS/MWCNT composite films increased by 137 and 49% respectively, compared to the undrawn PS film.

6. PROPERTIES OF CNT/POLYMER NANOCOMPOSITES

Incorporation of CNT in polymer matrix resulted in a significant change in mechanical, electrical and thermal properties of polymer matrices. Various factors that influence property modification are processing techniques, type of CNT, aspect ratio and CNT content. It is generally observed that a particular processing method which is good for one property may not be good for another. One such example is surface modification of CNT which generally enhances the mechanical properties but deteriorates the electronic properties. So it is very important to optimize the various conditions to obtain the nanocomposite with desired properties. A number of studies have been aimed at evaluating the mechanical, electrical and thermal properties of CNT/polymer nanocomposite under different conditions and filler loading.

6.1. Mechanical Properties Of Cnt/Polymer Nanocomposites

The excellent mechanical properties of CNTs, as discussed above, suggest that incorporation of very small amount of CNTs into a polymer matrix can lead to structural materials with significantly high modulus and strength. Significant advancement has been made in improving the mechanical properties of polymer matrix by mixing small fraction of CNTs. Qian et al. (2000) reported that adding 1 wt.% MWCNTs in the PS by solution-evaporation method, results in 36–42 and ~25% improvements in tensile modulus and tensile strength, respectively. Biercuk et al. (2002) have also reported increase of indentation resistance (Vickers hardness) by 3.5 times on adding 2 wt. % SWCNTs in epoxy resin.Cadek et al. (2002) also found significant improvement in the modulus and hardness (1.8 times and 1.6 times) on addition of 1 wt% MWCNTs in PVA matrix. Homogeneous dispersion and alignment of CNTs in polymer matrix had a significant effect on the properties of resulting composites. Velasco-Santos et al. (2003) reported that by enhancing the dispersion of CNT by using an in-situ polymerization, the storage modulus of PMMA/MWCNT nanocomposites at 1 wt.% of MWCNTs at 90º C increased by 1135%.

Although, addition of CNTs lead to enhancement of mechanical properties of the polymer matrix but the improvement is still well below the expected value. At current stage, the extraordinary properties of CNTs are still not fully utilized in polymer composites. Many research works have indicated that poor adhesion between the matrix and nanotube is the limiting factor in imparting the excellent mechanical properties of nanotubes in composites. As load transfer from matrix to CNTs play a key role in mechanical properties of the nanocomposite, good interfacial bonding is very important. Load transfer between the matrix and filler depends on the interfacial shear stress between the two (Schadler et al., 1998). A high interfacial shear stress will transfer the applied load to the filler over a short distance and a low interfacial shear stress will require a long distance. There are three main load transfer mechanisms operating between a matrix and filler.

a. Micro-mechanical interlocking

This is the amount of load transfer due to mechanical interlocking which is very poor in nanotube composites because of the atomically smooth surface of nanotubes. As CNTs has some surface defects like varying diameter and bends/twist due to non-hexagonal defects, along a CNT, mechanical interlocking do play a role in CNT–polymer adhesion.

b. Chemical bonding between filler and matrix

A chemical bond either ionic or covalent significantly improves the interfacial interaction between matrix and filler that enables a stress transfer.

c. Weak van der Waals bonding between filler and matrix

The van der Waals interaction arises from the molecular proximity and is the only mode of interaction between CNTs and the matrix in absence of chemical bonding.

Hence, formation of chemical bonding between CNT and polymer can significantly improve the mechanical properties of nanocomposites. Recently, Blake et al. (2004) developed butyl-lithium-functionalized MWCNTs which can be covalently bonded to chlorinated polypropylene (CPP). The CPP/MWCNT was then compounded with CPP/tetrahydrofuran (THF) solution to obtain CPP/MWCNT nanocomposites. They showed that on addition of 0.6 vol% MWCNT, the modulus increased by three times and both tensile strength and toughness (measured by the area under the stress–strain curve) increased by 3.8 times (from 13 to 49 MPa) and 4 times (from 27 to 108 J/g), respectively. Bal & Samal (2007) showed that the amine functionalized CNTs get completely dispersed in polymer matrix in comparison to unmodified CNTs. Telescopic pull-out was also observed in case of functionalized MWCNTs (Gojny et al., 2005). It was observed that although CNTs get pulled out from the matrix the outer wall still remained in the matrix. This is possible because only weak van der Waals forces are present between the various concentric tubes of the MWCNT where as the outer tube is covalently bonded to the matrix. Such a pull-out process suggests that efficient load transfer occurs from matrix to the outer tube, due to strong covalent bonding between epoxy matrix and CNT.

These observations suggest that the efficiency of property improvement depends on the type of CNT, processing techniques and the compatibility between CNT and host matrix. Although chemical functionalization of CNT can improve the compatibility between CNT and polymer which inturn improves the mechanical properties but it has a deteriorating effect on the other properties of nanocomposites such as electrical and thermal conductivity. However, the rapid growth of this field suggests the solution to these problems are not very far and in coming few years' desire of obtaining super strong polymer material will be realized.

6.2. Electrical properties of CNT/polymer nanocomposites

With exceptional mechanical properties, CNTs also possess very high intrinsic electrical conductivity. The electrical conductivity of individual CNTs ranged between 10^7 to 10^8 S/m that is comparable to metals (Ebbesen et al., 1996). Very high electrical conductivity of CNTs have helped to impart conductivity in highly insulating material like polymer by fabricating polymer nanocomposites. The enhancement in electrical conductivity of insulating polymer by several orders of magnitude has been achieved with a very small loading (0.021 wt %) of nanotubes in the polymer matrices, which helped in preserving other performance aspects of the polymers such as optical clarity, mechanical properties, low melt flow viscosities, etc. As a result, these conducting materials are in growing demand in different application area such as transparent conductive coatings, electrostatic dissipation, electrostatic painting and electromagnetic interference shielding applications.

The electrical conductivity of CNT reinforced polymer nanocomposites depends on many factors including type of CNTs, aspect ratio, surface functionalization and CNT content. The electrical conductivity of nanocomposite increases with increasing CNT loading till a critical filler concentration where a dramatic increase in conductivity is observed. This critical filler concentration is called electrical percolation threshold concentration. At percolation threshold concentration, filler forms a three-dimensional conductive network within the matrix, hence electron can tunnel from one filler to another, and in doing so it overcomes the high resistance offered by insulating polymer matrix. The percolation threshold is typically determined by plotting the electrical conductivity as a function of the reduced mass fraction of filler and fitting with a power law function (Fig. 8). As formation of percolating networks depends on both intrinsic conductivity and aspect ratio of the filler, the nanotube/polymer nanocomposites exhibit very low percolation threshold because of high conductivity and high aspect ratio of CNTs. Bryning et al. (2005b) studied the effect of aspect ratio on percolation threshold concentration by preparing SWCNT/epoxy nanocomposites with nanotubes from two different sources, HiPco and laser oven, having aspect ratios of ~150 and ~380 respectively. They reported a smaller percolation threshold with the higher aspect ratio nanotubes. Similar observation was also made by Bai & Allaoui (2003); they found more than 8-fold decrease in threshold concentration in MWCNT/epoxy composites when the MWCNT length was increased from 1 to 50 μm. In general, the minimum percolation threshold concentration for SWCNT/polymer nanocomposite is 0.0021wt % in epoxy matrix (Martin et al., 2004). For different polymer systems it ranged from 0.0021 to 15 wt% CNT loading (Bauhofer & Kovacs, 2009). These studies show that the percolation threshold concentration and nanocomposite conductivity also depends on polymer type, synthesis method, aspect ratio of CNTs, disentanglement of CNT agglomerates, uniform spatial distribution of individual CNTs and degree of alignment [Bryning, 2005b; Moniruzzaman, 2006b; Du, 2005].

Another factor which significantly affects the electrical conductivity of nanocomposite is the chemical functionalization of CNTs because it disrupts the extended π-conjugation of nanotubes and thereby reduces the electrical conductivity of isolated nanotubes. Sulong et al. (2009) showed that incorporation of acid and octadecylated functionalized MWCNT in epoxy matrix decreased the electrical conductivity of nanocomposites. Similar results have also been reported by Park et al (2009). Hence, it is important to optimize the modification condition or reagent to achieve minimum deterioration of electronic properties of CNTs.

Nevertheless, significant improvement in electrical conductivity of polymer on CNT addition lead to the development of CNT/polymer conductive nanocomposites for electronics, automotive and aerospace applications with uses such as electrostatic dissipation, electromagnetic interference (EMI) shielding, multilayer printed circuits, and conductive coatings (Baughman, 2002; Dresselhaus, 2004; Holt, 2006; Martel, 1998; Minoux, 2005;Zhang, 2005). The rapid development of electrical industry, demands fabrication of light weight and effective EMI shielding material for the protection of workspace and environment from radiation coming from computers and telecommunication equipment as well as for protection of sensitive circuits. Thus electrically conducting polymer nanocomposites have received much attention recently compared to conventional metal-based EMI shielding materials (Chung, 2001; Joo & Epstein, 1994;Kim, 2003; Yang, 2005a), because of their light weight, resistance to corrosion, flexibility and processing advantages. The use of CNT have significantly reduced the filler loading required to achieve desired EMI SE, thus reduced the cost and weight of the material (Chung, 2001; Micheli, 2009; Sundararaj & Al-Saleh, 2009; Yang, 2005b).

The effect of aspect ratio on EMI SE was shown by Sundararaj & Al-Saleh. They reported that SE of 1 mm thick shielding plate made of 7.5 vol% MWCNT/polypropylene (PP) nanocomposite was much higher (35 dB) than 7.5 vol% (HS-CB)/PP composite (18 dB) in the X-band frequency range. Yang et al. (2005b) studied the EMI shielding behavior of MWCNT/PS nanocomposites and achieved SE of ~20 dB at 7 wt % MWCNT loading. Although, lower value of SE of CNT composites have also been reported, e.g., Liu et al. (2007) achieved only ~17 dB at 20 wt% MWCNTs loading in polyurethane (PU) whereas Kim et al. (2004) reported ~27 dB SE at 40 wt% CNT loading for MWCNT/PMMA films. These observations suggest that SE of CNT filled polymer nanocomposite depends on many factors including fabrication techniques, purity of CNT, dispersion etc. Recently significant efforts have been made in understanding the EMI shielding mechanisms of polymer nanocomposite as it is very important for the best utilization of shielding capabilities of material and for designing nanocomposite at lowest possible filler loading and cost. Three types of EMI shielding mechanisms have been proposed, namely: reflection, absorption and multiple reflections [Sundararaj & Al-Saleh, 2009; Chung, 2001; Liu, 2007]. Some previous studies have shown that in MWCNT/polymer nanocomposites, SE is mainly absorption dominated where as SWCNT/polymer nanocomposites are mainly reflection dominated material. The application area of EMI shielding materials depends on the dominant shielding mechanism like absorption based EMI shielding materials are used in radar, microwave communication technology, stealth (self concealing) technology, microwave darkroom and anti-EMI coating application.

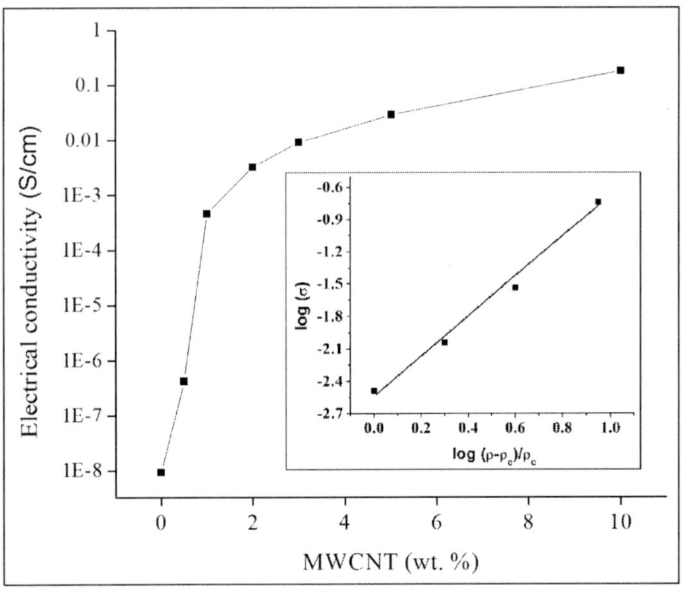

Figure 8. Plot of dc electrical conductivity (σ) vs weight fraction of MWCNT in PTT/MWCNT composites [inset shows the log-log plot of σ vs [(ρ- ρc)/ρc]

6.3. Thermal properties

Thermal properties of a composite are equally important as its mechanical and electrical properties because it gives more freedom in selecting processing conditions and also application area. It is observed that incorporation of CNTs in polymer matrices results in increase of glass transition, melting and thermal decomposition temperatures due to hindered chain and segmental mobility of the polymers. To improve the thermal endurance of CNT/polymer nanocomposites, surfactant as the wetting agent were also incorporated. Gong (2000) and Velasco-Santos et al. (2003) reported that addition of 1 wt.% CNTs with a surfactant in epoxy and PMMA matrix increased the glass transition temperature by ~25 and ~40º C respectively. Incorporation of CNTs in polymer matrices also enhances the rate of crystallization by acting as nucleating sites [Deshpande & Jape, 2009; Zhang, 2008] which in-turn reduces the processing time and enhances mechanical strength. There are reports according to which addition of CNTs in polymer matrix can hinder the crystallization of polymer matrices (Jin, 2007; Diez-Pascual, 2009). Our recent study on MWCNT/PTT nanocomposite suggested that presence of MWCNTs in PTT matrix delays the crystallization and lead to formation of bigger crystallites (Gupta & Choudhary, 2010).

Thermal stability and melting temperature of polymer matrices were improved in presence of CNTs.Kashiwagi et al. (2000) found that the thermal decomposition temperature of polypropylene in nitrogen increases by ~12º C on 2 vol. % MWCNTs loading. Because of the excellent thermal conductivity of CNTs, incorporation of CNTs significantly improves the thermal transport properties of polymer nanocomposites which lead to its usage as printed circuit boards, connectors, thermal interface materials, heat sinks, and other high-performance thermal management systems. Choi et al. (2003) reported 300% increase in thermal conductivity of epoxy matrix at room temperature on 3 wt % SWCNT loading and an additional increase (10%) when aligned magnetically. Biercuk et al. (2002) prepared an epoxy nanocomposite with 1 wt % raw (not purified) laser-oven SWCNT and showed a 125% increase in thermal conductivity at room temperature. Alignment of CNTs also plays an important role in improving the transport properties of a material. Choi et al. (2003) found 10% increase in thermal conductivity of epoxy composite with aligned MWCNTs in comparison to non-aligned MWCNTs.

7. APPLICATION OF CNT/POLYMER NANOCOMPOSITES

With their excellent range of properties, CNTs have opened up a new age of advanced multifunctional materials. Incorporation of CNTs in polymer matrices provides materials that could be used for many high performance engineering applications. Currently, the most widespread use of CNT nanocomposites is in electronics. These nanocomposites could be used to shield electromagnetic interference and as electrostatic-discharge components. The microwave-absorbing capability of nanotubes could be exploited to heat temporary housing structures and may have applications in space exploration. Thin layers of nanotubes on plastics might also be used in transparent conducting composites. High mechanical strength of these nanocomposites could be utilized to make some high-end sporting goods such as tennis rackets, baseball bat etc, and thus delivering superior performance. In short, the biggest market for CNT nanocomposites will undoubtedly be for high-value applications that can absorb the added costs, which includes commercial sectors such as electronics especially aerospace (which requires lightweight, high-strength, high-temperature-resistant composites) and energy (for example, in nanotube-reinforced rubber seals for large oil recovery platforms). Once the cost of nanotubes becomes comparable to that of carbon fibre (or even to that of the much cheaper reinforcing agent, carbon black), commodities such as nanotube-filled rubber tyres could become a reality.

8. CONCLUSION AND FUTURE SCOPE

Studies on CNT/polymer nanocomposite suggests that CNT has great potential in altering the properties of polymer matrices. The quality of CNT/polymer nanocomposites depends on many factors including type of CNTs, chirality, purity, defect density, aspect ratio, % loading, dispersion, alignment and interfacial adhesion between the nanotube and polymer matrix. A lot of research work has been carried out to improve the quality of CNTs and processing techniques. The biggest challenge in realizing the full potential of CNTs is to achieve homogeneous dispersion of CNTs so that maximum filler surface area will be available for load-transfer between filler and matrix. The functionalization of nanotubes provides a convenient route to improve dispersion and stress transfer between CNT and polymer matrix, but more improvement in this field is required to preserve the intrinsic properties of CNTs. It is important to focus on different methods of noncovalent functionalization of nanotubes and discover a route which can improve the dispersion and compatibility without negatively affecting the composite properties. The actual task of dispersing the CNTs in polymer matrix is performed during its manufacturing. The three major processing techniques namely, solution, melt and in-situ polymerization have their unique advantages in fabricating CNT/polymer nanocomposite. Although solution blending produces high quality composite but melt compounding is much simpler and provide option for large scale production. Recently, in-situ polymerization is also showing great potential in fabricating CNT/polymer nanocomposite. The success of processing technique is directly related to performance of composite. The maximum improvement in mechanical properties of polymer matrix is observed in case of in-situ polymerization which produces covalent bond between CNT and polymer matrix. However it negatively influences the electronic properties of the composite. The increase in electrical conductivity of polymer material on CNT addition is the biggest advantage of fabricating CNT/polymer nanocomposite. As significant improvement in electrical conductivity is observed at very low CNT loading, CNT/polymer nanocomposite is finding application as light weight, low cost and highly effective ESD and EMI shielding material. The thermal properties of the polymer matrix also modified by CNT addition like CNT increases the glass transition, melting and thermal decomposition temperatures. CNT also influences the crystallization rate and percentage crystallinity by acting as nucleating agent. Improvement in both crystallization rate and percentage crystallinity enhances its mechanical and processing properties. So finally we can conclude that CNT is ideal filler for fabricating polymer composite but some serious challenges need to be addressed before fully realizing the extraordinary properties of CNT in polymer nanocomposite.

REFERENCES

1. P. M. Ajayan, O. Stephan, C. Colliex, D. Trauth, 1994 Aligned carbon nanotube arrays formed by cutting a polymer resin-nanotube composite. Science, 265 12121214 , 0036-8075

2. P. M. Ajayan, L. S. Schadler, C. Giannaris, A. Rubio, 2000 Single-walled carbon nanotube-polymer composites: strength and weakness. Adv. Mater., 12 750753 , 0935-9648

3. K. Awasthi, A. Srivastava, O. N. Srivastava, 2005 Synthesis of carbon nanotubes. J Nanosci Nanotechnol, 5 16161636 , 1550-7033

4. S. Badaire, P. Poulin, M. Maugey, C. Zakri, 2004 In Situ Measurements of Nanotube Dimensions in Suspensions by Depolarized Dynamic Light Scattering. Langmuir, 20 1036710370 , 0743-7463

5. J. B. Bai, A. Allaoui, 2003 Effect of the length and the aggregate size of MWNTs on the improvement efficiency of the mechanical and electrical properties of composites- experimental investigation. Composites, Part A, 34A 689694 , 0135-9835X

6. S. Bal, S. S. Samal, 2007 Carbon nanotube reinforced polymer composites- a state of the art. Bull. Mater. Sci., 30 379386 , 0250-4707

7. R. Bandyopadhyaya, E. Nativ-Roth, O. Regev, R. Yerushalmi-Rozen, 2002 Stabilization of individual carbon nanotubes in aqueous solutions. Nano Lett., 2 2528 , 1530-6984

8. S. Banerjee, T. Hemraj-Benny, S. S. Wong, 2005 Covalent surface chemistry of single-walled carbon nanotubes. Adv. Mater., 17 1729 , 0935-9648

9. S. Barrau, P. Demont, E. Perez, A. Peigney, C. Laurent, C. Lacabanne, 2003 Effect of palmitic acid on the electrical conductivity of carbon nanotubes-epoxy resin composites. Macromolecules, 36 96789680 , 0024-9297

10. R. H. Baughman, A. A. Zakhidov, W. A. de Heer, 2002 Carbon nanotubes-the route toward applications. Science, 297 787792 , 0036-8075

11. W. Bauhofer, J. Z. Kovacs, 2009 A review and analysis of electrical percolation in carbon nanotube polymer composites. Compos. Sci. Technol., 69 14861498 , 0000-0266- 3538

12. J. M. Benoit, B. Corraze, S. Lefrant, W. J. Blau, P. Bernier, O. Chauvet, 2001 Transport properties of PMMA-Carbon Nanotubes composites. Synth. Met., 121 12151216 , 0379-6779

13. R. Bhattacharyya, T. V. Sreekumar, T. Liu, S. Kumar, L. M. Ericson, R. H. Hauge, 2003 Crystallization and orientation studies in polypropylene/single wall carbon nanotube composite. Polymer,44 23732377 , 0032-3861

14. M. J. Biercuk, M. C. Llaguno, M. Radosavljevic, J. K. Hyun, A. T. Johnson, J. E. Fischer, 2002 Carbon nanotube composites for thermal management. Appl. Phys. Lett., 80 27672769 , 0003-6951

15. R. Blake, Y. K. Gun'ko, J. Coleman, M. Cadek, A. Fonseca, J. B. Nagy, 2004 A Generic Organometallic Approach toward Ultra-Strong Carbon Nanotube Polymer Composites. J. Am. Chem. Soc., 126 1022610227 , 0002-7863

16. Bryning. M. B. (a, D. E. Milkie, M. F. Islam, J. M. Kikkawa, A. G. Yodh, 2005 Thermal conductivity and interfacial resistance in single-wall carbon nanotube epoxy composites. Appl. Phys. Lett., 87 161909 161901-161909/161903, 0003-6951

17. Bryning. M. B. (b, M. F. Islam, J. M. Kikkawa, A. G. Yodh, 2005 Very low conductivity threshold in bulk isotropic single-walled carbon nanotube-epoxy composites. Adv. Mater., 17 11861191 , 0935-9648

18. M. Cadek, J. N. Coleman, V. Barron, K. Hedicke, W. J. Blau, 2002 Morphological and mechanical properties of carbon-nanotube-reinforced semicrystalline and amorphous polymer composites. Appl. Phys. Lett., 81 51235125 , 0003-6951

19. L. Cai, H. Tabata, T. Kawai, 2000 Self-assembled DNA networks and their electrical conductivity. Appl. Phys. Lett., 77 31053106 , 0003-6951

20. J. C. Charlier, J. P. Issi, 1996 Electrical conductivity of novel forms of carbon. J. Phys. Chem. Solids, 57 957965 , 0022-3697

21. F. Cheng, P. Imin, C. Maunders, G. Botton, A. Adronov, 2008 Soluble, discrete supramolecular complexes of single-walled carbon nanotubes with fluorene-based conjugated polymers. Macromolecules, 41 23042308 , 0024-9297

22. E. S. Choi, J. S. Brooks, D. L. Eaton, M. S. Al-Haik, M. Y. Hussaini, H. Garmestani, 2003 Enhancement of thermal and electrical properties of carbon nanotube polymer composites by magnetic field processing. J. Appl. Phys., 94 60346039 , 0021-8979

23. D. D. L. Chung, 2001 Electromagnetic interference shielding effectiveness of carbon materials. Carbon, 39 279285 , 0008-6223

24. J. N. Coleman, A. B. Dalton, S. Curran, A. Rubio, A. P. Davey, A. Drury, 2000 Phase separation of carbon nanotubes and turbostratic graphite using a functional organic polymer. Adv. Mater., 12 213216 , 0935-9648

25. J. N. Coleman, U. Khan, W. J. Blau, Y. K. Gunko, 2006 Small but strong: A review of the mechanical properties of carbon nanotube-polymer composites. Carbon, 44 16241652 , 0008-6223

26. E. Camponeschi, R. Vance, M. Al-Haik, H. Garmestani, R. Tannenbaum, 2007 Properties of carbon nanotube-polymer composites aligned in a magnetic field. Carbon, 45 20372046 , 0008-6223

27. S. A. Curran, P. M. Ajayan, W. J. Blau, D. L. Carroll, J. N. Coleman, A. B. Dalton, 1998 A composite from poly(m-phenylenevinylene-co-2,5-dioctoxy-p-phenylenevinylene) and carbon nanotubes. A novel material for molecular optoelectronics. Adv. Mater., 10 10911093 , 0935-9648

28. H. Dai, E. W. Wong, C. M. Lieber, 1996 Probing electrical transport in nanomaterials: conductivity of individual carbon nanotubes. Science, 272 523526 , 0036-8075

29. V. D. Deshpande, S. Jape, 2009 Nonisothermal crystallization kinetics of nucleated poly(ethylene terephthalate). J. Appl. Polym. Sci., 111 13181327 , 0021-8995

30. M. Dıez-Pascual, M. Naffakh, M. A. Gomez, C. Marco, G. Ellis, B. Simard, 2009 2009). Development and characterization of PEEK/carbon nanotube composites. Vol. Carbon, 47 7, 30793090 , 0008-6223

31. M. S. Dresselhaus, 2004 Applied physics: Nanotube antennas. Nature, 432 959960 , 0028-0836

32. F. Du, J. E. Fischer, K. I. Winey, 2005 Effect of nanotube alignment on percolation conductivity in carbon nanotube/polymer composites. Phys. Rev. B: Condens. Matter Mater. Phys., 72 121404 121401-121404/121404, 1098-0121

33. F. Du, J. E. Fischer, K. I. Winey, 2003 Coagulation method for preparing single-walled carbon nanotube/poly(methyl methacrylate) composites and their modulus, electrical conductivity, and thermal stability. J. Polym. Sci., Part B: Polym. Phys., 41 33333338 , 0887-6266

34. P. Dubois, M. Alexandre, 2006 Performant clay/carbon nanotube polymer composites. Adv. Eng. Mater., 8 147154 , 1438-1656

35. T. W. Ebbesen, H. J. Lezec, H. Hiura, J. W. Bennett, H. F. Ghaemi, T. Thio, 1996 Electrical conductivity of individual carbon nanotubes. Nature, 382 54-56, 0028-0836

36. J. Feng, J. Sui, W. Cai, Z. Gao, 2008 Microstructure and mechanical properties of carboxylated carbon nanotubes/poly(L-lactic acid) composite. J. Compos. Mater., 42 15871595 , 0021-9983

37. W. Feng, X. D. Bai, Y. Q. Lian, J. Liang, X. G. Wang, K. Yoshino, 2003 Well-aligned polyaniline/carbon-nanotube composite films grown by in-situ aniline polymerization. Carbon, 41 15511557 , 0008-6223

38. G. B. M. Fiege, A. Altes, R. Heiderhoff, L. J. Balk, 1999 Quantitative thermal conductivity measurements with nanometer resolution. J. Phys. D: Appl. Phys., 32 L13L17 , 0022-3727

39. J. E. Fischer, H. Dai, A. Thess, R. Lee, N. M. Hanjani, D. L. Dehaas, 1997 Metallic resistivity in crystalline ropes of single-wall carbon nanotubes. Phys. Rev. B: Condens. Matter, 55 R4921R4924 , 1098-0121

40. K. Fu, W. Huang, Y. Lin, L. A. Riddle, D. L. Carroll, Y. Sun, P. , 2001 Defunctionalization of Functionalized Carbon Nanotubes. Nano Lett., 1 439441 , 1530-6984

41. J. Gao, B. Zhao, M. E. Itkis, E. Bekyarova, H. Hu, V. Kranak, 2006 Chemical engineering of the single-walled carbon nanotube-nylon 6 interface. J. Am. Chem. Soc., 128 74927496 , 0002-7863

42. J. Gao, M. E. Itkis, A. Yu, E. Bekyarova, B. Zhao, R. C. Haddon, 2005 Continuous Spinning of a Single-Walled Carbon Nanotube-Nylon Composite Fiber. J. Am. Chem. Soc., 127 38473854 , 0002-7863

43. J. Gao, A. Yu, M. E. Itkis, E. Bekyarova, B. Zhao, S. Niyogi, 2004 Large-scale fabrication of aligned single-walled carbon nanotube array and hierarchical single-walled carbon nanotube assembly. J. Am. Chem. Soc., 126 1669816699 , 0002-7863

44. F. H. Gojny, M. H. G. Wichmann, B. Fiedler, W. Bauhofer, K. Schulte, 2005 Influence of nano-modification on the mechanical and electrical properties of conventional fibre-reinforced composites. Composites, Part A, 36A 15251535 , 0135-9835X

45. S. Gomes, N. Trannoy, P. Grossel, 1999 DC thermal microscopy: study of the thermal exchange between a probe and a sample. Meas. Sci. Technol., 10 805811 , 0957-0233

46. X. Gong, J. Liu, S. Baskaran, R. D. Voise, J. S. Young, 2000 Surfactant-assisted processing of carbon nanotube/polymer composites. Chem. Mater., 12 10491052 , 0897-4756

47. N. Grossiord, J. Loos, C. E. Koning, 2005 Strategies for dispersing carbon nanotubes in highly viscous polymers. J. Mater. Chem., 15 23492352 , 0959-9428

48. Gupta, V. Choudhary, 2010 Isothermal and non-Isothermal crystallization kinetics and morphology of poly(trimethylene terephthalate)/multiwalled carbon nanotube composites. Macromole. Symp., 290 5669 , 1022-1360

49. R. Haggenmueller, J. E. Fischer, K. I. Winey, 2006 Single wall carbon nanotube/polyethylene composites: nucleating and templating polyethylene crystallites. Macromolecules, 39 29642971 , 0024-9297

50. R. Haggenmueller, H. H. Gommans, A. G. Rinzler, J. E. Fischer, K. I. Winey, 2000 Aligned single-wall carbon nanotubes in composites by melt processing methods. Chem. Phys. Lett., 330 219225 , 0009-2614

51. R. Haggenmueller, W. Zhou, J. E. Fischer, K. I. Winey, 2003 Production and characterization of highly aligned single-walled carbon nanotubes in polymer composites. J. Nanosci.Nanotechnol., 3 105110 , 1550-7033

52. N. Hamada, S. Sawada, A. Oshiyama, 1992 New one-dimensional conductors: graphitic microtubules. Phys. Rev. Lett., 68 15791581 , 0031-9007

53. M. C. Hersam, A. C. F. Hoole, S. J. O'Shea, M. E. Welland, 1998 Potentiometry and repair of electrically stressed nanowires using atomic force microscopy. Appl. Phys. Lett., 72 915917 , 0003-6951

54. Hirsch, 2002 Functionalization of single-walled carbon nanotubes. Angewandte Chemie International Edition, 41 18531859 , 0570-0833

55. J. K. Holt, H. G. Park, Y. Wang, M. Stadermann, A. B. Artyukhin, C. P. Grigoropoulos, 2006 Fast mass transport through sub-2-nanometer carbon nanotubes. Science, 312 10341037 , 0036-8075

56. J. Hone, M. Whitney, A. Zettl, 1999 Thermal conductivity of single-walled carbon nanotubes. Synth. Met., 103 24982499 , 0379-6779

57. H. Hou, J. J. Ge, J. Zeng, Q. Li, D. H. Reneker, A. Greiner, 2005 Electrospun polyacrylonitrile nanofibers containing a high concentration of well-aligned multiwall carbon nanotubes. Chem. Mater., 17 967973 , 0897-4756

58. G. L. Hwang, Y. Shieh, T. , K. C. Hwang, 2004 Efficient load transfer to polymer-grafted multiwalled carbon nanotubes in polymer composites. Adv. Funct. Mater., 14 487491 , 0161-6301X

59. M. F. Islam, E. Rojas, D. M. Bergey, A. T. Johnson, A. G. Yodh, 2003 High weight fraction surfactant solubilization of single-wall carbon nanotubes in water. Nano Lett., 3 269273 , 1530-6984

60. S. Iijima, 1991 Helical microtubules of graphitic carbon. Nature, 354 5658 , 0028-0836

61. H. Ishikawa, S. Fudetani, M. Hirohashi, 2001 Mechanical properties of thin films measured by nanoindenters. Appl. Surf. Sci., 178 5662 , 0169-4332

62. Z. Jia, Z. Wang, C. Xu, J. Liang, B. Wei, D. Wu, 1999 Study on poly(methyl methacrylate)/carbon nanotube composites. Mater. Sci. Eng., A, A271 395400 , 0921-5093

63. J. Jin, M. Song, F. Pan, 2007 A DSC study of effect of carbon nanotubes on crystallization behaviour of poly(ethylene oxide). Thermo. Acta, 45 2531 , 0040-6031

64. Z. Jin, K. P. Pramoda, S. H. Goh, G. Xu, 2002 Poly(vinylidene fluoride)-assisted melt-blending of multi-walled carbon nanotube/poly(methyl methacrylate) composites. Mater. Res. Bull., 37 271278 , 0025-5408

65. L. Jin, C. Bower, O. Zhoua, 1998 Alignment of carbon nanotubes in a polymer matrix by mechanical stretching. Appl. Phys. Lett., 73 11971202 , 0003-6951

66. J. Joo, A. J. Epstein, 1994 Electromagnetic radiation shielding by intrinsically conducting polymers. Appl. Phys. Lett., 65 22782280 , 0003-6951

67. T. Kashiwagi, J. W. Gilman, K. Butler, R. H. Harris, J. R. Shields, A. Asano, 2000 Flame retardant mechanism of silica gel/silica. Fire Mater., 24 277289 , 0308-0501

68. C. L. Kane, E. J. Mele, 1997 Size, shape, and low energy electronic structure of carbon nanotubes. Phys. Rev. Lett., 78 19321935 , 0031-9007

69. P. Kim, L. Shi, A. Majumdar, P. L. Mc Euen, 2001 Thermal transport measurements of individual multiwalled nanotubes. Phys. Rev. Lett., 87 215502 215501-215502/215504, 0031-9007

70. H. M. Kim, K. Kim, C. Y. Lee, J. Joo, S. J. Cho, H. S. Yoon, 2004 Electrical conductivity and electromagnetic interference shielding of multiwalled carbon nanotube composites containing Fe catalyst. Appl. Phys. Lett., 84 589591 , 0003-6951

71. S. H. Kim, S. H. Jang, S. W. Byun, J. Y. Lee, J. S. Joo, S. H. Jeong, 2003 Electrical properties and EMI shielding characteristics of polypyrrole-nylon 6 composite fabrics. J. Appl. Polym. Sci., 87 19691974 , 0021-8995

72. F. Ko, Y. Gogotsi, A. Ali, N. Naguib, H. Ye, G. Yang, 2003 Electrospinning of continuous carbon nanotube-filled nanofiber yarns. Adv. Mater., 15 11611165 , 0935-9648

73. B. Kracke, B. Damaschke, 2000 Measurement of nanohardness and nanoelasticity of thin gold films with scanning force microscope. Appl. Phys. Lett., 77 361363 , 0003-6951

74. Chapelle. M. Lamy la, C. Stephan, T. P. Nguyen, S. Lefrant, C. Journet, P. Bernier, 1999 Raman characterization of singlewalled carbon nanotubes-PMMA composites. Synth. Met., 103 25102512 , 0379-6779

75. J. Luo, T. , H. Wen, C. , W. Wu, F. , C. Chou, P. , 2008 Mechanical research of carbon nanotubes/PMMA composite films. Polym. Compos., 29 12851290 , 0272-8397

76. P. Liu, 2005 Modifications of carbon nanotubes with polymers. Eur. Polym. J., 41 26932703 , 0014-3057

77. Z. Liu, G. Bai, Y. Huang, Y. , F. Du, F. Li, 2007 Reflection and absorption contributions to the electromagnetic interference shielding of single-walled carbon nanotube/polyurethane composites. Carbon, 45 821827 , 0008-6223

78. T. Liu, I. Y. Phang, L. Shen, S. Y. Chow, W. Zhang-D, 2004 Morphology and Mechanical Properties of Multiwalled Carbon Nanotubes Reinforced Nylon-6 Composites. Macromolecules, 37 72147222 , 0024-9297

79. C. , W. Zhang, Y. Zhu, L. Ji, R. Zhang, N.. Koratkar, J. Liang, 2006 Alignment and dispersion of functionalized carbon nanotubes in polymer composites induced by an electric field. Carbon, 46 706720 , 0008-6223

80. R. Martel, T. Schmidt, H. R. Shea, T. Hertel, P. Avouris, 1998 Single- and multi-wall carbon nanotube field-effect transistors. Appl. Phys. Lett., 73 24472449 , 0003-6951

81. C. A. Martin, J. K. W. Sandler, M. S. P. Shaffer, M. K. Schwarz, W. Bauhofer, K. Schulte, 2004 Formation of percolating networks in multi-wall carbon-nanotube-epoxy composites. Compos. Sci. Technol., 64 23092316 , 0000-0266- 3538

82. O. Matarredona, H. Rhoads, Z. Li, J. H. Harwell, L. Balzano, D. E. Resasco, 2003 Dispersion of single-walled carbon nanotubes in aqueous solutions of the anionic surfactant NaDDBS. J. Phys. Chem. B, 107 1335713367 , 1089-5647

83. D. Micheli, R. Pastore, C. Apollo, M. Marchetti, G. Gradoni, F. Moglie, 2009 Carbon based nanomaterial composites in RAM and microwave shielding applications, 9th IEEE Conf. Nanotechnol., 226235 , 978-1-42444-832-6 Genoa, 26-30 July, 2009

84. E. Minoux, O. Groening, K. B. K. Teo, S. H. Dalal, L. Gangloff, J. Schnell, P. , 2005 Achieving high-current carbon nanotube emitters. Nano Lett., 5 21352138 , 1530-6984

85. J. W. Mintmire, B. I. Dunlap, C. T. White, 1992 Are fullerene tubules metallic? Phys. Rev. Lett., 68 631634 , 0031-9007

86. Moniruzzaman. M. (a, F. Du, N. Romero, K. I. Winey, 2006 Increased flexural modulus and strength in SWNT/epoxy composites by a new fabrication method. Polymer, 47 293298 , 0032-3861

87. Moniruzzaman. M. (b, K. I. Winey, 2006 Polymer composites containing carbon nanotubes. Macromolecules, 39 51945205 , 0024-9297

88. M. Monthioux, 2002 Filling single-wall carbon nanotubes. Carbon, 40 18091823 , 0008-6223

89. V. C. Moore, M. S. Strano, E. H. Haroz, R. H. Hauge, R. E. Smalley, J. Schmidt, 2003 Individually suspended single-walled carbon nanotubes in various surfactants. Nano Lett., 3 13791382 , 1530-6984

90. M. J. O'Connell, P. Boul, L. M. Ericson, C. Huffman, Y. Wang, E. Haroz, 2001 Reversible water-solubilization of single-walled carbon nanotubes by polymer wrapping. Chem. Phys. Lett., 342 265271 , 0009-2614

91. T. W. Odom, J. Huang, P. Kim, C. M. Lieber, 1998 Atomic structure and electronic properties of single-walled carbon nanotubes. Nature, 391 6264 , 0028-0836

92. C. Park, Z. Ounaies, K. A. Watson, R. E. Crooks, J. Smith, S. E. Lowther, 2002 Dispersion of single wall carbon nanotubes by in-situ polymerization under sonication. Chem. Phys. Lett., 364 303308 , 0009-2614

93. O. Park-K, T. Jeevananda, N. H. Kim, S.-i. Kim, J. H. Lee, 2009 Effects of surface modification on the dispersion and electrical conductivity of carbon nanotube/polyaniline composites. Scr. Mater., 60 551554 , 1359-6462

94. X. Pei, L. Hu, W. Liu, J. Hao, 2008 Synthesis of water-soluble carbon nanotubes via surface initiated redox polymerization and their tribological properties as water-based lubricant additive. Eur. Polym. J., 44 24582464 , 0014-3057

95. P. Poetschke, A. R. Bhattacharyya, A. Janke, H. Goering, 2003 Melt mixing of polycarbonate/multi-wall carbon nanotube composites. Compos. Interfaces, 389404 , 1568-5543

96. K. W. Putz, C. A. Mitchell, R. Krishnamoorti, P. F. Green, 2004 Elastic modulus of single-walled carbon nanotube/poly(methyl methacrylate) composites. J. Polym. Sci., Part B: Polym. Phys., 42 22862293 , 0887-6266

97. D. Qian, E. C. Dickey, R. Andrews, T. Rantell, 2000 Load transfer and deformation mechanisms in carbon nanotube-polystyrene composites. Appl. Phys. Lett., 76 28682870 , 0003-6951

98. S. Qin, D. Qin, W. T. Ford, D. E. Resasco, J. E. Herrera, 2004 Functionalization of single-walled carbon nanotubes with polystyrene via grafting to and grafting from methods. Macromolecules, 37 752757 , 0024-9297

99. N. R. Raravikar, L. S. Schadler, A. Vijayaraghavan, Y. Zhao, B. Wei, P. M. Ajayan, 2005 Synthesis and characterization of thickness-aligned carbon nanotube-polymer composite films. Chem. Mater., 17 974983 , 0897-4756

100. Rasheed. A. (a, M. D. Dadmun, P. F. Britt, 2006 Polymer-nanofiber composites: enhancing composite properties by nanofiber oxidation. J. Polym. Sci., Part B: Polym. Phys., 44 30533061 , 0887-6266

101. Rasheed. A. (b, H. G. Chae, S. Kumar, M. D. Dadmun, 2006 Polymer nanotube composites: Correlating intermolecular interaction to ultimate properties. Polymer, 47 47344741 , 0032-3861

102. O. Regev, P. N. B. El Kati, J. Loos, C. E. Koning, 2004 Preparation of conductive nanotube-polymer composites using latex technology. Adv. Mater., 16 248251 , 0935-9648

103. F. Ruiz, W. D. Sun, F. H. Pollak, C. Venkatraman, 1998 Determination of the thermal conductivity of diamond-like composite films using a scanning thermal microscope. Appl. Phys. Lett., 73 18021804 , 0003-6951

104. B. Safadi, R. Andrews, E. A. Grulke, 2002 Multiwalled carbon nanotube polymer composites: synthesis and characterization of thin films. J. Appl. Polym. Sci., 84 26602669 , 0021-8995

105. N. G. Sahoo, Y. C. Jung, H. J. Yoo, J. W. Cho, 2006 Effect of functionalized carbon nanotubes on molecular interaction and properties of polyurethane composites. Macromol. Chem. Phys., 207 17731780 , 1521-3935

106. R. Saito, M. Fujita, G. Dresselhaus, M. S. Dresselhaus, 1992 Electronic structure of chiral graphene tubules. Appl. Phys. Lett., 60 22042206 , 0003-6951

107. L. S. Schadler, S. C. Giannaris, P. M. Ajayan, 1998 Load transfer in carbon nanotube epoxy composites. Appl. Phys. Lett., 73 38423844 , 0003-6951

108. P. Singh, S. Campidelli, S. Giordani, D. Bonifazi, A. Bianco, M. Prato, 2009 Organic functionalisation and characterisation of single-walled carbon nanotubes. Chem Soc Rev, 38 22142230 , 0306-0012

109. E. J. Siochi, D. C. Working, C. Park, P. T. Lillehei, J. H. Rouse, C. C. Topping, 2004 Melt processing of SWCNT-polyimide composite fibers. Composites, Part B, 35B 439446 , 0000-1359- 8368

110. G. Sui, W. H. Zhong, X. P. Yang, Y. H. Yu, S. H. Zhao, 2008 Preparation and properties of natural rubber composites reinforced with pretreated carbon nanotubes. Polym. Adv. Technol., 19 15431549 , 0000-1042- 7147

111. B. Sulong, N. Muhamad, J. Sahari, R. Ramli, B. Deros, J. Park, 2009 Electrical conductivity behavior of chemical functionalized MWCNTs epoxy composites. European Journal of Scientific Research. 29 1321 , 0145-0216X

112. L. Sun, G. L. Warren, J. Y. O'Reilly, W. N. Everett, S. M. Lee, D. Davis, 2008 Mechanical properties of surface-functionalized SWCNT/epoxy composites. Carbon, 46 320328 , 0008-6223

113. U. Sundararaj, M. H. Al-Saleh, 2009 Electromagnetic interference shielding mechanisms of CNT/polymer composites. Carbon, 47 17381746 , 0008-6223

114. S. J. Tans, M. H. Devoret, H. Dal, A. Thess, R. E. Smalley, L. J. Geerligs, 1997 Individual single-wall carbon nanotubes as quantum wires. Nature, 386 474477 , 0028-0836

115. E. T. Thostenson, Z. Ren, T. W. Chou, 2001 Advances in the science and technology of carbon nanotubes and their composites: a review. Compos. Sci. Technol., 61 18991912 , 0000-0266- 3538

116. E. T. Thostenson, T. Chou, W. , 2002 Aligned multi-walled carbon nanotube-reinforced composites: processing and mechanical characterization. J. Phys. D: Appl. Phys., 35 L77L80 , 0022-3727

117. C. Velasco-Santos, A. L. Martinez-Hernandez, F. T. Fisher, R. Ruoff, V. M. Castano, 2003 Improvement of thermal and mechanical properties of carbon nanotube composites through chemical functionalization. Chem. Mater., 15 44704475 , 0897-4756

118. B. Vigolo, A. Penicaud, C. Coulon, C. Sauder, R. Pailler, C. Journet, 2000 Macroscopic fibers and ribbons of oriented carbon nanotubes. Science, 290 13311334 , 0036-8075

119. K. Wong, K. , S. Shi, Q. , K. Lau, T. , 2007 Mechanical and thermal behavior of a polymer composite reinforced with functionalized carbon nanotubes. Key Eng. Mater., 334-335 , 705708 , 1013-9826

120. C. Wu, S. , 2009 Synthesis and characterization of poly(trimethylene terephthalate) composites incorporating multi-walled carbon nanotubes. J. Appl. Polym. Sci., 114 16331642 , 0021-8995

121. H. Wu, X. , X. Qiu, Q. , W. Cao, M. , Y. Lin, H. , R. Cai, F. , S. Qian, X. , 2007 Polymer-wrapped multiwalled carbon nanotubes synthesized via microwave-assisted in situ emulsion polymerization and their optical limiting properties. Carbon, 45 28662872 , 0008-6223

122. X. Xie, L. , Y. Mai, W. , X. Zhou, P. , 2005 Dispersion and alignment of carbon nanotubes in polymer matrix: A review. Mater. Sci. Eng., R, R49 89112 , 0092-7796X

123. Yang. Y. (a, M. C. Gupta, K. L. Dudley, R. W. Lawrence, 2005 Conductive carbon nanofiber-polymer foam structures. Adv. Mater., 17 19992003 , 0935-9648

124. Yang. Y. (b, M. C. Gupta, K. L. Dudley, R. W. Lawrence, 2005 Novel carbon nanotube-polystyrene foam composites for electromagnetic interference shielding. Nano Lett., 5 21312134 , 1530-6984

125. Yu. M. (a, F. , B. S. Files, S. Arepalli, R. S. Ruoff, 2000 Tensile loading of ropes of single wall carbon nanotubes and their mechanical properties. Phys. Rev. Lett., 84 55525555 , 0031-9007

126. Yu. M. (b, F. , O. Lourie, M. J. Dyer, K. Moloni, T. F. Kelly, R. S. Ruoff, 2000 Strength and breaking mechanism of multiwalled carbon nanotubes under tensile load. Science, 287 637640 , 0036-8075

127. Yuen. S. (a, M. , C. , C. M. , 2008 Morphological, electrical, and mechanical properties of multiwall carbon nanotube/polysilsesquioxane composite. J. Appl. Polym. Sci., 109 20002007 , 0021-8995

128. Yuen. S. (b, M. , C. , C. M. , C. Chuang, Y. , K. Yu, C. , S. Wu, Y. , C. Yang, C. , 2008 Effect of processing method on the shielding effectiveness of electromagnetic interference of MWCNT/PMMA composites. Compos. Sci. Technol., 68 963968 , 0000-0266- 3538

129. Yuen. S. (c, M. , C. , C. M. , C. Teng, C. , H. Wu, H. , H. Kuan, C. , C. Chiang, L., 2008 Molecular motion, morphology, and thermal properties of multiwall

carbon nanotube/polysilsesquioxane composite. J. Polym. Sci., Part B: Polym. Phys., 46 472482 , 0887-6266

130. W. D. Zhang, L. Shen, I. Y. Phang, T. Liu, 2004 Carbon nanotubes reinforced nylon-6 composite prepared by simple melt-compounding. Macromolecules, 37 256259 , 0024-9297

131. M. Zhang, S. Fang, A. A. Zakhidov, S. B. Lee, A. E. Aliev, C. D. Williams, 2005 Strong, transparent, multifunctional, carbon nanotube sheets. Science, 309 12151219 , 0036-8075

132. S. Zhang, M. L. Minus, L. Zhu, C. Wong, P. , S. Kumar, 2008 Polymer transcrystallinity induced by carbon nanotubes. Polymer, 49 13561364 , 0032-3861

133. B. Zhao, J. Wang, Z. Li, P. Liu, D. Chen, Y. Zhang, 2008 Mechanical strength improvement of polypropylene threads modified by PVA/CNT composite coatings. Mater. Lett., 62 43804382 , 0016-7577X

134. J. Zhu, J. Kim, H. Peng, J. L. Margrave, V. N. Khabashesku, E. V. Barrera, 2003 Improving the dispersion and integration of single-walled carbon nanotubes in epoxy composites through functionalization. Nano Lett., 3 11071113 , 1530-6984

135. J. Zhu, H. Peng, F. Rodriguez-Macias, J. L. Margrave, V. N. Khabashesku, A. M. Imam, 2004 Reinforcing epoxy polymer composites through covalent integration of functionalized nanotubes. Adv. Funct. Mater., 14 643648 , 0161-6301X

Chapter 3

Stimuli-Responsive Polymer-Clay Nanocomposites under Electric Fields

Shang Hao Piao, Seung Hyuk Kwon and Hyoung Jin Choi *

Department of Polymer Science and Engineering, Inha University, Incheon 402-751, Korea

ABSTRACT

This short Feature Article reviews electric stimuli-responsive polymer/clay nanocomposites with respect to their fabrication, physical characteristics and electrorheological (ER) behaviors under applied electric fields when dispersed in oil. Their structural characteristics, morphological features and thermal degradation behavior were examined by X-ray diffraction pattern, scanning electron microscopy and transmission electron microscopy, and thermogravimetric analysis, respectively. Particular focus is given to the electro-responsive ER characteristics of the polymer/clay nanocomposites in terms of the yield stress and viscoelastic properties along with their applications.

KEYWORDS

polymer; clay; nanocomposite; electrorheological; conducting polymer

1. INTRODUCTION

Nanoscale materials have attracted considerable interest for their potential technological applications in a range of areas because their chemical and physical characteristics can be altered drastically from a macroscopic bulk to the molecular level. In particular, polymeric materials are being improved by forming polymer nanocomposite systems for desired engineering applications because of their synergistic effects on the molecular nanoscale from both pure polymers and inorganic materials [1,2,3]. Among the various inorganics for nanocomposites, there has been particular interest in polymer/clay nanocomposites consisting of a polymer and clay, in which the clay has been selected conventionally as fillers in polymer compound research and industrial applications [4,5]. Particles form internal interfaces with large specific surfaces in dispersions, making it possible to stabilize different director configurations, in which clay minerals are introduced to the structures, either intercalated or exfoliated, because of their small particle size, simple chemical modification mainly using cationic surfactants, layer expanding capabilities with various treatments, and low cost. On the other hand, it is well known that using molecular or nanoscale strengthening techniques instead of conventional particulate filled composites, polymer nanocomposites have the ability to extend their usage, through which they offer unique characteristics that are drastically different from their bulk counterparts [6,7,8,9]. Note that in conventional polymer/clay composites, the clay particles are simply dispersed as a filler in a polymer matrix, mainly to improve the mechanical and thermal properties, such as the heat-distortion temperature [10]. Nevertheless, the characteristics of a polymer-based nanocomposite immediately after its manufacture may be rather different from those of the same material after conversion to a final useful shape by some processing technique. In the case of polymer/clay nanocomposites, Wang *et al.* [11] reported that for a polypropylene/organic modified montmorillonite (MMT) nanocomposite processed via dynamic packing injection molding, the morphological change in the shear-induced morphology with a core in the center, an oriented zone surrounding the core and a skin layer in the cross-section areas of the samples was observed. Another processing technique of ultrasonication also affects the dispersion of clay in a polymer matrix [12].

Concurrently, polymer nanocomposites with conducting polymers with conjugated repeating units have facilitated the new development of enhanced thermal, electrical and optical properties [13,14], via the appraisal of important parameters, such as electron affinities, ionization potential, band gap, and bandwidths [15,16,17], ensuring their potential engineering adoptability with better mechanical strength, dispersion stability or physical properties than pure conducting polymers [18,19,20,21,22,23,24]. In particular, the electrical conductivity can be controlled more easily, and the thermal or mechanical stability can be improved by the fabrication of the nanocomposites [25].

On the other hand, an electrorheological (ER) fluid is a type of smart and intelligent material that possesses extraordinary transition characteristics from a liquid-like to a solid-like state under an applied external electric field [26,27,28,29,30]. In general, a typical ER fluid is formed by electrically polarizable particles dispersed in a non-conducting medium, such as silicone oil or mineral oil [31,32,33,34]. The particles are dispersed randomly in the medium phase, exhibiting Newtonian fluid-like behavior when the external electric field is absent. On the other hand, when an electric field is applied to the fluid, all particles initially dispersed will become rapidly polarized and attract neighboring particles to form a chain-like structure with a robust dipole-dipole interaction aligned along the applied field. When the electric field strength is increased, the resulting chains agglomerate to produce stronger columns, indicating that the ER fluid is a more-like an elastic solid. During this electrically controllable and immediate process, the rheological property of the ER fluids can generally be expressed using a Bingham fluid model. Compared to the liquid-like state of ER fluids without an electric field, the solid-like state induced by an external electric field is provided with a non-vanishing yield stress, higher shear viscosity and dynamic moduli. Therefore, similar to the case of the magnetorheological (MR) fluids under a magnetic field [35,36,37], the ER fluids have a considerable potential in a variety of engineering applications, such as dampers, torque transducers, optical finishing, and haptic devices [38,39,40].

Various electro-responsive materials, including biopolymers, such as cellulose and starch, inorganic particles, such as titania, zeolite, mesoporous material, and silica, and conducting polymers demonstrate typical ER behaviors, while hybrids of organic/inorganic materials have been introduced because of their enhanced characteristics [25].

This paper reviews the fabrication methods and characteristics of smart polymer/clay nanocomposites used as ER materials prepared from a range of different fabrication methods. The ER performance of the polymer/clay nanocomposites was observed by either optical microscopy or rotational rheometry.

1.1. Fabrication of Electro-Responsive Polymer/Clay Nanocomposites

As a simple and interesting way to construct organic-inorganic hybrid systems and provide synergistic properties, which cannot be achieved from individual materials, such as easily controlled conductivity and higher mechanical stability, several methods to prepare electro-responsive polymer/clay nanocomposites have been reported [25,40].

1.1.1. *In-Situ* Chemical Oxidation Polymerization

Yeh *et al.* [41] introduced a polyaniline (PANI)/clay nanocomposite by effectively dispersing the inorganic nanolayers of montmorillonite (MMT) clay in an organic PANI matrix via *in-situ* polymerization. Initially, an appropriate amount of organophilic clay was added to an aqueous HCl solution, and the organic aniline monomers were added to the solution for intercalation into the interlayer regions of the organophilic clay hosts. Upon the addition of ammonium persulfate (APS) in HCl, HCl-doped lamellar nanocomposite precipitates were obtained. The final nanocomposite products were obtained by immersing the HCl-doped nanocomposites into aqueous an NH_4OH solution, followed by filtration and drying. PANI/clay nanocomposites were prepared in the form of coatings with a low clay loading from 0.75 to 3 wt %. Scheme 1 presents a diagram of the experimental procedure.

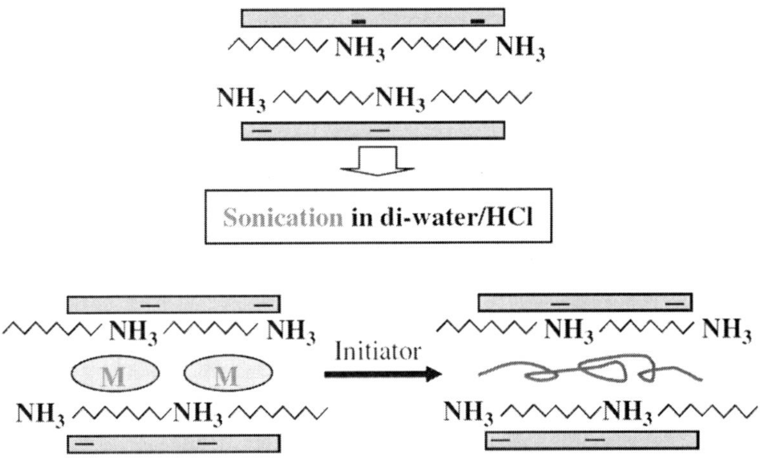

Scheme 1. Preparation of the polyaniline (PANI)/clay nanocomposite via an *in-situ*polymerization (Reprinted from Reference [25] with permission).

Furthermore, using a copolyaniline of poly(o-ethoxyaniline) (PEA), Yeh *et al.* fabricated PEA/MMT nanocomposites with different clay loadings up to 5 wt %, finding that the nanocomposites at low clay loadings up to 3 wt % exhibited a much superior corrosion inhibition effect compared to pristine PEA [42].

Concurrently, using other types of tube-like clay particles, the PANI/halloysite (HNT) nanocomposites have been also reported [43,44,45]. Zhang *et al.* [45] reported the facile synthesis of organic/inorganic PANI-wrapped HNT composite by the*in situ* polymerization of aniline in a HNT dispersion using APS as an initiator, and adopted it as an ER potential material.

Chea *et al.* [46] fabricated palygorskite (Pal) clay coated with semiconducting PANI nanocomposite particles by an oxidative polymerization process, as shown in Scheme 2 and used them as ER materials without a post-treatment step except for the doping process. Sulfuric acid was added drop-wise to initiate aniline polymerization. The sulfuric acid-modified aniline monomer was then polymerized with the aid of APS at the surface of Pal, resulting in Pal/PANI composite particles.

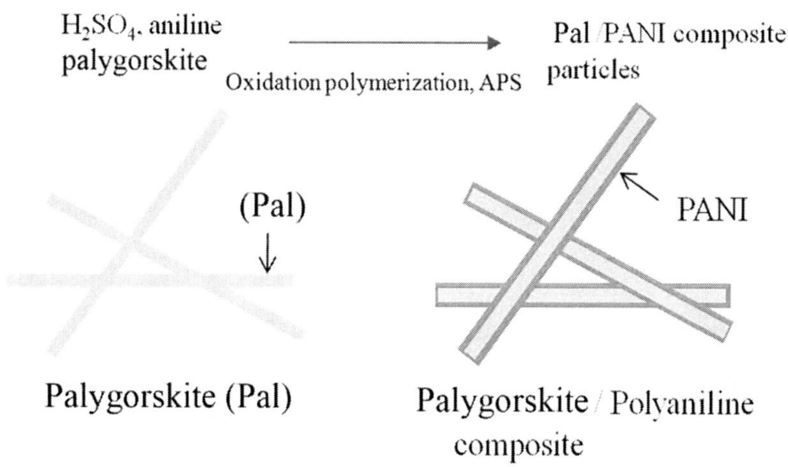

Scheme 2. Schematic diagram of the experimental route to synthesize the palygorskite (Pal)/PANI composite particles.

1.1.2. Suspension Polymerization

Compared to other polymerization processes in electro-responsive polymer/clay nanocomposites, suspension polymerization was seldom adopted. Jun and Suh [47] synthesized poly(urethane acrylate) (PUA)/clay nanocomposite particles by suspension polymerization, with a continuous phase comprised of an aqueous solution of poly(vinyl alcohol) and sodium nitrite, and the dispersed phase comprised of monomer, toluene as an organic diluent, and the oil-soluble initiator with a clay content of 5 wt %. They then examined their ER characteristics.

1.1.3. Emulsion Polymerization

The emulsion polymerization process is considered to be a very useful fabrication technique for synthesizing polymer/clay nanocomposites using glassy polymers with high glass-transition temperatures, such as poly(methyl methacrylate) (PMMA), polystyrene (PS) and styrene-acrylonitrile (SAN), and epoxy and rubbery polymers, such as poly(ethyl acrylate) [25,48]. Kim *et al.* [49] synthesized SAN copolymer-Na^+-MMT clay nanocomposite particles by emulsion polymerization and examined their ER performance. For the emulsion polymerization of polymer nanocomposites, the Na^+-MMT clay was introduced to synthesize a polymer/clay nanocomposite with SAN in the presence of potassium persulfate as an initiator and sodium lauryl sulfate as an emulsifier, and the product was then coagulated by the addition of aluminum sulfate solution. The final weight fraction of the Na^+-MMT in the SAN/clay nanocomposite was 4.76 wt %. ER fluids composed of SAN-clay composite exhibited typical ER behavior and possessed "pseudo-Newtonian" behavior at high shear rates.

1.1.4. Pickering Emulsion Polymerization

Recently, the Pickering emulsion polymerization process has attracted considerable attention as a new method for the fabrication of smart nanocomposites, in which the emulsion droplets prior to polymerization are being stabilized by various solid particles instead of conventional organic surfactants or stabilizers. Therefore, Pickering emulsions impart better stability against coalescence and, in many cases, are biologically compatible and environmentally friendly [50]. Although Pickering emulsions have huge industrial potential applications in the areas of petroleum, food, biomedicine, pharmaceuticals, and cosmetics, Pickering emulsion polymerized particles have recently been adopted for both ER and MR materials [51].

Fang *et al.* [52] fabricated PANI/clay nanoparticles with a special core-shell structure via Pickering emulsion in a toluene phase by employing an exfoliated clay sheet as a stabilizer, using organophilically modified MMT (OMMT). The synthesized PANI nanospheres, which were initialized by oil-soluble benzoyl peroxide, possessed a polydisperse size distribution of particles, ranging from 200 nm to 1 μm.

Kim *et al.* [53] introduced polystyrene (PS)/laponite composite nanoparticles fabricated using Pickering emulsion polymerization. The hydrophilic laponite modified with cetyltrimethylammonium bromide was used as a stabilizer, in which emulsions of styrene were dispersed in water. Scheme 3 outlines the mechanism for preparing PS/laponite core-shell particles by surfactant-free Pickering emulsion polymerization. The modified laponite was adsorbed on the surface of the styrene monomer droplet to stabilize the system. After adding a water soluble initiator, the mixture became milky white and polymerization occurred in the styrene droplets with laponite adsorbed at the boundary surface. The weight ratio of laponite in the PS/laponite nanoparticles was approximately 8.85%.

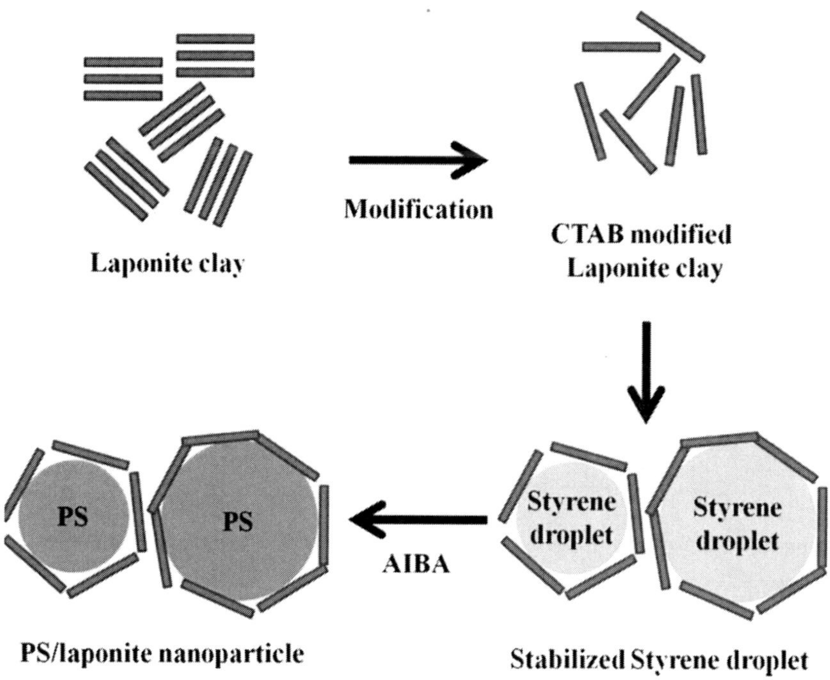

Scheme 3. Mechanism of surfactant-free Pickering emulsion polymerization for polystyrene (PS)/laponite nanoparticles (Reprinted from Reference [53] with permission).

1.2. Melt Processing

Based on the effects of the applied electric field on the structural evolution of poly(propylene) (PP)/clay nanocomposites, showing a tendency toward exfoliation [54], Kim *et al.* [55] reported a novel method to produce poly(propylene)/clay nanocomposites continuously with a clay content of 5 wt % using an electric melt pipe equipped with a twin-screw extruder, as shown in Scheme 4. In their study, partial intercalation was obtained by continuous processing, showing the possibility to produce nanocomposites using this method. As this physical process can be appropriate for conventional extrusion, the approach may also be used in other polymer/clay nanocomposite systems [55].

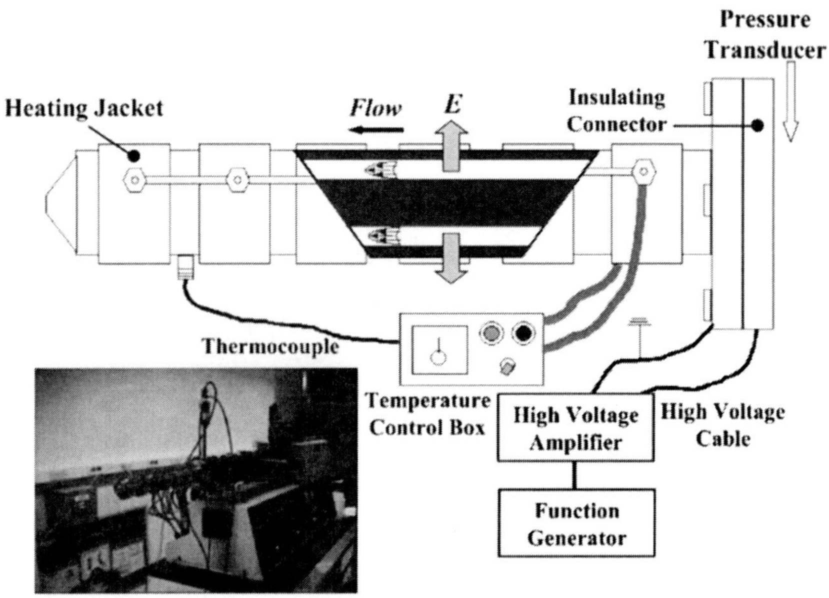

Scheme 4. Schematic diagram of an electric melt pipe equipped with a twin-screw extruder. (Reprinted from Reference [55] with permission).

2. CHARACTERIZATION OF POLYMER-CLAY NANOCOMPOSITES

2.1. Morphology

Once the electro-responsive smart polymer/clay nanocomposites were synthesized, their morphology was examined to determine if the nanocomposite had been synthesized successfully. These can be answered by either scanning electron microscopy (SEM) or transmission electron microscopy (TEM).

Figure 1 provides direct evidence of the synthesis of a PANI-coated HNT surface provided by the SEM images, in which the neat nanotubes of the raw HNT can be seen. Compared to the pure HNT, the distinctive tubular shape of the PANI/HNT composite in Figure 1b disappeared, indicating the formation of PANI.

Regarding the morphology of the PANI/OMMT nanocomposite confirmed by SEM (Figure 2a), pure clay exhibits a lamellar structure with a nanosized thickness and huge surface area in each layer. After the polymerization of aniline, a large number of nano-granules were observed with a broad particle size distribution, as shown in Figure 2b. The granular surface became rough due to the adsorbed OMMT sheets [56,57,58,59]. Therefore, the PANI/OMMT nanocomposite particles have a core-shell structure, in which the core material is an organic PANI and the shell material is the exfoliated clay sheet.

Figure 1. Scanning electron microscopy (SEM) images of (**a**) halloysites (HNTs) and (**b**) PANI/HNT composites (Reprinted from Reference [45] with permission).

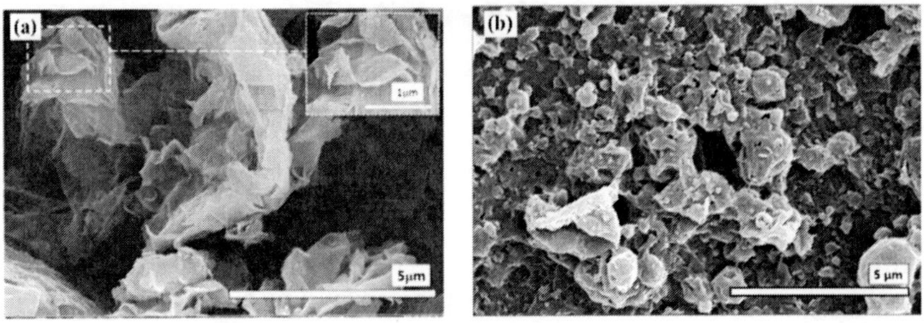

Figure 2. SEM images of pristine clay (**a**) and fabricated PANI/OMMT nanocomposite particles (**b**) (Reprinted from Reference [52] with permission).

The morphologies of both Pal/PANI and Pal were examined by TEM to characterize the surface morphology of the pure Pal (Figure 3a) and Pal/PANI particles (Figure 3b). Figure 3a showed that the surface of Pal was quite smooth, in which Pal has a highly fibrous morphology that forms bundles. The length of each fiber was varied from the sub-micrometer to micrometer range with a mean diameter of approximately 20 nm. In contrast, the Pal/PANI composite particles had a much rougher surface due to the wrapping of PANI (Figure 3b), meaning that the aniline had been polymerized onto the Pal template by a chemical oxidation method, altering the outside surface.

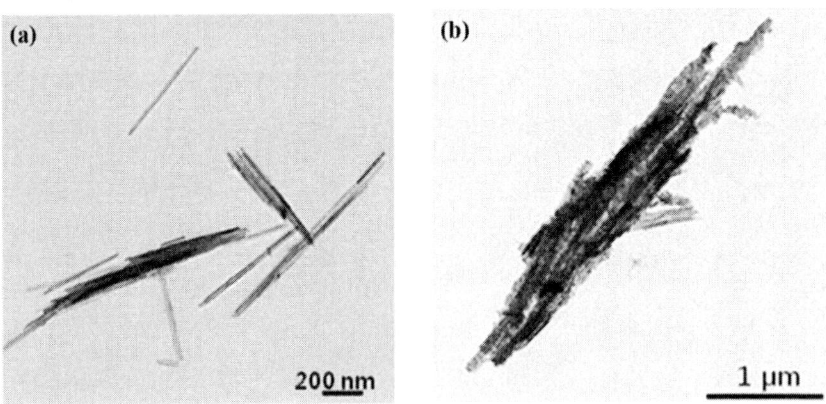

Figure 3. Transmission electron microscopy (TEM) images of Pal (**a**) and Pal/PANI composite particles (**b**), respectively (Reprinted from Reference [46] with permission).

Fang *et al.* [60] published a TEM image of a cross-sectional view of the synthesized nano-sized laponite stabilized poly(methyl methacrylate) (PMMA) spheres, as shown in Figure 4. The PMMA nanospheres were synthesized by surfactant-free Pickering emulsion polymerization, the emulsions of methyl methacrylate (MMA) monomer were dispersed in water stabilized by the hydrophilic laponite clay. The grey spherical regions were considered to be PMMA cores and the dark strips were laponite plates. The grey cores were surrounded by multitudinous densely stacked laponite plates, demonstrating the role of laponite plates as a stabilizer.

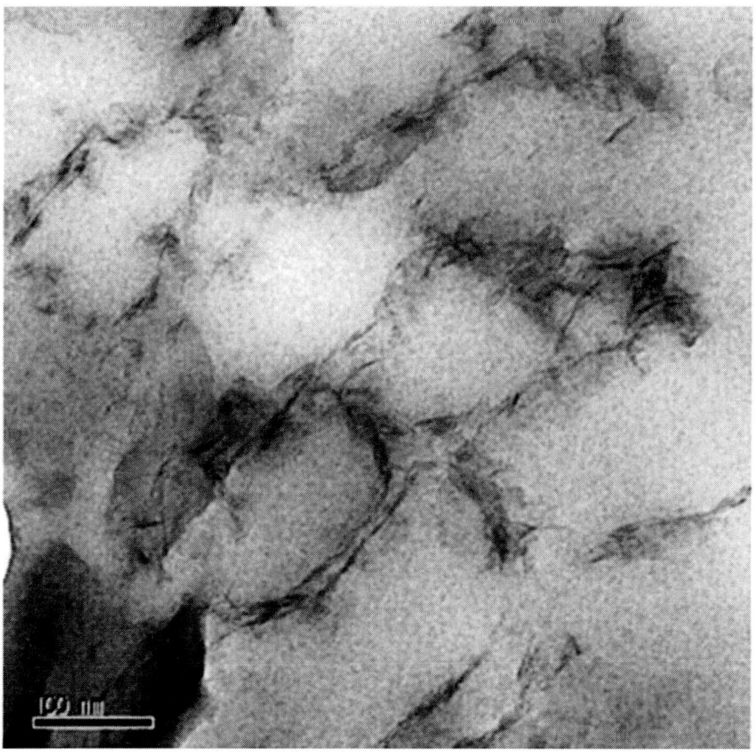

Figure 4. TEM image of a poly(methyl methacrylate) (PMMA) nanoparticle stabilized by laponite (Reprinted from Reference [60] with permission).

2.2. Crystalline State

In the case of crystalline clay, the delicate layer-layer structure and the changes in the d-spacing between adjacent layers were examined by X-ray diffraction

(XRD). The initial XRD patterns were different from the final XRD patterns, in which the characteristic peak for clay showed a shift in intensity and position.

From the XRD pattern of pure clay and PANI/clay nanoparticles [61] (Figure 5), the salient peak (d001) equivalent to the basal spacing of MMT was calculated to be 3.25 nm using the Bragg Equation: $\lambda = 2d \cdot \sin\theta$ (λ = 0.154 nm) [62]. The PANI/clay nanoparticles in Figure 5 showed no distinct sharp peak but a wide plateau, proving the disappearance of the layered structure. Nevertheless, it can be deduced that a few clay particles were not exfoliated, and the entire exfoliation of clay remains a difficult task [63].

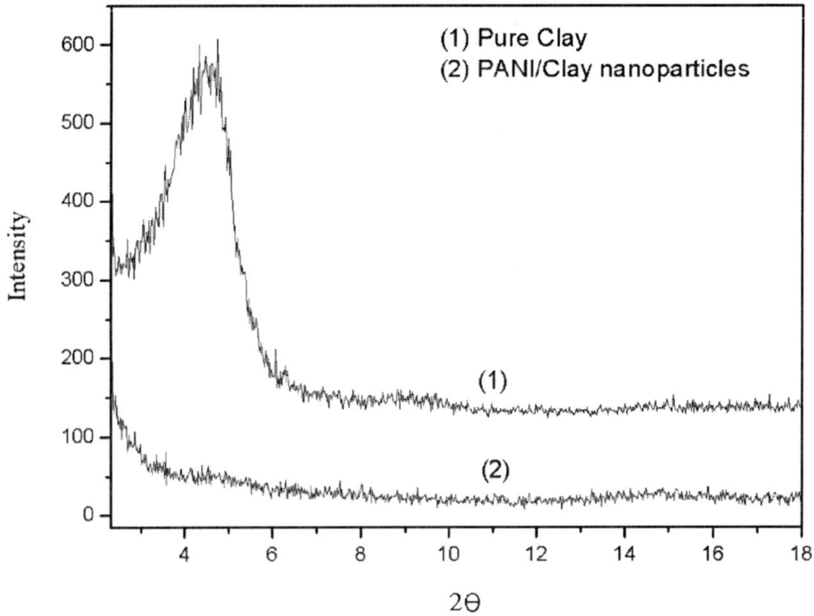

Figure 5. X-ray diffraction (XRD) patterns of pure clay (1) and clay sheet-stabilized polyaniline granules (2) (Reprinted from Reference [61] with permission).

Figure 6 presents XRD patterns of Pal, PANI and synthesized Pal/PANI nanocomposite particles. The crystalline PANI possessed reflections at 18.4° and 25.7° 2θ. The reflection centered at 18.4° 2θ was due to the periodicity in the orientation parallel to the polymer chain, while the peak at 25.7° 2θ was assigned to the periodicity in the orientation perpendicular to the polymer chain [64]. For Pal, the representative reflections were observed at 8.3°, 13.6°,

19.7° and 26.6° 2θ, which were assigned to the (1 1 0), (2 0 0), (0 4 0) and (4 0 0) planes of Pal, respectively [65]. Pal/PANI showed a similar set of characteristic reflections, which means that the crystal structure of Pal has been maintained during its polymerization reaction. The PANI reflection in Pal/PANI is almost invisible because Pal/PANI has a relative thin layer of amorphous PANI synthesized by this polymerization reaction [65].

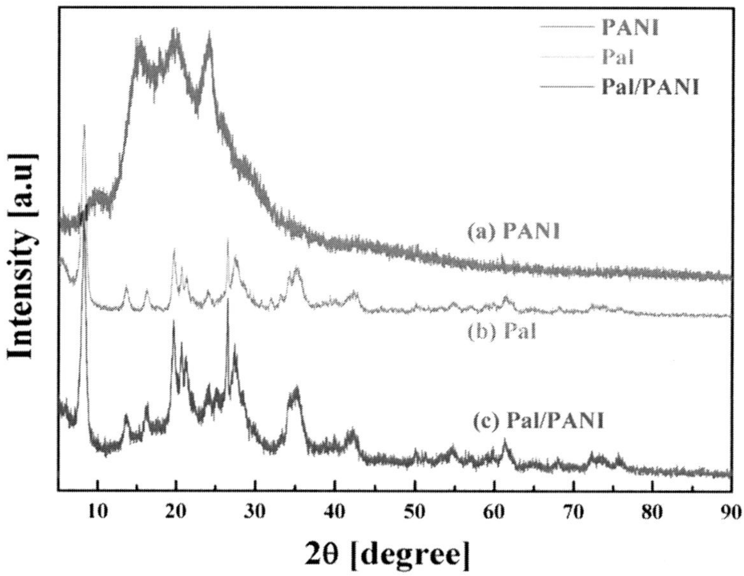

Figure 6. XRD patterns for pristine PANI (**a**), Pal (**b**) and Pal/PANI (**c**) (Reprinted from Reference [46] with permission).

2.3. Thermal Properties

The nanoscopic combination of conducting polymers with clay affects the thermal performance, in which the enhanced thermal properties are due to the significant enhancements in the interfacial conglutination between the polymer and clay [66]. Generally, better interfacial bonding imparts better properties to a polymer nanocomposite, such as tensile strength, hardness and high modulus, as well as resistance to fatigue, tear, corrosion and cracking [67]. Thermogravimetric analysis (TGA) shows the changed thermal behaviors.

Figure 7 presents the weight composition and thermal stability of the HNT and polypyrrole (PPy)/HNT nanocomposite [68]. The PPy/HNT nanocomposites exhibited two-step thermal degradation behavior with the first weight loss from 300 °C and the second thermal degradation from 500 °C, indicating that they were thermally stable up to 300 °C. A sharp loss in mass was observed at 300 and 500 °C, possibly due to the large scale thermal degradation of the PPy chains [69] and the dehydroxylation of HNT [70]. On the other hand, this char formation temperature indicates the increased thermal stability of the polymer/clay nanocomposite [71].

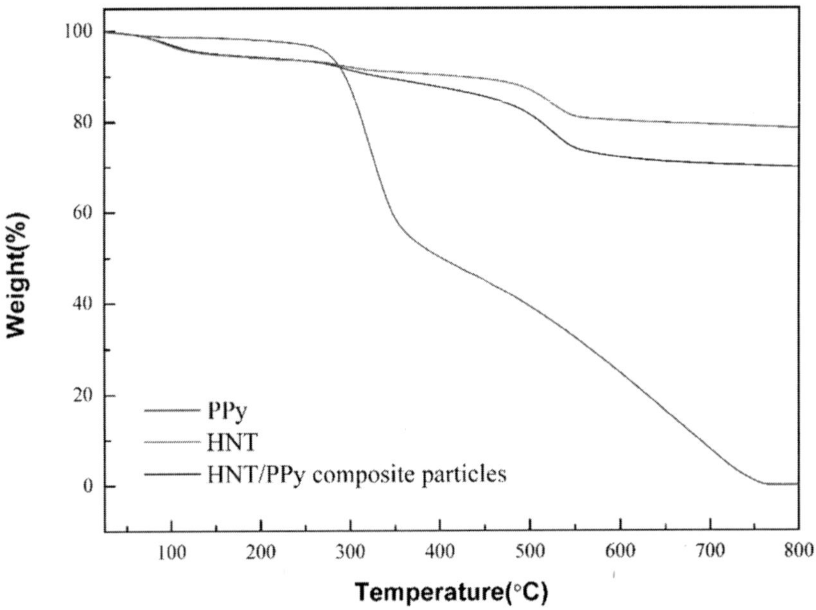

Figure 7. Thermogravimetric analysis (TGA) curve of HNT and PPy/HNT composite (Reprinted from Reference [68] with permission).

3. ELECTRORHEOLOGICAL (ER) CHARACTERISTICS

Kim *et al.* [53] prepared the an ER fluid by dispersing the PS/laponite nanoparticles in silicone oil and observed its structural change directly by optical microscope (OM) under an external electric field with a Direct current (DC) high voltage source. The PS/laponite nanoparticle-based ER fluid exhibited a typical ER chain structure. In the absence of an electric field, the particles were dispersed randomly in silicone oil, indicating a liquid-like state (Figure 8a). In an applied electric field, the particles moved immediately and formed a chain structure aligned along the orientation of the applied electric field (Figure

8b). Normally, this phenomenon of chain formation under an external applied electric field can be maintained as long as the electric field is applied.

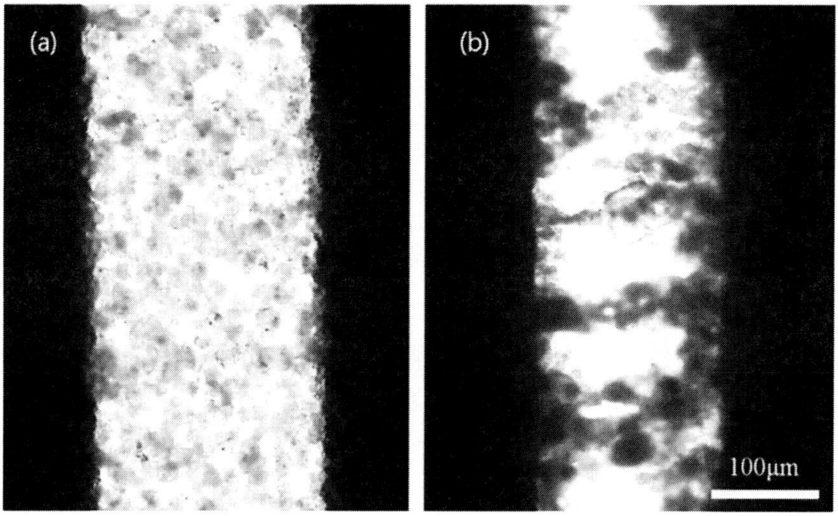

Figure 8. Optical microscope (OM) images of the electrorheological (ER) fluid based on PS/laponite nanoparticles in the absence of an electric field (**a**) and in an electric field (**b**) (Reprinted from Reference [53] with permission).

Owing to the dielectric polarization [72,73] of ER fluids arising from the mismatch between the dielectric constants of the medium oil and the dispersed nanocomposites, particulate materials with a higher dielectric constant [74], are expected to be beneficial for the superior ER effects. Figure 9 shows the variation of the dielectric constants as a function of the applied electrical frequency for the PUA and PUA/clay composite particles [47]. The clay amalgamated PUA particles showed significantly improved dielectric constant values than the bare PUA particles.

Regarding the flow curve, however, the Pal/PANI nanocomposite particle-based ER fluid exhibited an unusual decreasing trend in shear stress at a low shear rate region, and an increased shear stress with an increasing shear rate. The chain structures began to deteriorate with hydrodynamic shear deformation, and the damaged structures tended to reform the chains repeatedly due to the applied electric field, depending on the magnitude of the applied shear and the particle to particle interactions in the fibrils. In the low shear rate region, the electrostatic interactions were dominant [75,76]. At a high shear rate region, where the hydrodynamic interaction was greater, the broken chain particles had less chance to reform and the ER fluid behaved like a pseudo-Newtonian fluid [75]. Therefore, the Bingham fluid model and Cho–Choi–Jhon (CCJ) model were used to explain the shear stress behavior and yield stress [76].

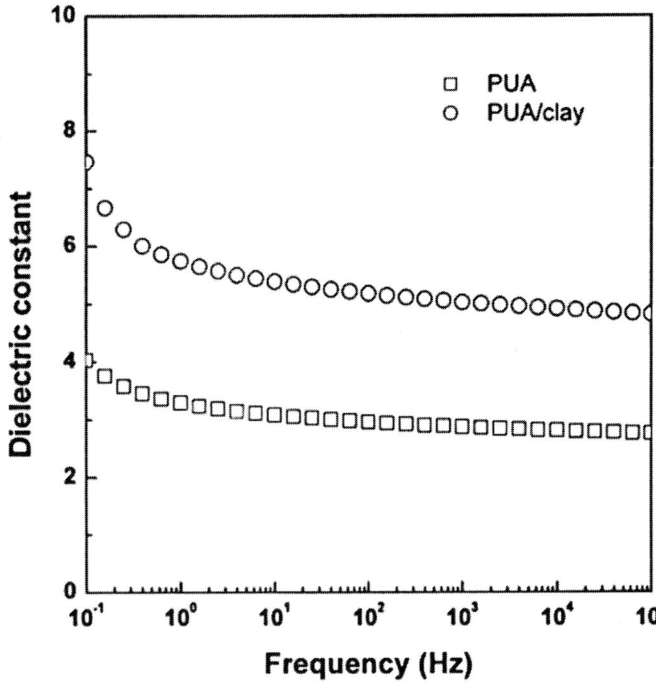

Figure 9. Dielectric constants as a function of the applied electrical frequency for bare poly(urethane acrylate) (PUA) and PUA/clay composite particles (Reprinted from Reference [47] with permission).

The Bingham fluid model shown in Equation (1), which is the simplest model with two parameters originating from Newtonian viscosity (η_0) and yield stress (τ_0), is used widely to describe the shear stress behavior of ER suspensions and conventional suspension systems.

$$\begin{aligned} \tau &= \tau_0 + \eta_0 \dot{\gamma} \ (\tau \geqslant \tau_0) \\ \dot{\gamma} &= 0 \ (\tau < \tau_0) \end{aligned} \quad (1)$$

where τ represents the shear stress and $\dot{\gamma}$ is the shear rate. The dotted lines in Figure 10a were from Equation (1). The simple Bingham model, however, could not be fitted to the flow curve of Pal/PANI ER fluid. Therefore, the CCJ model shown in Equation (2) was suggested to re-plot the shear stress behavior by fitting the curves using six parameters [75].

$$\tau = \frac{\tau_0}{1+(t_1\dot{\gamma})^{\alpha}} + \eta_{\infty}\left(1+\frac{1}{(t_2\dot{\gamma})^{\beta}}\right)\dot{\gamma} \quad (2)$$

where η_{∞} represents the viscosity at the infinite shear rate that is interpreted as the viscosity in the absence of an external electric field. The parameters, t_1 and t_2, are time constants, the exponents α and β are defined as the decrease and increase in shear stress; the exponents β has the range $0 < \beta \leq 1$, due to $d\tau\dot{\gamma} \geq 0$ [68]. Figure 10a shows the fitting of the two model equations for the Pal/PANI composite-based ER fluid. The solid lines from the CCJ model showed a better fit to the flow curves than the dotted lines generated by fitting the Bingham model in both the low and high shear rate regions [75].

As shown in Figure 10b, the shear viscosity exhibited shear thinning behavior as a function of the shear rate as similar to that in various polymeric systems [77]. Generally, the non-Newtonian behavior in the absence of an electric field is due to the particle dispersed state in a high concentrated ER fluid.

The dynamic yield stress is a characteristic factor of an ER fluid. In Figure 11, the dynamic yield stress of the Pal/PANI composite particle-based ER fluid was plotted as a function of the electric field strength (E) in log–log scale curves. The yield stress (τ_y) of an ER fluid is related to the electric field strength, and can be described as a power law relationship:

$$\tau_y \propto E^{m} \quad (3)$$

where the exponent m was obtained by fitting the yield stress over a broad electric field range. The dependence of the dynamic yield stress could be expressed as $\tau_y \propto E^2$, which is a polarization model of the ER mechanism [78].

Volume fraction of electro-responsive ER particles dispersed in the suspension is one of the major factors that affect the electric field dependent shear viscosity [79,80,81,82,83]. Guzel *et al.* [81] fabricated polyindene (PIN) and three volume fractions of OMMT nanocomposites namely K1 (5.5%), K2 (7.2%) and K3 (12.1%) to study their ER characteristics. Figure 12shows the change of the electric field-dependent shear viscosity with clay volume fraction at constant conditions (E = 3 kV/mm, $\dot{\gamma}$ = $1 \cdot s^{-1}$, T = 25 °C). The higher particle concentration leads to the intensive particle chains formed by the influence of an external electric field, resulting in a higher resistance to flow.

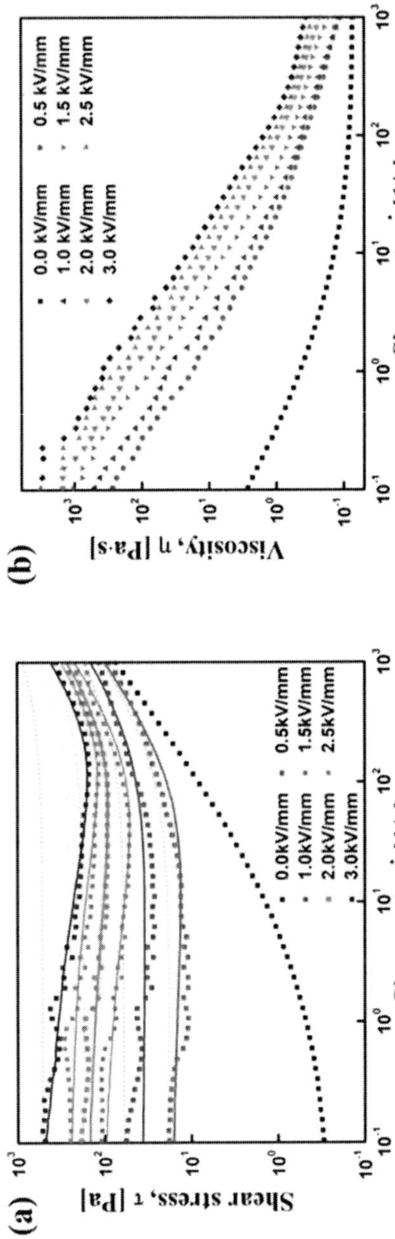

Figure 10. Shear stress (**a**) and shear viscosity (**b**) curves *vs.* shear rate for Pal/PANI based on ER fluids (10 vol. %) under a range of electric field strengths, the dashed lines were fitted using a conventional Bingham model, the solid lines were fitted via a suggested CCJ model (Reprinted from Reference [46] with permission).

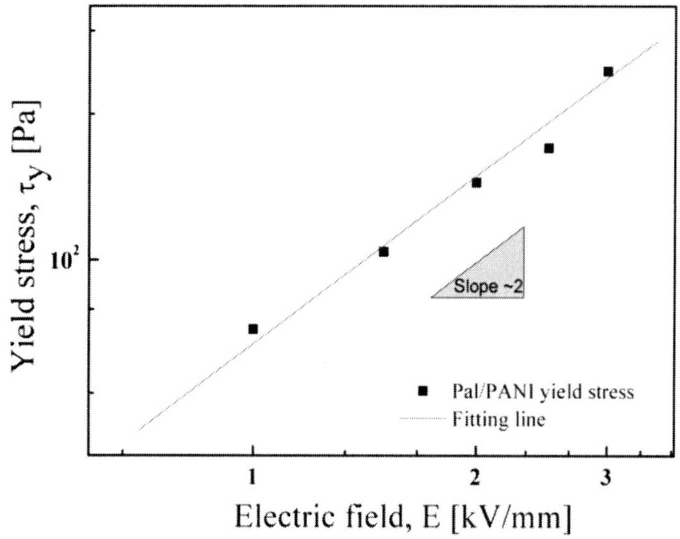

Figure 11. Dynamic yield stress as a function of the electric field strength of the Pal/PANI composite particles based ER fluid. The solid line was fitted using the equation, $\tau_y \propto E^2$ (Reprinted from Reference [46] with permission).

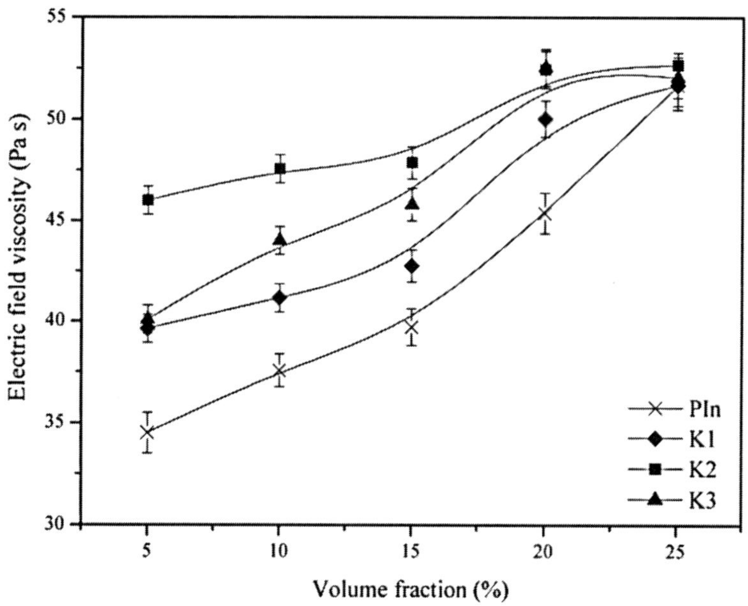

Figure 12. Effect of volume fraction on electric field viscosity (E = 3 kV/mm, $\dot{\gamma}$ = $1{\cdot}s^{-1}$, T= 25 °C). (Reprinted from Reference [81] with permission).

Furthermore, Eristi *et al.* [80] synthesized PIN and five PIN/kaolinite composites containing different amounts of kaolinite of K1 (78%), K2 (63%), K3 (47%), K4 (25%) and K5 (15%). Figure 13 shows that the electric field-dependent shear viscosity increased with enhanced electric field strength. However at a given electric field applied of 3 kV/mm, it was observed that the electric field viscosity reduced with increasing clay content.

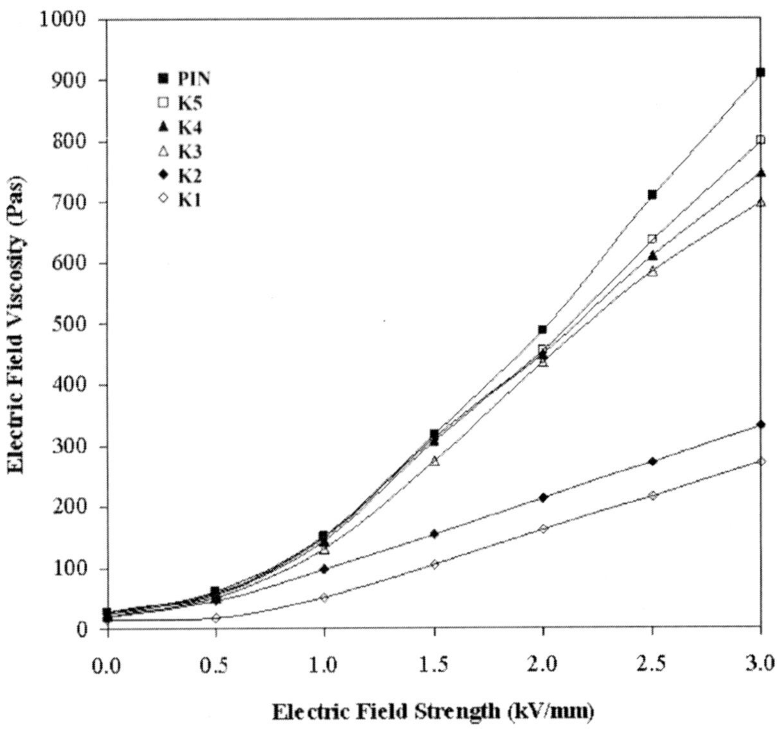

Figure 13. The change of viscosity with electric field strength, (c = 15 (m/m, %), $\dot{\gamma}$ = 1.0·s^{-1}, T = 25 °C). (Reprinted from Reference [80] with permission).

Figure 14 compares the frequency dependence of the storage modulus (G') and loss modulus (G'') of a PPy/HNT fluid, the measured frequency range was 1–100 rad/s. In the absence of an electric field, the storage modulus increased linearly with frequency, showing liquid-like characteristics. In an applied external electric field, the G' and G'' values increased in proportion to the electric field. The G' values representing an elastic response were higher than those of G'', representing viscous property indicating that the ER fluid has very strong solid-like behavior, which is the dominant factor of the elastic property over the viscous one.

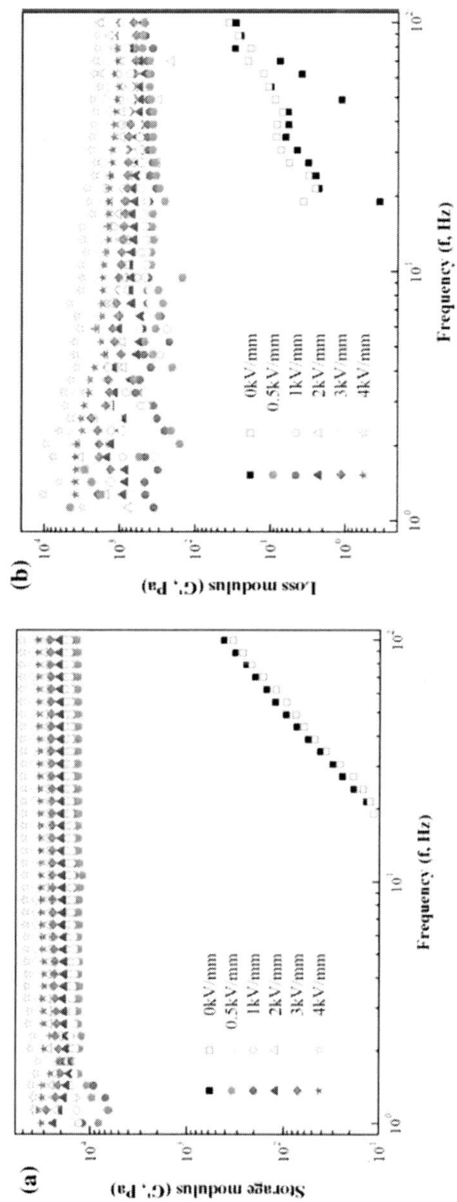

Figure 14. Storage modulus (**a**) and loss modulus (**b**) *versus* frequency for PPy/HNT composite-based ER fluid (close symbol 10 vol. % and open symbol 15 vol. % particle concentration) under various electric field strengths (Reprinted from Reference [68] with permission).

To further examine the ER characteristics of the PS/laponite nanoparticle-based ER fluid, the relationship between dielectric and ER properties were examined using an inductance capacitance resistance (LCR) meter. Figure 15 shows the dielectric spectra and Cole–Cole plot, respectively [84]. Both the permittivity (ε') and loss factor (ε'') were measured as a function of frequency (ω). The model is represented in terms of the complex dielectric constant as follows:

$$\varepsilon^* = \varepsilon' - \varepsilon'' = \varepsilon_\infty + \frac{\Delta\varepsilon}{(1 + i\omega\lambda)^{1-\alpha}} \tag{4}$$

In Equation (4), ε^* is a complex dielectric constant and ε_0 is the dielectric constant when ω approaches 0. $\Delta\varepsilon$ is the difference between the dielectric constant at 0 and infinite frequency (ε_0 and ε_∞). They are the distribution curves over a broad frequency range. λ is the dielectric relaxation time at the frequency of which the dielectric loss arrives the maximum value. The exponent (1–α) represents the broadness of the relaxation time distribution and α is a value in the range 0–1. When α is zero, Equation (4) reduces to the Debye's single relaxation time model. $\Delta\varepsilon$ is the achievable polarizability in the ER fluids, which equals 0.546. Owing to the laponite content in the PS/laponite particles, the value of $\Delta\varepsilon$ was much lower than those reported elsewhere [85,86,87]. A similar correlation between the dielectric properties and ER performance was also reported for the silica nanoparticle decorated polyaniline nanofiber-based ER fluid [88].

As shown in Figure 16, observing the relaxation behavior is one way of inspecting the phase change from a liquid-like to solid-like phase. The relaxation modulus $G(t)$ was calculated from the G' and G'' values using the typical formula known as the Schwarzl Equation, as given in Equation (5). Note that $G(t)$ is difficult to measure experimentally due to the intrinsic properties of the materials and the limitation of the mechanical measurements [89]. $G(t)$, as a function of time showed a linear increase with increasing electric field strength, which confirmed the strong interaction among PANI/HNT particles.

$$G(t) \cong G'(\omega) - 0.560G''\left(\frac{\omega}{2}\right) + 0.200G''(\omega) \tag{5}$$

In addition, Table 1 summarizes fabrication method, clay content and slope of dynamic yield stress of most electro-responsive polymer/clay nanocomposites covered in this review. The slope of dynamic yield stress of the ER fluids ranged from 1.2 to 2, implying that the mechanism seems to be dependent on not only different materials but also different fabrication methods.

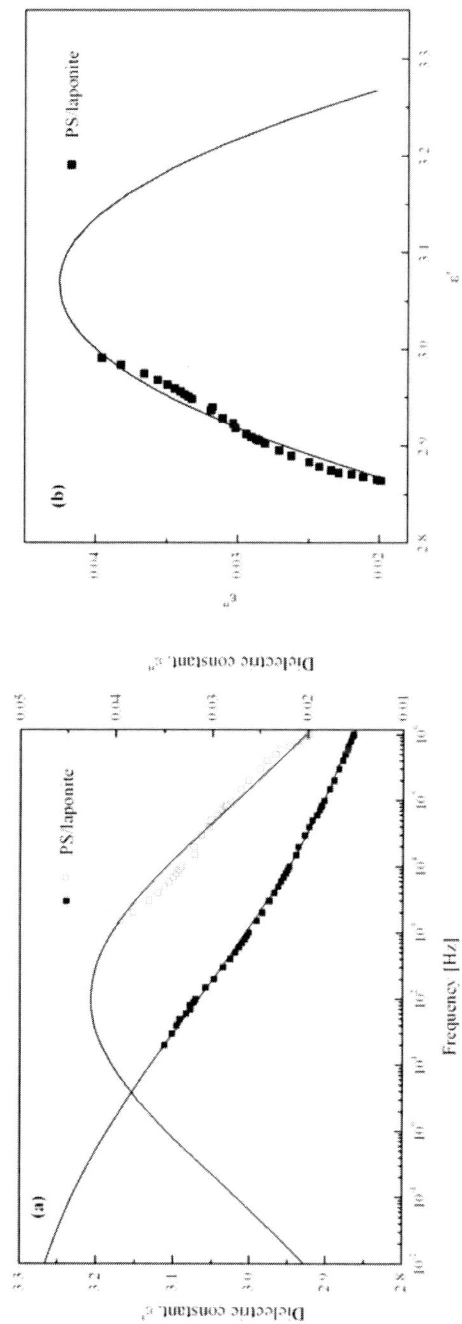

Figure 15. (**a**) Dielectric spectra (ε′: closed symbols; ε″: open symbols) and (**b**) Cole–Cole plot of the ER fluids. The fitting lines are generated from Equation (4) (Reprinted from Reference [53] with permission).

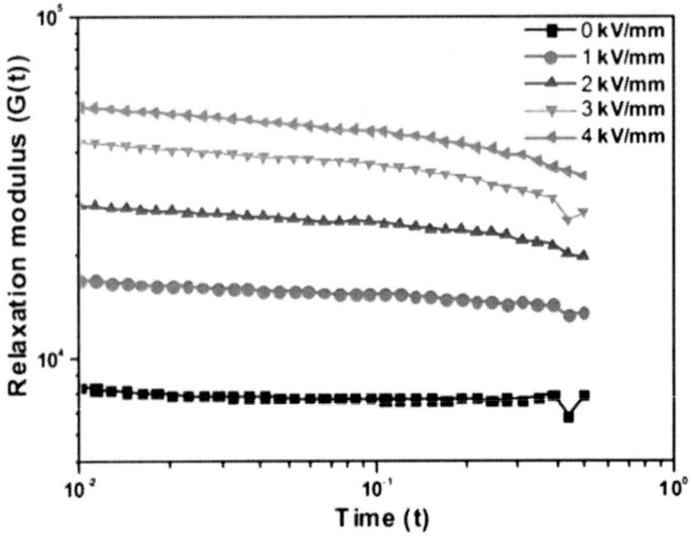

Figure 16. Relaxation modulus *G*(*t*) of PANI/HNT composite-based ER fluid as calculated from *G*′(ω) and *G*″(ω) (Reprinted from Reference [45] with permission).

4. CONCLUSIONS

Various electric stimuli-responsive polymer-clay nanocomposites synthesized by a range of methods were reviewed. SEM, TEM and XRD confirmed their successful intercalation. The thermal stability of the conducting polymer chain was also enhanced due to the shielding role of the clay. In addition, the conducting polymer/clay nanocomposite-based ER fluids also found showed excellent ER behaviors and follow the previously proposed CCJ model.

ACKNOWLEDGMENTS

This work was supported by a research grant from National Research Foundation, Korea (NRF-2013R1A1A2057955).

Table 1. Summary of electro-responsive polymer/clay nanocomposites with fabrication method, clay content and slope of dynamic yield stress.

Polymer/Clay Nanocomposites	Fabrication Method	Clay Content (wt %)	Slope of Dynamic Yield Stress	Reference
PANI/MMT	*In-situ* polymerization	0.75–3	-	[41]
PEA/MMT	*In-situ* polymerization	0.5–5	-	[42]
PUA/MMT	Suspension polymerization	5	-	[47]
SAN/MMT	Emulsion polymerization	4.76	-	[49]
PANI/OMMT	Pickering emulsion polymerization	-	1.5	[52]
PIN/OMMT	*In-situ* polymerization	5.5–12.1	-	[81]
PIN/kaolinite	cationic radical polymerization	15–78	-	[80]
PANI/HNT	*In-situ* polymerization	-	2	[45]
PPy/HNT	*In-situ* polymerization	-	1.5	[68]
PANI/PAL	*In-situ* polymerization	-	2	[46]
PS/laponite	Pickering emulsion polymerization	8.85	1.2	[53]

AUTHOR CONTRIBUTIONS

The whole process of preparation of this review has been initiated and guided by Hyoung Jin Choi, while Shang Hao Piao and Seung Hyuk Kwon wrote the original version of the manuscript.

CONFLICTS OF INTEREST

The authors declare no conflict of interest.

REFERENCES

1. Abranyi, A.; Szazdi, L.; Pukanszky, B.; Vancso, G.J.; Pukansky, B. Formation and detection of clay network structure in Poly(propylene)/layered silicate nanocomposites. Macromol. Rapid Commun. 2006, 27, 132–165.

2. Messersmith, P.B.; Giannelis, E.P. Synthesis and characterization of layered silicate-expoxy nanocomposites. Chem. Mater. 1994, 6, 1719–1725.

3. Ray, S.S.; Okamoto, M. Biodegradable polylactide and its nanocomposites: Opening a new dimension for plastics and composites. Macromol. Rapid Commun. 2003, 24, 815–840.

4. Kuksenok, O.V.; Ruhwandl, R.W.; Shiyanovskii, S.V.; Terentjev, E.M. Director structure around a colloid particle suspended in a nematic liquid crystal. Phys. Rev. 1996, 54, 5198–5203.

5. Chang, Y.M.; Tsai, Y.Y.; Huang, Y.P.; Chen, W.S.; Lee, W. Electrical and electro-optical properties of nematic-liquid-crystal-montmorillonite-clay nanocomposites. Jpn. J. Appl. Phys. 2007, 46, 7368–7370.

6. Lim, S.T.; Hyun, Y.H.; Choi, H.J.; Jhon, M.S. Synthetic biodegradable aliphatic polyester/montmorillonite nanocomposites. Chem. Mater. 2002, 14, 1839–1844.

7. Alexandre, M.; Beyer, G.; Henrist, C.; Cloots, R.; Rulmant, A.; Jerome, R.; Dubois, P. Preparation and properties of layered silicate nanocomposites based on ethylene vinyl acetate copolymers. Macromol. Rapid Commun. 2001, 22, 643–646.

8. Hyun, Y.H.; Lim, S.T.; Choi, H.J.; Jhon, M.S. Rheology of poly(ethylene oxide)/organoclay nanocomposites.Macromolecules 2001, 34, 8084–8093.

9. Fong, H.; Vaia, R.A.; Sanders, J.H.; Lincoln, D.; Vreugdenhil, A.J.; Liu, W.D.; Bultman, J.; Chen, C.G. Self-passivation of polymer-layered silicate nanocomposites. Chem. Mater. 2001, 13, 4123–4129.

10. Chen, B.Q.; Evans, J.R.G.; Greenwell, H.C.; Boulet, P.; Coveney, P.V.; Bowden, A.A.; Whiting, A. A critical appraisal of polymer-clay nanocomposites. Chem. Soc. Rev. 2008, 37, 568–594.

11. Chen, G.M.; Liu, S.H.; Zhang, S.F.; Qi, Z.N. Self-assembly in a polystyrene/montmorillonite nanocomposite. Macromol. Rapid Commun. 2008, 21, 746–749.

12. Wang, K.; Liang, S.; Du, R.N.; Zhang, Q.; Fu, Q. The interplay of thermodynamics and shear on the dispersion of polymer nanocomposite. Polymer 2004, 45, 7953–7960.

13. Faguy, P.W.; Lucas, R.A.; Ma, W.L. An FT-IR-ATR spectroscopic study of the spontaneous polymerization of pyrrole in iron-exchanged montmorillonite. Colloid. Surf. Sci. A 1995, 105, 105–112.

14. Carrado, K.A.; Xu, L.Q. In situ synthesis of polymer-clay nanocomposites from silicate gels. Chem. Mater. 1998, 10, 1440–1445.

15. Ballav, N.; Sardar, P.S.; Ghosh, S.; Biswas, M. Polyaniline and polypyrrole modified conductive nanocomposites of polyfuran and polythiophene with montmorillonite clay. J. Ind. Eng. Chem. 2006, 41, 2959–2964.

16. Masdarolomoor, F.; Innis, P.C.; Ashraf, S.; Kaner, R.B.; Walace, G.G. Nanocomposites of polyaniline/poly (2-methoxyaniline-5-sulfonic acid). Macromol. Rapid Commun. 2006, 27, 1995–2000.

17. Yeh, J.M.; Chin, C.P.; Chang, S. Enhanced corrosion protection coatings prepared from soluble electronically conductive polypyrrole-clay nanocomposite materials. J. Appl. Polym. Sci. 2003, 88, 3264–3272.

18. Choi, H.J.; Kim, S.G.; Hyun, Y.H.; Jhon, M.S. Preparation and rheological characteristics of solvent-cast poly(ethylene oxide)-montmorillonite nanocomposites. Macromol. Rapid Commun. 2001, 22, 320–325.

19. Ogawa, M.; Ishii, T.; Miyamoto, N.; Kuroda, K. Intercalation of a cationic azobenzene into montmorillonite. Appl. Clay Sci. 2003, 22, 179–185.

20. Ray, S.S.; Yamada, K.; Okamoto, M.; Ueda, K. Control of biodegradability of polylactide via nanocomposite technilogy.Macromol. Mater. Eng. 2003, 288, 203–208.

21. Tjong, S.C.; Meng, Y.Z.; Hay, A.S. Novel preparation and properties of polypropylene-vermiculite nanocomposites.Chem. Mater. 2002, 14, 44–51.

22. Manias, E.; Touny, A.; Wu, L.; Strawhecker, K.; Lu, B.; Chung, T.C. Polypropylene/montmorillonite nanocomposites. Review of the synthetic routes and materials properties. Chem. Mater. 2001, 13, 3516–3523.

23. Caseri, W. Nanocomposites of polymers and metals or semiconductors: Historical background and optical properties.Macromol. Rapid Commun. 2000, 21, 705–722.

24. Leroux, F.; Besse, J.P. Polymer interleaved layered double hydroxide: A new emerging class of nanocomosites. Chem. Mater. 2001, 13, 3507–3515.

25. Fang, F.F.; Choi, H.J.; Joo, J. Conducting polymer/clay nanocomposites and their applications. J. Nanosci. Nanotechnol.2008, 8, 1559–1581.

26. Zhao, X.P.; Yin, J.B. Preparation and electrorheological characteristics of rare-earth-doped TiO2 suspensions. Chem. Mater. 2002, 14, 2258–2263.

27. Loudet, J.C.; Poulin, P. Application of an electric field to colloidal particles suspended in a liquid-crystal solvent. Phys. Rev. Lett. 2001, 87, 165503.

28. Oz, K.; Yavuz, M.; Yilmaz, H.; Unal, H.I.; Sari, B. Electrorheological properties and creep behavior of polyindole/poly(vinyl acetate) composite suspensions. J. Mater. Sci. 2008, 43, 1451–1459.

29. Cheng, Y.C.; Guo, J.J.; Xu, G.J.; Cui, P.; Liu, X.H.; Liu, F.H.; Wu, J.H. Electrorheological property and microstructure of acetamide-modified TiO2 nanoparticles. Colloid Polym. Sci. 2008, 286, 1493–1497.

30. Chin, B.D.; Lee, Y.S.; Park, O.O. Dielectric and rheological behaviors of semiconductive polymer suspension as an electrorheological fluid. J. Mod. Phys. B 1999, 13, 1852–1859.

31. Kim, S.G.; Kim, J.W.; Jang, W.H.; Choi, H.J.; Jhon, M.S. Electrorheological characteristics of phosphate cellulose-based suspensions. Polymer 2001, 42, 5005–5012.

32. Kim, J.W.; Liu, F.; Choi, H.J. Polypyrrole/clay nanocomposite and its electrorheological characteristics. J. Ind. Eng. Chem. 2002, 8, 399–403.

33. Wen, W.J.; Huang, X.X.; Sheng, P. Electrorheological fluids, structures and mechanisms. Soft Matter 2008, 4, 200–210.

34. Lei, X.P.; Liu, Y.S.; Su, Z.X. Synthesis and characterization of organo-attapulgite/polyaniline-dodecylbenzenesulfonic acid based on emulsion polymerization method. Polym. Compos. 2008, 29, 239–244.

35. Chin, B.D.; Park, J.H.; Kwon, M.H.; Park, O.O. Rheological properties and dispersion stability of magnetorheological (MR) suspensions. Rheol. Acta 2001, 40, 211–219.

36. Park, J.H.; Chin, B.D.; Park, O.O. Rheological properties and stabilization of magnetorheological fluids in a water-in-oil emulsion. J. Colloid Interf. Sci. 2001, 240, 349–354.

37. Lim, S.T.; Cho, M.S.; Jang, I.B.; Choi, H.J. Magnetorheological characterization of carbonyl iron based suspension stabilized by fumed silica. J. Magn. Magn. Mater. 2004, 282, 170–173.

38. Bossis, G.; Abbo, C.; Cutillas, S.; Lacis, S.; Métayer, C. Electroactive and electrostructured elastomers. Int. J. Mod. Phys. B 2001, 15, 564–573.

39. Kim, J.W.; Kim, C.A.; Choi, H.J.; Choi, S.B. Role of surfactant on damping performance of polyaniline based electrorheological suspension. Korea Aust. Rheol. J. 2006, 18, 25–30.

40. Ruiz-Hitzky, E.; Aranda, P.; Casal, B.; Galván, J.C. Nanocomposite materials with controlled ion mobility. Adv. Mater.1995, 7, 180–184.

41. Yeh, J.M.; Liou, S.J.; Lai, C.Y.; Wu, P.C. Enhancement of corrosion protection effect in polyaniline via the formation of polyaniline-clay nanocomposite materials. Chem. Mater. 2001, 13, 1131–1136.

42. Yeh, J.M.; Chen, C.L.; Chen, Y.C.; Ma, C.Y.; Lee, K.R.; Wei, Y.; Li, S.X. Enhancement of corrosion protection effect of poly(o-ethoxyaniline) via the formation of poly (o-ethoxyaniline)-clay nanocomposite materials. Polymer 2002, 43, 2729–2736.

43. Zhou, C.; Du, X.; Liu, Z.; Mai, Y.W.; Ringer, S.P. Multi-holed clay nanotubes and their modification with a polyaniline nanolayer. J. Mater. Sci. 2011, 46, 446–450.

44. Zhang, L.; Wang, T.; Liu, P. Polyaniline-coated halloysite nanotubes via in-situ chemical polymerization. Appl. Surf. Sci. 2008, 255, 2091–2097.

45. Zhang, W.L.; Choi, H.J. Fabrication of semiconducting polyaniline-wrapped halloysite nanotube composite and its electrorheology. Colloid Polym. Sci. 2012, 290, 1743–1748.

46. Chae, H.S.; Zhang, W.L.; Piao, S.H.; Choi, H.J. Synthesized palygorskite/polyaniline nanocomposite particles by oxidative polymerization and their electrorheology. Appl. Clay Sci. 2015, 107, 165–172.

47. Jun, J.B.; Suh, K.D. Preparation and electrorheological characterization of suspensions of poly(urethane acrylate)/clay nanocomposite particles. J. Appl. Polym. Sci. 2003, 90, 458–464.

48. Tong, X.; Zhao, H.; Tang, T.; Feng, Z.; Huang, B. Preparation and characterization of poly(ethyl acrylate)/bentonite nanocomposites by in situ emulsion polymerization. J. Polym. Sci. A Polym. Chem. 2002, 40, 1706–1711.

49. Kim, J.W.; Jang, L.W.; Choi, H.J.; Jhon, M.S. Physical and electroresponsive characteristics of the intercalated styrene-acrylonitrile copolymer/clay nanocomposite under applied electric fields. J. Appl. Polym. Sci. 2003, 89, 821–827.

50. Tang, J.; Quinlan, P.J.; Tam, K.C. Stimuli-responsive Pickering emulsions: Recent advances and potential applications.Soft Matter 2015, 11, 3512–3529.

51. Piao, S.H.; Kwon, S.H.; Zhang, W.L.; Choi, H.J. Celebrating soft matter's 10th anniversary: Stimuli-responsive Pickering emulsion polymerized smart fluids. Soft Matter 2015, 11, 646–654.

52. Fang, F.F.; Liu, Y.D.; Choi, H.J. Synthesis and electrorheological characteristics of polyaniline/organoclay nanoparticles via Pickering emulsion polymerization. Smart Mater. Struct. 2010, 19, 124002.

53. Kim, Y.J.; Liu, Y.D.; Choi, H.J.; Park, S.J. Facile fabrication of Pickering emulsion polymerized polystyrene/laponite composite nanoparticles and their electrorheology. J. Colloid Interf. Sci. 2013, 394, 108–114.

54. Park, J.U.; Choi, Y.S.; Cho, K.S.; Kim, D.H.; Ahn, K.H.; Lee, S.J. Time–electric field superposition in electrically activated polypropylene/layered silicate nanocomposites. Polymer 2006, 47, 5145–5153.

55. Kim, D.H.; Park, J.U.; Cho, K.S.; Ahn, K.H.; Lee, S.J. A novel fabrication method for poly(propylene)/clay nanocomposites by continuous processing. Macromol. Mater. Eng. 2006, 291, 1127–1135.

56. Strounina, E.V.; Shepherd, R.; Kane-Maguire, L.A.P.; Wallace, G.G. Conformational changes in sulfonated polyaniline caused by metal salts and OH−. Synth. Met. 2003, 135/136, 289–290.

57. Bon, S.A.F.; Chen, T. Pickering stabilization as a tool in the fabrication of complex nanopatterned silica microcapsules.Langmuir 2007, 23, 9527–9530.

58. Bon, S.A.F.; Chen, T. Pickering miniemulsion polymerization using laponite clay as a stabilizer. Langmuir 2007, 23, 8316–8322.

59. Negrete-Herrera, N.; Putaux, J.L.; David, L.; Bourgeat-Lami, E. Polymer/laponite composite colloids through emulsion polymerization: Influence of the clay modification level on particle morphology. Macromolecules 2006, 39, 9177–9184.

60. Fang, F.F.; Kim, J.H.; Choi, H.J.; Kim, C.A. Synthesis and electrorheological response of nano-sized laponite stabilized poly(methyl methacrylate) spheres. Colloid Polym. Sci. 2009, 287, 745–749.

61. Fang, F.F.; Choi, H.J.; Seo, Y.S. Novel fabrication of polyaniline particles wrapped by exfoliated clay sheets and their electrorheology. J. Nanosci. Nanotechnol. 2009, 9, 1–5.

62. Vaia, R.A.; Giannelis, E.P. Polymer melt intercalation in organically-modified layered silicates: Model predictions and experiment. Macromolecules 1997, 30, 8000–8009.

63. Lee, D.K.; Lee, S.H.; Char, K.; Kim, J. Expansion distribution of basal spacing of the silicate layers in polyaniline/Na-montmorillonite nanocomposites monitored with X-ray diffraction. Macromol. Rapid Commun. 2000, 21, 1136–1139.

64. Jin, E.; Lu, X.F.; Bian, X.J.; Kong, L.R.; Zhang, W.J.; Wang, C. Unique tetragonal starlike polyanilinemicrostructure and its application in electrochemical biosensing. J. Mater. Chem. 2010, 20, 3079–3083.

65. Wang, J.H.; Han, X.J.; Ma, H.R.; Ji, Y.F.; Bi, L.J. Adsorptive removal of humic acid from aqueous solution on polyaniline/attapulgite composite. Chem. Eng. J. 2011, 173, 171–177.

66. Agag, T.; Koga, T.; Takeichi, T. Thermal and mechanical properties of polyimide-clay nanocomposites. Polymer 2001,42, 3399–3408.

67. Jang, B.Z. Control of interfacial adhesion in continuous carbon and Kevlar fiber reinforced polymer composites.Compos. Sci. Technol. 1992, 44, 333–349.

68. Jang, D.S.; Zhang, W.L.; Choi, H.J. Polypyrrole-wrapped halloysite nanocomposite and its rheological response under electric fields. J. Mater. Sci. 2014, 49, 7309–7316.

69. Jang, J.S.; Yoon, H.S. Novel fabrication of size-tunable silica nanotubes using a reverse-microemulsion-mediated sol–gel method. Adv. Mater. 2004, 16, 799–802.

70. Nicolini, K.P.; Fukamachi, C.R.B.; Wypych, F.; Mangrich, A.S. Dehydrated halloysite intercalated mechanochemically with urea: Thermal behavior and structural aspects. J. Colloid Interface Sci. 2009, 338, 474–479.

71. Park, D.P.; Lim, S.T.; Lim, J.Y.; Choi, H.J.; Choi, S.B. Electrorheological characteristics of solvent-cast polypyrrole/clay nanocomposite. J. Appl. Polym. Sci. 2009, 112, 1365–1371.

72. Parthasarathy, M.; Klingenberg, D.J. Electrorheology: Mechanisms and models. Mater. Sci. Eng. 1996, 17, 57–103.

73. Davis, L.C. Polarization forces and conductivity effects in electrorheological fluids. J. Appl. Phys. 1992, 72, 1334–1340.

74. Hao, T.; Kawai, A.; Ikazaki, F. Mechanism of the electrorheological effect—Evidence from the conductive, dielectric, and surface characteristics of water-free electrorheological fluids. Langmuir 1998, 14, 1256–1262.

75. Cho, M.S.; Cho, Y.H.; Choi, H.J.; Jhon, M.S. Synthesis and electrorheological characteristics of polyaniline-coated poly(methyl methacrylate) microsphere: Size effect. Langmuir 2003, 19, 5875–5881.

76. Cho, M.S.; Choi, H.J.; Jhon, M.S. Shear stress analysis of a semiconducting polymer based electrorheological fluid system. Polymer 2005, 46, 11484–11488.

77. Choi, H.J.; Park, S.H.; Yoon, J.S.; Lee, H.S.; Choi, S.J. Rheological study on poly-D(-)(3-hydroxybutyrate) and its blends with poly(ethylene oxide). Polym. Eng. Sci. 1995, 35, 1636–1642.

78. Fang, F.F.; Liu, Y.D.; Lee, I.S.; Choi, H.J. Well controlled core/shell type polymeric microspheres coated with conducting polyaniline: Fabrication and electrorheology. RSC Adv. 2011, 1, 1026–1032.

79. Erol, O.; Karakisla, M.; Unal, H.I.; Sacak, M. Electrorheological properties of polyaniline/K-feldspar conducting composite. J. Compos. Mater. 2012, 46, 1295–1304.

80. Eristi, C.; Yavus, M.; Yilmaz, H.; Sari, B.; Unal, H.I. Synthesis, characterization and electrorheological properties of polyindene/kaolinite composites. J. Macromol. Sci. A Pure Appl. Chem. 2007, 44, 759–767.

81. Guzel, S.; Erol, O.; Unal, H.I. Polyindene/Organo-montmorillonite conducting nanocomposites. II. Electrorheological properties. J. Appl. Polym. Sci. 2012, 124, 4935–4944.

82. Lu, J.; Zhao, X.P. Electrorheological properties of suspensions based on polyaniline-montmorillonite clay nanocomposite. J. Mater. Res. 2002, 17, 1513–1519.

83. Erol, O.; Unal, H.I.; Sari, B. Synthesis, electrorheology, and creep behaviors of in situ intercalated polyindole/organo-montmorillonite conducting nanocomposite. Polym. Compos. 2010, 31, 471–481.

84. Yin, J.B.; Xia, X.; Zhao, X.P. The electrorheological effect and dielectric properties of suspensions containing polyaniline@tutania nanocable-like particles. Soft Matter 2011, 7, 10978–10986.

85. Liu, C.; Li, C.; Chen, P.; He, J.; Fan, Q. Influence of long-chain branching on linear viscoelastic flow properties and dielectric relaxation of polycarbonates. Polymer 2004, 45, 2803–2812.

86. Pluta, M.; Jeszka, J.K.; Boiteux, G. Polylactide/montmorillonite nanocomposites: Structure, dielectric, viscoelastic and thermal properties. Eur. Polym. J. 2007, 43, 2819–2835.

87. Xu, J.; Wong, C.P. Characterization and properties of an organic-inorganic dielectric nanocomposite for embedded decoupling capacitor applications. Compos. A 2007, 38, 13–19.

88. Liu, Y.D.; Fang, F.F.; Choi, H.J. Silica nanoparticle decorated polyaniline nanofiber and its electrorheological response.Soft Matter 2011, 7, 2782–2789.

89. Schwarzl, F.L. Numerical calculation of stress relaxation modulus from dynamic data for linear viscoelastic materials.Rheol. Acta 1975, 14, 581–590.

Chapter 4

Nanoindentation of functionally graded polymer nanocomposites: assessment of the strengthening parameters through experiments and modeling

Tommaso Nardi[1], **Chad Hammerquist**[2], **John A. Nairn**[2], **Ayat Karimi**[3], **Jan-Anders E. Månson**[1] **and Yves Leterrier**[1]*

[1]Laboratoire de Technologie des Composites et Polymères (LTC), École Polytechnique Fédérale de Lausanne (EPFL), Lausanne, Switzerland

[2]Department of Wood Science and Engineering, Oregon State University, Corvallis, OR, USA

[3]Laboratoire de Physique de la Matière Complexe (LPMC), École Polytechnique Fédérale de Lausanne (EPFL), Lausanne, Switzerland

Nanoindentation tests were carried out on the surface of polymer nanocomposites exhibiting either graded or homogeneous distributions of Fe_3O_4@silica core-shell nanoparticles in a photocurable polymeric matrix. The results reveal a complex interplay between graded morphology, indentation depth, and calculated modulus and hardness values, which was elucidated through numerical simulations. First, it was experimentally shown how for small (1 µm) indentations, large increases in modulus (up to +40%) and hardness (up to +93%) were obtained for graded composites with respect to their homogeneous counterparts, whereas at a larger indentation depth (20 µm), the modulus and hardness of the graded and homogeneous composites did not substantially differ from each other and from those of the pure polymer. Then, through a material point method approach, experimental nanoindentation tests were successfully simulated, confirming the importance of the indentation depth and of the associated plastic zone as key factors for a more accurate design of graded polymer nanocomposites whose mechanical properties are able to fulfill the requirements encountered during operational life.

1. INTRODUCTION

Nanoindentation is a well-established characterization technique widely applied to metallic (Nair et al., 2011), ceramic (Woirgard et al., 1998), and polymeric materials (Chen and Diebels, 2014). Although challenging, the extraction of material properties by this method has been successfully carried out for macroscopically homogeneous samples, such as polycrystalline solids (Dao et al., 2001), amorphous solids (Rodríguez et al., 2012), and even nanocomposite structures (Penumadu et al., 2011; Díez-Pascual et al., 2015). On the contrary, less attention has been paid to the analysis of graded structures. Available studies were devoted to graded metal–polymer films (Nunes and Piedade, 2013) and graded bioactive ceramic coatings (Roop Kumar and Wang, 2002a,b). However, these works lack systematic studies of the effect of the gradient morphology on the obtained properties, when it was made clear that the properties directly reflect the concentration gradients of the constitutive elements (Nardi et al., 2014a). The attractiveness of such materials derives from their intrinsic ability to combine different and apparently incompatible features in a single body, making them appealing candidates for applications, such as coatings for either oxidation (Smeacetto et al., 2002), corrosion (Alegría-Ortega et al., 2012), or abrasion protection (Qureshi et al., 2013). Although a general increment of the mechanical properties (e.g., hardness) is desirable for many applications, a more precise design of the gradient morphology could lead to materials that perform better under loading conditions encountered during their operational life. In other words, the performance of graded structures synthesized via concentration of reinforcing elements in a soft matrix [e.g., nanoparticles (NPs) in a polymer] is strongly affected by the loads to which these materials are subjected, both at their final application stage and during testing (Reshetnyak and Kübarsepp, 1997; Mann and Arya, 2002; Jeong et al., 2003; Sevim and Barlas Eryurek, 2006).

In this work, two graded structures synthesized through the application of an external magnetic field gradient to a suspension of Fe_3O_4@SiO_2-MPS core-shell NPs in a photocurable polymeric matrix (Nardi et al., 2014b) were analyzed by means of nanoindentation, alongside with their two homogeneous counterparts and the pure polymeric matrix. The employment of core-shell NPs was motivated by the possibility to tune the properties of the nanocomposite depending on the characteristics of the shell, while maintaining the magnetic properties typical of the core. In this particular study, silica was chosen as the constituent of the shell for its remarkable hardness [2.8 GPa (Nardi et al., 2014a)] and stiffness [20 GPa (Nardi et al., 2014a)] in order to generate considerable gradients in mechanical properties within the final graded nanocomposite. Silica is also transparent to light contrary to magnetite, enabling photopolymerization of the graded nanocomposite (Nardi et al., 2013). Moreover, the surface of the employed core-shell NPs was functionalized with an appropriate acrylated ligand (3-methacryloyloxy propyltrimethoxysilane, MPS) that had the specific role of reducing the viscosity of the nanosuspension and speed-up the magnetophoretic process, as explained in a previous publication (Nardi et al., 2014a). The key question, which is sketched in Figure 1, is whether there exist an optimal gradient, which maximizes the hardness of the graded composite for a given average particle volume fraction and a given indentation load. Indentations performed on the NP-enriched surfaces revealed how the obtained elastic moduli and hardnesses do not always reflect the local increment of reinforcing elements, but are strongly affected by the indentation depth. Despite the approximations given by the utilized approach (Oliver and Pharr, 1992), experimental results have been successfully reproduced by material point method (MPM) simulations. Unlike previous publications that mostly deal either with the nanoindentation of homogeneous amorphous materials (Rodríguez et al., 2012) or with the effect of the substrate during the tests (Nunes and Piedade, 2013), the present work clarifies the relationship between the mechanical properties obtained by indentation and the morphology of graded polymer composites, constituting a useful tool for a smarter design of functionally graded nanocomposites showing improved performances under the loading conditions of interest.

2. EXPERIMENTAL

2.1 Materials

Iron (III) acetyacetonate (Fe(acac)3, 99+%) and benzyl ether (99%) were purchased from Acros. Oleylamine (Tech.70%), polyoxyethylene(5)non-ylphenyl ether (Igepal CO-520), 3-(Methacryloyloxy)propyltrimethoxysilane (MPS, 98%), ethanol (≥99.5%), and the epoxy embedding medium kit were purchased from Aldrich. Cyclohexane (p.A.) was purchased from Applichem. Ammonia solution (25% min) and tetraethylorthosilicate (TEOS, ≥99%) were purchased from VWR and Merck, respectively. The hyperbranched polyester acrylated oligomer

(HBP, commercial name CN2302) was purchased from Sartomer, whereas the photo-initiator (Lucirin TPO) was purchased from BASF.

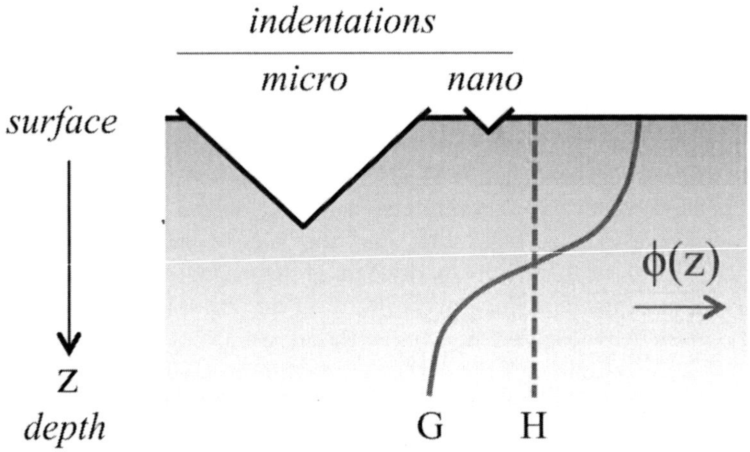

Figure 1. For a given average particle volume fraction φ and a given indentation load (micro and nano), which composite (homogeneous particle fraction, H, or graded particle fraction, G) is the hardest?

2.2 Synthesis of Fe_3O_4@SiO_2-MPS Nanoparticles

The synthetic procedure for MPS-functionalized Fe_3O_4@SiO_2 NPs has been reported in previous publications (Zhichuan et al., 2009; Nardi et al., 2014b). Briefly, Fe_3O_4 NPs were synthesized dissolving Fe(acac)$_3$ in a mixture of benzyl ether and oleylamine (Zhichuan et al., 2009). The solution was heated to 260°C and aged at this temperature for 1 h. After that, ethanol was added to precipitate the nanocrystals, which were then separated by centrifugation and dispersed in cyclohexane. In a typical synthesis of Fe_3O_4@SiO_2-MPS NPs, the dispersion of oleylamine-coated Fe_3O_4 NPs in cyclohexane was added to cyclohexane and mixed with Igepal CO-520. The pH was adjusted to 10 through the addition of ammonium hydroxide and the formation of a transparent brown microemulsion was ensured with the aid of a sonicator. Then, TEOS was added and the reaction was stirred at room temperature for 24 h, when the desired amount of MPS was also added. Hence, the reaction was continued for 24 additional hours.

2.3 Preparation of Nanocomposites for Indentation

As described earlier (Nardi et al., 2013), the photo-initiator was first dissolved in the HBP in a 6 wt% fraction and subsequently the selected amount of MPS-functionalized $Fe_3O_4@SiO_2$ core-shell NPs was added. A certain amount of as-prepared nanocomposites was placed between two UV-transparent polystyrene sheets (16 mm × 16 mm × 0.3 mm, made by hot pressing polystyrene beads) kept at a distance of 150 μm by two plastic spacers (16 mm × 3 mm × 0.15 mm), which ensured a constant thickness of the final polymer composite films. Graded composites were synthesized placing the 150 μm thick nanoparticulate films containing 1 or 8 vol% of homogeneously distributed $Fe_3O_4@SiO_2$-MPS NPs under the effect of the magnetic field gradient generated by two magnets in repulsion configuration (i.e., with the magnetizations opposing each other) (Nardi et al., 2014a). The polymerization process was then performed on the homogeneous and graded samples irradiating them from both sides using two identical 200 W high-pressure mercury UV lamps (OmniCure 2000, Exfo, Canada). Using a calibrated radiometer (Silver Line, CON-TROL-CURE, Germany), the measured UV-A light intensity on the sample surface ensured by each lamp was around 190 $mW \cdot cm^{-2}$. The films were then detached from the two polystyrene sheets with a razor blade and glued to a metallic support that fit the sampling slots of the nanoindenter.

3. METHODS

The morphology of the silica-coated magnetic NPs was characterized using a Philips FEI CM12 microscope at an accelerating voltage of 120 kV. Typically, few drops of a diluted suspension of the NPs in ethanol were deposited on carbon filmed copper grids (200 mesh, Plano GmbH). The concentration gradient was investigated analyzing the Si content (i.e., from the SiO_2 shell of the particles) on cross-sections of composite coatings by energy-dispersive X-ray spectroscopy in a scanning electron microscope (FEI XLF-30 FEG) at an accelerating voltage of 13 kV and constant working distance of 11 mm and spot size of 4. The elemental composition of the samples throughout the 150 μm thickness of the coating was obtained by data averaging over a 120 μm wide window.

Quasi-static tensile tests were performed on the pure HBP matrix and on the nanocomposites, after curing, using a Linkam TST 350 tensile testing machine at a temperature of 24 ± 0.1°C and at a constant tensile strain rate of 3.3 10^{-4} s^{-1}. The yield stress was determined from the intersect between the tangents to the stress–strain curves at zero strain and at failure.

Nanoindentation measurements were performed on the particle-enriched surfaces of the samples using a Nano Indenter XP™(MTS Nano Instruments) with a three-sided diamond Berkovich indenter. For all tests, the indenter tip approached the surface from a distance of 1000 nm at a rate of 10 nm/s, and the maximum penetration depth into the sample was set either to 1 or 20 μm. The possible influence of the metallic substrate was disregarded since the maximum indentation depth of 20 μm was only 13% of the coating thickness of 150 μm. A minimum of five indentations were performed on each type of material. Hardness and Young's modulus were determined from the unloading part of the test using the Oliver and Pharr's analysis method (Oliver and Pharr, 1992).

The numerical modeling was done using the MPM. This particle-based method has advantages dealing with large deformations (Sulsky et al., 1994) that might occur under the indenter top. Furthermore, MPM works well for problems with contact that were needed here for modeling contact forces between the tip and the films (Bardenhagen et al., 2001; Lemiale et al., 2010; Nairn, 2013). Figure 2 shows the geometry for an axisymmetric model of the nanoindentation experiment. Both axisymmetric (with a conical indenter, Figure 2B) and 3D (with a Berkovich, three-side pyramidal indenter, Figure 2C) simulations were performed, revealing sufficiently close results. Being faster and having the possibility to be run at higher resolution, focus was hence laid on axisymmetric simulations. For the best analogy between axisymmetric and 3D, an effective cone angle was determined by equating contact area as a function of depth between conical and Berkovich indenters. For a Berkovich indenter with θ_b = 65.3° tip angle, the equivalent cone has a tip half-angle of θ_c = 70.2°(Oliver and Pharr, 1992).

The indenter was modeled as a rigid material that moved at constant velocity up to a specified displacement (or force) and then retracted at the same speed. The material points in the tip of the indenter were placed on the grid to match the surface of the indenter (see Figure 2A). The bulk material was modeled as a large-deformation neo-Hookean solid with elastic properties E and ν, that plastically deforms by non-linear hardening implemented by J_2 plasticity with hardening law (Aimene and Nairn, 2015):

$$\sigma_y = \sigma_{y0}(1 + K\varepsilon)^n \tag{1}$$

where σ_{y0} is initial yield stress, ε is the strain and K and n are dimensionless hardening parameters. For modeling of gradient films, all properties were adjusted as a function of depth to represent homogenized properties for filled polymer given the filler concentration as a function of depth. The determination of material properties is discussed in the modeling results section. The first tasks were to select size of the modeled region (r and h in

Figure 2A), determine required particle resolution, and choose the indenter speed. The experimental film was 150 μm thick, but to save analysis time, some modeling was restricted to a smaller region around the indenter tip. We varied r and h until we found dimensions such that the results were insensitive to model size ($r = h = 50$ μm for the 1 μm indents and $r = h = 150$ μm for the 20 μm indents). In MPM, the particles are placed on a background grid of regular cells using four particles per cell (see Figure 2A). Because MPM contact calculations are done on the grid, the size of the grid cells can affect those calculations. The loading portion of the indentation curves in Figure 3 for a 900 nm indentation shows results at two different resolutions. At low resolutions (1 μm cells), the curves showed periodic ridges corresponding to grid dimension. These ridges were significantly reduced at higher resolutions (0.15 μm cells). All 1 μm indentation calculations used this higher resolution, which corresponded to 75 nm diameter material points. Because of the larger total displacement for 20 μm indentations, the cell size could be larger; for these simulations, we used 2 μm cells corresponding to 1 μm material points. The indenter speed was set to minimize inertial effects such that we could run simulations without needing artificial damping. The unloading curves in Figure 3 show results at two different speeds. At high speed (10 m/s), the reversal of load after reaching maximum displacement acted as an impact event causing an abrupt jump in the results. At low loading rates (1 m/s), the results showed only very small inertial effects. Thus, all simulations were performed at 1 m/s. Although this speed is much higher than actual experiments, the material model used had no time dependence. In other words, elimination of inertial effects implies that the results when using a rate-independent material model are equivalent to quasi-static loading.

The output of the simulation was given as total force on the indenter, calculated in the contact algorithm, as a function of time (or depth). On each time step, the multimaterial MPM algorithm separately updates position of the rigid particles and both momentum and position of the bulk material (Bardenhagen et al., 2001). Moreover, for any nodes containing information from both types of materials, the contact methods change the momentum of the bulk material to be consistent with position of the rigid particles (Lemiale et al., 2010; Nairn, 2013). Although various contact laws can be implanted, in this work all simulations used frictionless contact. Contact force was derived by the imposed change in momentum (per unit time) and the total force of the indenter on the object was obtained by summing these forces over all nodes. Using the constant loading and unloading rates, the original force vs. time output could be converted into a force vs. displacement output. Modulus and hardness were then determined from the simulated force vs. displacement curves by the same nanoindentation methods described in Oliver and Pharr (1992) to treat experimental results.

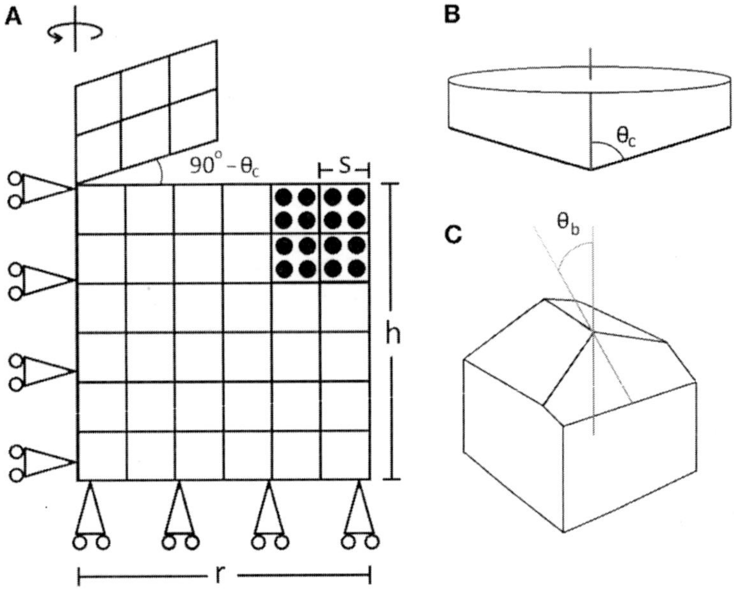

Figure 2. (A) Schematic view of the axisymmetric MPM model. The bulk object has background grid of regular cells with four particles per cell. **(B)** A conical indenter. **(C)** Tip of a Berkovich indenter.

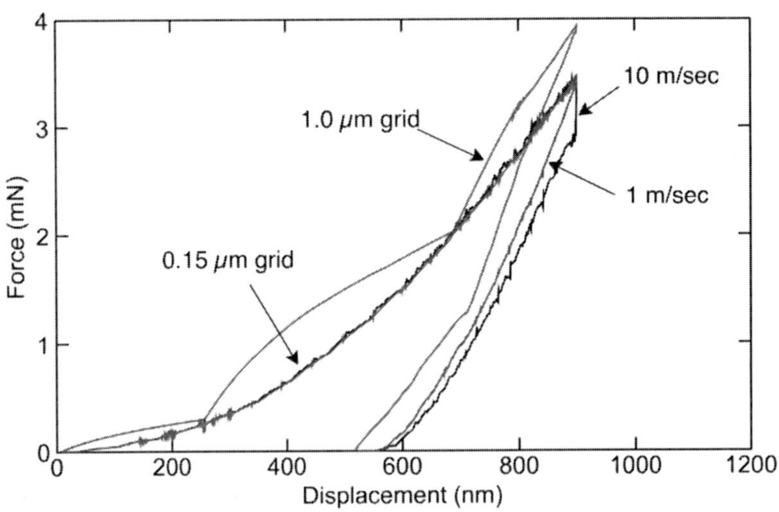

Figure 3. Simulated nanoindentation curves for force as a function of indentation depth showing the effects of MPM resolution and of indenter loading rate.

4. RESULTS AND DISCUSSION

4.1 Tensile Behavior of Homogeneous Nanocomposites

Figure 4 shows the tensile behavior of the HBP and the homogeneous composites with 1 and 8 vol% of particles. The polymer was rather ductile with a yield stress equal to 26 MPa and a strain at failure of 9%. The addition of 1 vol% of particles slightly increased the yield stress to 28 MPa and the composite failed at 6% strain. The composite with 8%vol of particles was brittle with a stress and strain at break of 22 MPa and 1.5%, respectively. The increase in stress after yielding could be fit well using the hardening law in Eq. 1.

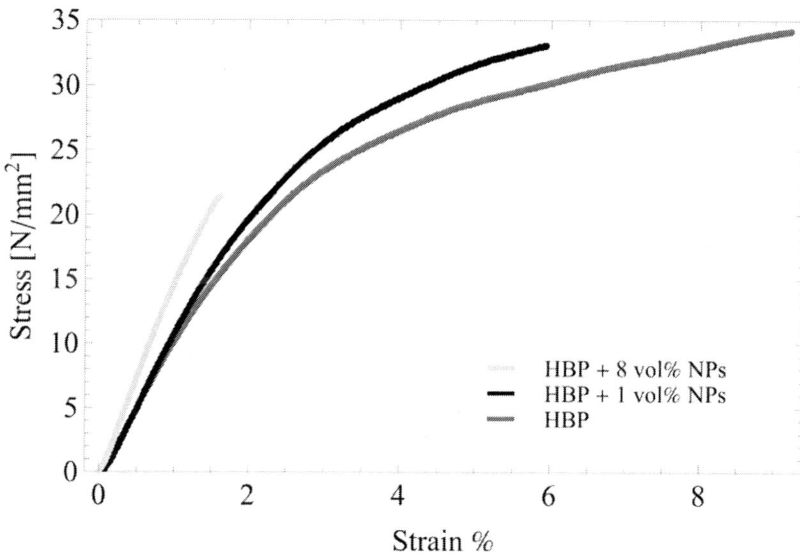

Figure 4. Tensile behavior of the HBP and the homogeneous composites with 1 and 8 vol% of particles.

4.2 Graded Nanocomposite Morphology

TEM analysis on MPS-functionalized Fe_3O_4@SiO_2 core-shell NPs shown in Figure 5 evidenced a log-normal particle size distribution with a mean diameter of 23.7 nm and SD 6.2 nm. Experimental and numerical nanoindentation tests were performed on the pure HBP, on the homogeneous composites containing 1 and 8 vol% Fe_3O_4@SiO_2-MPS NPs, and on the graded structures synthesized through the application of an external magnetic field gradient on the above-mentioned homogeneous particulate suspensions (Nardi et al., 2014a). The NP distribution of the four analyzed nanocomposite materials was determined from the cross-sectional SEM–EDX analysis of the composite coatings showing the Si content (from the SiO_2 shell) of the analyzed materials. The results shown in Figure 6 are compared with the indentation depths, represented by the red (1 µm) and the blue (20 µm) lines, and demonstrate the good dispersion of the NPs in the polymeric matrix. Notice that the lines for the average particle concentrations are not exactly at 1 and 8 vol% and these slight variations correspond to the experimental average distribution of a finite volume element of the samples, deduced from the SEM–EDX analysis.

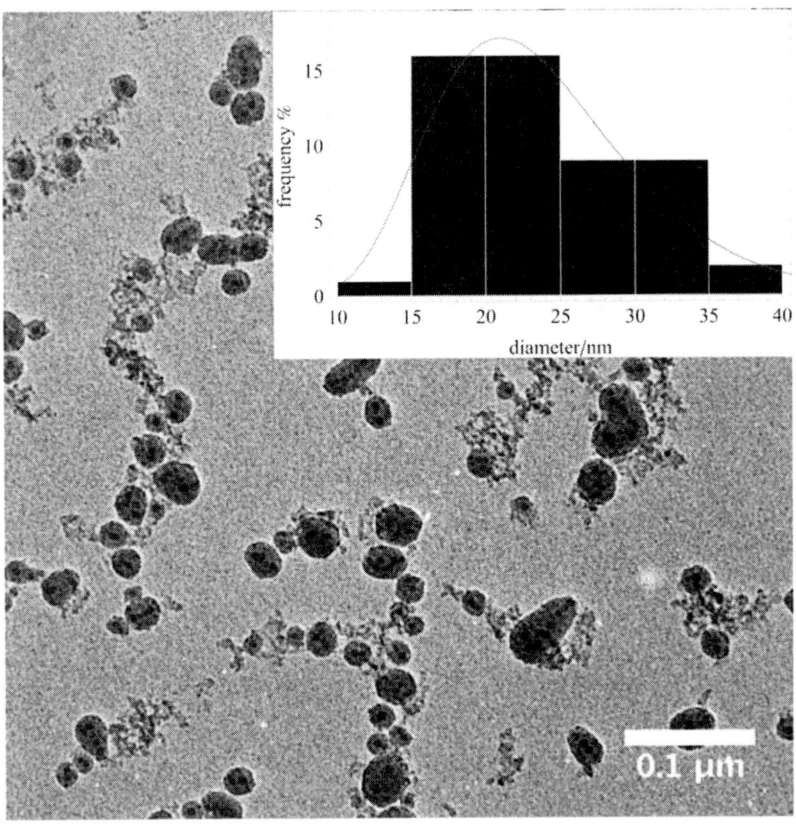

Figure 5. TEM micrograph of Fe_3O_4@SiO_2-MPS core-shell nanoparticles with the corresponding particle size distribution.

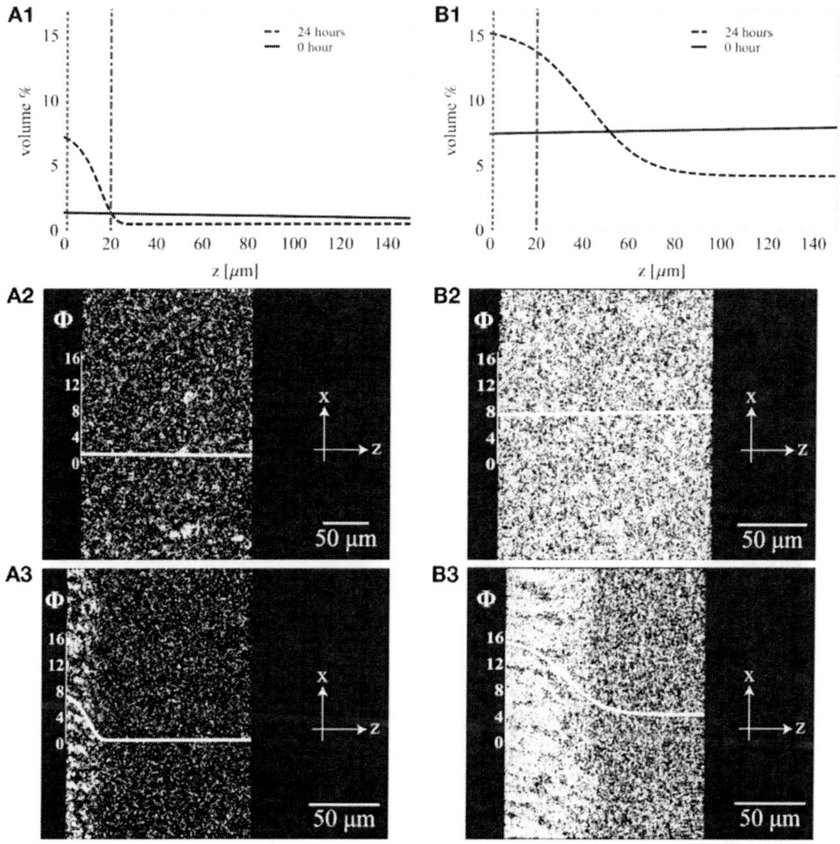

Figure 6. Nanoparticle volume fraction as a function of the position *z* along the thickness of the film for nanocomposites with average particle concentrations of 1 vol% (A1) and 8 vol% (B1). For both nanocomposite systems, the homogeneous case ("0 h") and the gradient morphology after 24 h of magnetic field application are shown. The two vertical lines represent the two nanoindentation depths (1 and 20 μm) examined in this study. The corresponding cross-sectional SEM–EDX spectral images showing the Si content are also reported for the nanocomposite containing 1 vol% of nanoparticles **(A2,A3)** and 8 vol% of nanoparticles **(B2,B3)**.

4.3 Numerical Modeling

This section describes numerical modeling results for the properties of gradient films. First, the parameters E, σ_{y0}, K, and n (see Eq. 1) were determined from homogeneous control films at constant NP fraction. Although the tensile behavior of the films (Figure 4) was accurately fitted by Eq. 1, the derived properties did not correspond well with those found from the nanoindentation experiments. Our hypothesis was that the effective properties under the indenter tip are different than those from thin-film tests. Hence, the following approach made use of Eq. 1, but determined the parameters by fits to nanoindentation experiments for three different particle volume fractions – 0, 1, and 8%. The fitting process used a derivative-free optimization algorithm provided by optim function in the R software package in order to minimize the sum-squared cost function between load curves of the experiments and the simulations. Optimization was performed over elasticity *E*, strain hardening exponent *n*, plastic modulus *K*, and yield stress σ_{y0}. Due to the high-colinearity between plasticity parameters, the yield stress was fixed to σ_{y0} = 20 MPa and the optimization process was performed over *E*, *n*, and *K*. The fitting results by particle volume fraction V_f are given in Table 1. Because of possible indenter effects on effective properties, Table 1 includes fitting results derived separately from 1 to 20 μm nanoindentation experiments on homogeneous films. A sample comparison between 1 and 20 μm depth experiments and nanoindentation modeling for the homogeneous 8 vol% composite is given in Figure 7. It is evident that the experiments were stable with smooth increase of the force during loading, whereas the simulations appeared to be noisy because they were reaction forces at contact points from an explicit, dynamic code. The steps and "noise" at 1 μm were due to use of contact physics to get force and relation between total depth of penetration and the background grid. Both greatly reduce at the deeper depth of penetration. Such "noise" in numerical results can be removed by damping, be we preferred to avoid dealing with damping terms. The jumps in simulation curves are tied to resolution and particles crossing between elements (see Figure 3 for larger effects).

Table 1. Homogenized properties of the film as determined by fitting nanoindentation simulations to experimental results as a function of filler content and depth of nanoindentation.

V_f (%)	1 μm nanoindentations			20 μm nanoindentations		
	***E* (MPa)**	***n***	***K***	***E* (MPa)**	***n***	***K***
0	2302	0.362	371	2031	0.342	412
1	2714	0.459	426	1899	0.346	369
8	3064	0.572	420	2371	0.400	419

The yield stress for all was fixed at $\sigma_{y0} = 20$ *MPa.*

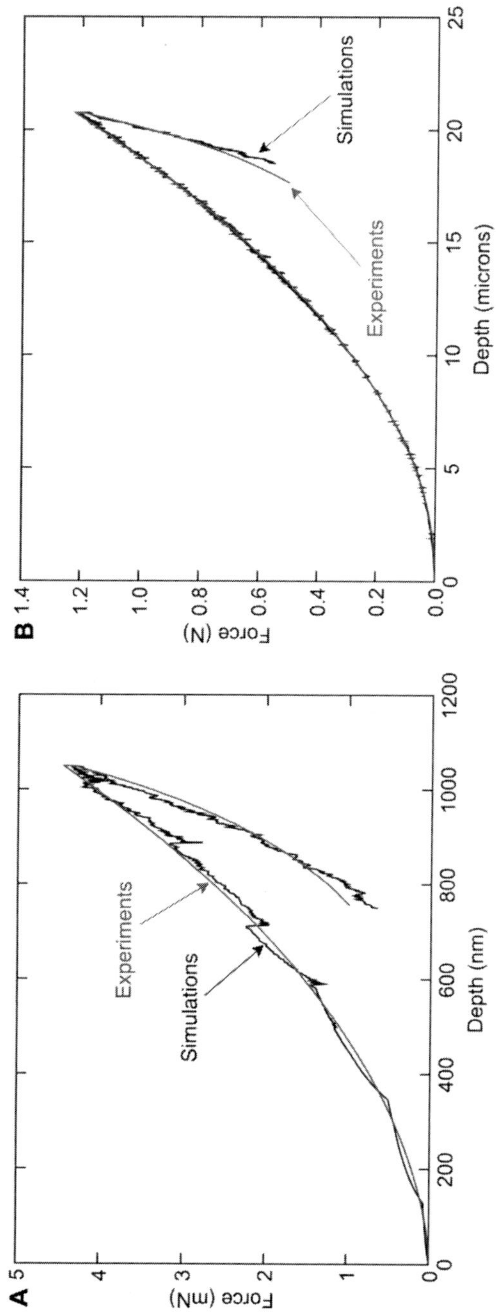

Figure 7. Comparison of simulated nanoindentation to experimental results for 8 vol% homogeneous control films and indentation depth of 1 μm (A) and 20 μm (B) using properties in Table **1.**

To model gradient composites, the vertical axis of the film was divided into 40 layers, each 1.25 μm thick. For each layer, the volume percent of filler particles was determined from experimental measurements shown in Figure 6. To determine material properties as a function of filler content, the fitting parameters in Table 1 were linearly extrapolated to local filler content (i.e., each property was set to $a V_f + b$, where a and b were determined from properties in Table 1). In multimaterial MPM, each material uses its own velocity field (Bardenhagen et al., 2001), but for our purpose only contact between the indenter and all other materials combined was needed. As a refinement to standard MPM, all materials in the bulk film were set to share the same velocity field. In this way, the film acts as a bulk continuum, albeit with position-dependent properties, while the indenter interacts with that bulk by contact mechanics.

4.4 Experimental and Numerical Properties

Figure 8 compares the experimental and numerical modulus and hardness values for the HBP and the homogeneous and graded composites. The fitting procedure described in the previous section for the properties of the pure HBP led to an excellent match between experimental and modeled data, with modulus comprised between 2.2 and 2.5 GPa and hardness between 0.14 and 0.15 GPa.

For the homogeneous composites (Figures 8A,C), the modulus derived from the unloading part of the load–displacement curves was substantially independent of the indentation depth, and equal to ~2.7 GPa (1 vol%) and around 3 GPa (8 vol%). Comparable increments in modulus with increasing NP content were indeed measured at both indentations depths, for both experimental and numerical tests. Experimental hardness values, successfully reproduced by numerical simulations, were higher when the indentation depth was kept at 1 μm (0.24 GPa) rather than 20 μm (0.16 GPa). Similar increases were reported for aluminum and titanium, which was attributed to surface hardening, whereas the hardness of fragile materials, such as quartz and glass, were found to be independent of indentation load (Oliver and Pharr, 1992). The plastic energy fields for the 1 and 20 μm indentations in HBP graded composites with 8 vol% of particles were considerably different, as shown in Figure 9. The present finding reflects the intrinsic softening effect arising from correspondingly large deformations (Figure 4) (Lam and Chong, 1999) and specific strain gradients associated with the Berkovich geometry (Alisafaei et al., 2014), rather than an indentation size effect, which usually manifests itself for shallow, sub-micron penetration depths (Pharr et al., 2010). The observed hardness increases at small indentation depth, however, were negligible compared to the considerable variations in mechanical properties encountered for graded composites, especially when indented at different depths.

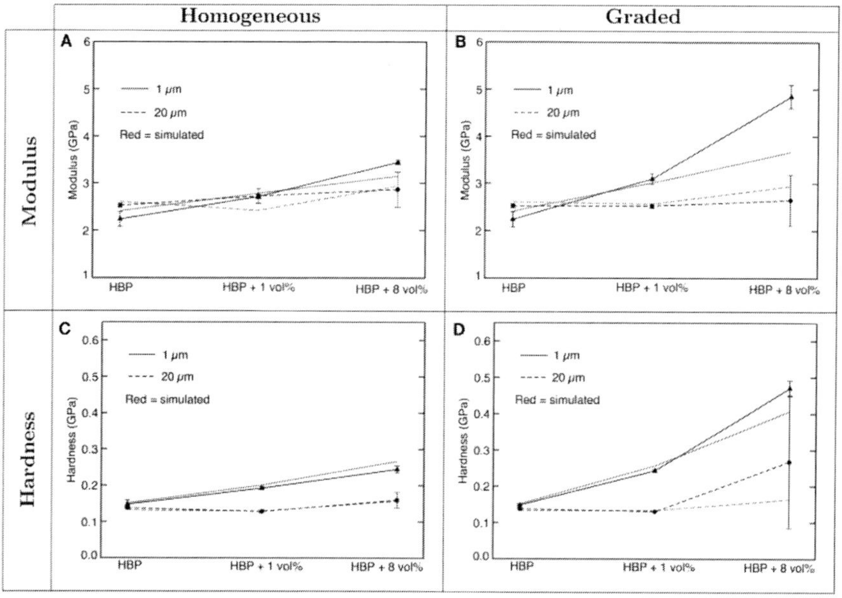

Figure 8. Modulus (A,B) and hardness (C,D) derived by the unloading part of the load-displacement curves for experimental (black) and numerical (red) tests. The results for the homogeneous **(A,C)** and graded **(B,D)** materials are shown as a function of the nanoparticle content for 1 μm (solid lines) and 20 μm (dotted lines) indentation depths.

Indeed, for the graded composite containing 1 vol% of NPs and for 1 μm indentations, the modulus and hardness increased by as much as 14 and 27%, respectively, compared with its homogeneous analog, whereas for 20 μm indentations, these two properties were equal to those of the pure HBP within experimental scatter (Figures 8A,B). Notice that, in this case, the indentation depth was comparable with the size of the gradient (Figure 6). A similar response was obtained for the graded composite with 8 vol% of NPs, with rather small increases for 20 μm indentations, but very large increases for 1 μm indentations, as high as 40% for the modulus and 93% for the hardness compared with the homogeneous composite. These latter results are interesting: for a 1 μm indentation on the graded composite, the plastic zone extended to ~5 μm below the indenter tip (Figure 9), which hence only probed the material with about 15 vol% particles (Figure 6B1). By contrast, for a 20 μm indentation the plastic zone extended more than 60 μm below the surface, hence far beyond the gradient area. Therefore, the properties at this level of indentation would be some average of the homogeneous material properties over the particle concentration range of 5–15 vol%.

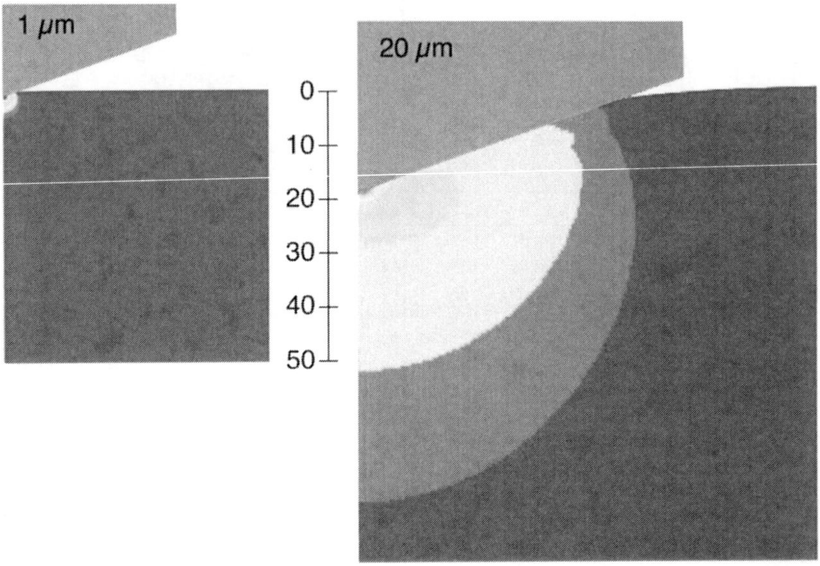

Figure 9. Plastic energy fields for 1 and 20 μm indentations in HBP with 8 vol% filler and graded composition. The plastic energy ranges from 0 (blue) to 3 kJ/m^3 (red).

Clearly, already at 1 vol% of NPs, smaller indentation depths led to both stiffening, going from 2.52 GPa at 20 μm to 3.10 GPa at 1 μm, and hardening, going from 0.13 GPa at 20 μm to 0.24 GPa at 1 μm. At 8 vol% of NPs, this effect became even larger, producing variations on the order of 83% (4.84 vs. 2.64 GPa) and 81% (0.47 vs. 0.26 GPa) in terms of modulus and hardness, respectively. Simulations run on the same graded structures evidenced similar trends at the two different indentation depths. Considerable increments in both modulus (3.67 vs. 2.64 GPa) and hardness (0.41 vs. 0.16 GPa) were indeed obtained when films filled with 8 vol% NPs were indented at 1 μm rather than 20 μm (Figures 8B,D). Moreover, one should point out the very high hardness values of the graded nanocomposites under 1 μm indentations, which were close to 500 MPa.

In light of the good agreement between experiments and simulations (Figure 8), virtual experiments on films with different gradients but with a total NP content of 8 vol% were performed. Figure 10 shows the four investigated morphologies together with the Boltzmann sigmoidal function utilized to represent them. For the graded structures, the parameters of the sigmoidal function were adjusted to either fit experimentally developed gradients ("Grade 1"), or to generate hypothetical graded morphologies ("Grade 2," "Grade 3") with higher superficial NP volume fractions but with a constant global NP content. From a qualitative standpoint, an indentation performed on the blue gradient with a depth of 20 μm would almost entirely penetrate the part of the material with enhanced filler content, and the associated plastic zone would affect a large amount of unfilled polymer.

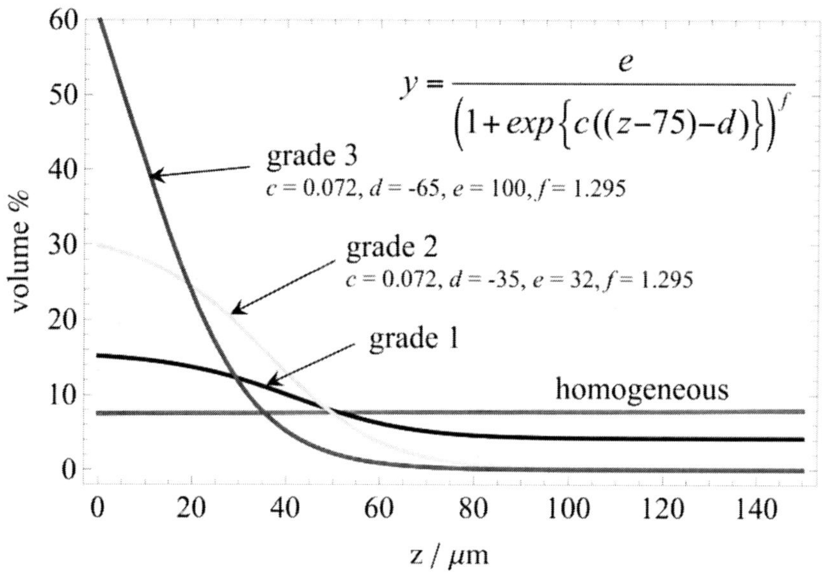

Figure 10. Homogeneous and graded morphologies with average nanoparticle volume fraction of 8%. "Grade 1" is the experimental gradient already shown in Figure 5, whereas "Grade 2" and "Grade 3" were created for numerical simulation purposes using the equation and parameters shown in the figure. "Grade 2" goes from a minimum particle fraction of 0 at depths $z > 100$ μm to a maximum particle fraction of 30% on the surface ($z = 0$). "Grade 3" goes from a minimum particle fraction of 0 at depths $z > 80$ μm to a maximum particle fraction of 60% on the surface ($z = 0$), close to the random packing fraction for monodispersed spheres of ~64% (Desmond and Weeks, 2009).

We ran nanoindentation simulations for all four gradients for both 1 and 20 μm indentation depths. When we extrapolated the 1 μm mechanical properties from Table 1 to 60 vol% on the surface for the blue Grade 3, it gave plasticity properties that were likely unrealistic (e.g., hardening power law $n > 1$). The properties from 20 μm experiments, however, gave more reasonable extrapolations. We therefore ran all these virtual experiments using extrapolated 20 μm properties from Table 1. The results are in Figure 11. Note that the results at 1 μm depth in Figure 11 differ from corresponding results in Figure 8 because of the use of different mechanical properties.

The simulations clarified the role of the gradient on the properties measured at different indentation depths. For 1 μm indentations, both modulus and hardness continued to increase as the surface layer was enriched with NPs, confirming that at this indentation depth results are sensitive only to the properties near the surface. On the contrary, the modulus derived from 20 μm indentations peaked for experimental "Grade 1" and then decreased. Similarly, the hardness stopped increasing after "Grade 2." These results point out that the optimum gradient for enhanced modulus and toughness strongly depends on the indentation depth experienced in a given application, and more specifically on the size of the plastic zone below the indenter. The present analysis suggests that there exists an optimal concentration gradient that maximizes the stiffness and hardness of polymer nanocomposite for a given indentation load, although additional work would be useful to improve the design of such optimal gradients.

5. CONCLUSION

The modulus and hardness of graded polymer nanocomposites were found to depend on enhanced concentrations of particles near the surface and on the indentation depth. For small (1 μm) indentations, large increases in modulus (+40%) and hardness (+93% to very high values close to 500 MPa) were obtained for graded composites with an average 8%vol of NPs, compared with their homogeneous counterparts. At larger indentation depth (20 μm, comparable with the size of the composition gradient), however, the modulus and hardness of the graded and homogeneous composites were almost the same, and equal to those of the pure polymer.

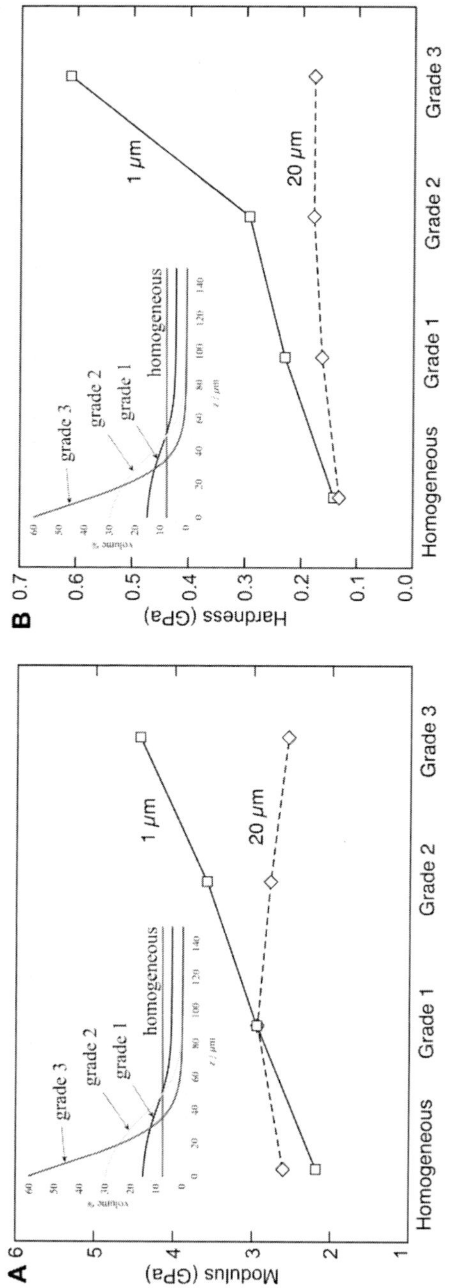

Figure 11. Simulated nanoindentation experiments for modulus (A) and hardness (B) at both 1 and 20 μm indentation depths. The "Homogeneous," "Grade 1," "Grade 2," and "Grade 3" refer to the black, red, green, and blue gradients in the inset, respectively.

Through a MPM approach, a tool for the interpretation and the prediction of the mechanical properties obtained from nanoindentation tests was developed. With this tool, experimental nanoindentation tests were successfully simulated, confirming the importance of the indentation depth and associated plastic zone as key factors toward the determination of the material properties. It turned out that very high-surface concentrations of hard NPs do not necessary lead to the hardest nanocomposites. On the contrary, the performance of a certain concentration gradient should also be defined in conjunction with the indentation load to which it is subjected. The present approach is thus particularly effective to design and prepare hard polymer nanocomposites through the generation of a controlled concentration gradient from a dilute suspension of particles.

CONFLICT OF INTEREST STATEMENT

The authors declare that the research was conducted in the absence of any commercial or financial relationships that could be construed as a potential conflict of interest.

ACKNOWLEDGMENTS

The authors acknowledge the Swiss National Science Foundation (project 200020_155888) for financial support. Two authors (CH and JN) acknowledge support by a National Institute of Food and Agriculture (NIFA) award #2013-34638-21483, of the United States Department of Agriculture (USDA).

REFERENCES

1. Aimene, Y. E., and Nairn, J. A. (2015). Simulation of transverse wood compression using a large-deformation, hyperelastic-plastic material model. *Wood Sci. Technol.* 49, 21–39.

2. Alegría-Ortega, J. A., Ocampo-Carmona, L. M., Suárez-Bustamante, F. A., and Olaya-Flórez, J. J. (2012). Erosion-corrosion wear of Cr/CrN multi-layer coating deposited on AISI-304 stainless steel using the unbalanced magnetron (UBM) sputtering system. *Wear* 29, 149–153.

3. Alisafaei, F., Han, C.-S., and Lakhera, N. (2014). Characterization of indentation size effects in epoxy. *Polym. Test.* 40, 70–78.

4. Bardenhagen, S. G., Guilkey, J. E., Roessig, K. M., Brackbill, J. U., Witzel, W. M., and Foster, J. C. (2001). An improved contact algorithm for the material point method and application to stress propagation in granular material. *CMES* 2, 509–522.

5. Chen, Z., and Diebels, S. (2014). Nanoindentation of soft polymers: modeling, experiments and parameter identification. *Tech. Mech.* 34, 166–189.

6. Dao, M., Chollacoop, N., Van Vliet, K. J., Venkatesh, T. A., and Suresh, S. (2001). Computational modeling of the forward and reverse problems in instrumented sharp indentation. *Acta Mater.* 49, 3899–3918.

7. Desmond, K. W., and Weeks, E. R. (2009). Random close packing of disks and spheres in confined geometries. *Phys. Rev. E Stat. Nonlin. Soft Matter Phys.* 80, 051305.

8. Díez-Pascual, A. M., Gómez-Fatou, M. A., Ania, F., and Flores, A. (2015). Nanoindentation in polymer nanocomposites. *Prog. Mater. Sci.*67, 1–94.

9. Jeong, D. H., Erb, U., Aust, K. T., and Palumbo, G. (2003). The relationship between hardness and abrasive wear resistance of electrodeposited nanocrystalline Ni-P coatings. *Scr. Mater.* 48, 1067–1072.

10. Lam, D. C. C., and Chong, A. C. M. (1999). Indentation model and strain gradient plasticity law for glassy polymers. *J. Mater. Res.* 14, 3784–3788.

11. Lemiale, V., Hurmane, A., and Nairn, J. A. (2010). Material point method simulation of equal channel angular pressing involving large plastic strain and contact through sharp corners. *Comput. Model. Eng. Sci.* 70, 41–66.

12. Mann, B. S., and Arya, V. (2002). An experimental study to correlate water jet impingement erosion resistance and properties of metallic materials and coatings. *Wear* 253, 650–661.

13. Nair, A. K., Cordill, M. J., Farkas, D., and Gerberich, W. W. (2011). Nanoindentation of thin films: simulations and experiments. *J. Mater. Res.* 24, 1135–1141.

14. Nairn, J. A. (2013). Modeling of imperfect interfaces in the material point method using multimaterial methods. *Comput. Model. Eng. Sci.* 92, 271–299.

15. Nardi, T., Leterrier, Y., Karimi, A., and Månson, J.-A. E. (2014a). A novel synthetic strategy for bioinspired functionally graded nanocomposites employing magnetic field gradients. *RSC Adv.* 4, 7246–7255.

16. Nardi, T., Leterrier, Y., and Månson, J.-A. E. (2014b). Bioinspired functionally graded nanocomposites synthesized through magnetophoretic processes for tailored stress reduction. *MRS Online Proc. Libr.* 1685.

17. Nardi, T., Sangermano, M., Leterrier, Y., Allia, P., Tiberto, P., and Månson, J.-A. E. (2013). UV-cured transparent magnetic polymer nanocomposites. *Polymer* 54, 4472–4479.

18. Nunes, J., and Piedade, A. P. (2013). Nanoindentation of functionally graded hybrid polymer/metal thin films. *Appl. Surf. Sci.* 284, 792–797.

19. Oliver, W. C., and Pharr, G. M. (1992). An improved technique for determining hardness and elastic modulus using load and displacement sensing indentation experiments. *J. Mater. Res.* 7, 1564–1583.

20. Penumadu, D., Dutta, A., Pharr, G. M., and Files, B. (2011). Mechanical properties of blended single-wall carbon nanotube composites. *J. Mater. Res.* 18, 1849–1853.

21. Pharr, G. M., Herbert, E. G., and Gao, Y. F. (2010). The indentation size effect: a critical examination of experimental observations and mechanistic interpretations. *Annu. Rev. Mater. Res.* 40, 271–292.

22. Qureshi, S. S., Zheng, Z., Sarwar, M. I., Félix, O., and Decher, G. (2013). Nanoprotective layer-by-layer coatings with epoxy components for enhancing abrasion resistance: toward robust multimaterial nanoscale films. *ACS Nano* 7, 9336–9344.

23. Reshetnyak, H., and Kübarsepp, J. (1997). Structure sensitivity of wear resistance of hardmetals. *Int. J. Refract. Metals Hard Mater.* 15, 89–95.

24. Rodríguez, M., Molina-Aldareguía, J. M., González, C., and Llorca, J. (2012). Determination of the mechanical properties of amorphous materials through instrumented nanoindentation. *Acta Mater.* 60, 3953–3964.

25. Roop Kumar, R., and Wang, M. (2002a). Modulus and hardness evaluations of sintered bioceramic powders and functionally graded bioactive composites by nano-indentation technique. *Mater. Sci. Eng. A* 338, 230–236.

26. Roop Kumar, R., and Wang, M. (2002b). Functionally graded bioactive coatings of hydroxyapatite/titanium oxide composite system.*Mater. Lett.* 55, 133–137.

27. Sevim, I., and Barlas Eryurek, I. (2006). Materials & design effect of abrasive particle size on wear resistance in steels. *Mater. Des.* 27, 173–181. doi:10.1016/j.matdes.2005.03.009

28. Smeacetto, F., Salvo, M., and Ferraris, M. (2002). Oxidation protective multilayer coatings for carbon-carbon composites. *Carbon N. Y.*40, 583–587.

29. Sulsky, D., Chen, Z., and Schreyer, H. L. (1994). A particle method for history-dependent materials. *Comput. Methods Appl. Mech. Eng.*118, 179–186.

30. Woirgard, J., Tromas, C., Girard, J. C., and Audurier, V. (1998). Study of the mechanical properties of ceramic materials by the nanoindentation technique. *J. Eur. Ceram. Soc.* 18, 2297–2305.

31. Zhichuan, X., Chengmin, S., Yanglong, H., Hongjun, G., and Shouheng, S. (2009). Oleylamine as both reducing agent and stabilizer in a facile synthesis of magnetite NPs. *Chem. Mater.* 21, 1778–1780.

Chapter 5

Nanocomposites Based on Conducting Polymers and Functionalized Carbon Nanotubes with Different Dopants Obtained by Electropolymerization

Florina Branzoi[1*], Viorel Branzoi[2]

[1]Institute of Physical Chemistry "Ilie Murgulescu" 202 Splaiul Independenţei, Bucharest, Romania

[2]Department of Applied Physical Chemistry and Electrochemistry, University Politehnica of Bucharest, Bucharest, Romania

ABSTRACT

Nanocomposite films based on poly (3,4-ethylenedioxy thiophene) (PEDOT), functionalized singlewalled carbon nanotubes and different dopants were studied. It was fabricated by a simple oxidative electropolymerization method. The dopant substances used were SDS (Sodium dodecyl sulfate) and tiron (1,2-Dihydroxybenzene-3,5-disulfonic acid disodium salt hydrate). These nanocomposite films were grown electrochemically from aqueous solutions such that constituents were deposited simultaneously onto substrate electrode. The synthetic, morphological and electrical properties of the nanocomposite films obtained were compared. Scanning electron microscopy (SEM) revealed that the composite films consisted of nanoporous networks of

SWCNTS (single-walled carbon nanotubes) coated with polymeric film. Cyclic voltammetry (CV), electro-chemical impedance spectroscopy (EIS) and FT-IR spectroscopy demonstrated that these composite films had similar electrochemical response rates to pure polymeric films but a lower resistance and much improved mechanical integrity. The negatively charged functionalized carbon nanotubes (CNTSF) served as anionic dopant during the electropolymerization to synthesize polymer/CNTSF composite films. The specific electrochemical capacitance of the composite films is a significantly higher value than that for pure polymer films prepared similarly. Using these composite films, the modified electrodes with improved properties were obtained. (In this paper, for simplicity, the SWANTs-COOH group will be noted CNTsF which means functionalized carbon nanotubes.)

KEYWORDS

Nanocomposite Films, Functionalized Carbon Nanotubes, SEM, Capacitance, Real Impedance

1. INTRODUCTION

As a result of the first report concerning the preparation of carbon nanotubes (CNTs)/polymer composite like that many efforts have been made to combine CNTs and polymers to produce functional composite materials with superior properties [1] -[4] . Apart from possible improvements in the mechanical and electrical properties of polymers, the formation of CNTs/polymer composites have been and still are explored as a promising approach for an effective incorporation of CNTs into practical devices [5] -[9] . Composite material based on the coupling of conducting organic polymers (CPs) and CNTs have shown that they possess properties of the individual components with a synergistic effect [9] . In this context, a special attention has been paid to the following CPs: polyaniline (PANI), polypyrrole (PPY), polythiophene (PTh) [10] -[13] etc. Among the numerous materials devised, development of the polythiophene derivative, polyethylenedioxythiophene (PEDOT), has shown significant promise for the challenge at hand [14].

The combination of CNTs with CPs offers an attractive route to reinforce the polymer as well as to introduce electronic properties based on morphological modification or electronic interaction between the two components [4] -[9] [15] -[18] . However, it is difficult to process CNTs and insoluble in most solvents. In order to broaden their applications, it is necessary to tailor their solubility properties. For this reason, in this study, SWCNTs (singlewalled carbon nanotubes) covalently functionalized with polycarboxylic acids, were used. The SWCNTs graft copolymer has excellent solubilities in water and some organic solvents and it also exhibits an order of magnitude increase in electrical conductivity. It was also demonstrated that SWCNTs-COOH (single walled carbon nanotubes functionalized with polycarboxylic acids) showed an improved sensor performance compared to unfunctionalized SWCNTs because polycarboxylic acid is an conductive organic compound in its own right. Additionally, the presence of numerous functional groups in SWCNTs-COOH means that there is potential for covalent immobilization of various big molecules and specially biomolecules.

The chemical functionalization of CNTs (carbon nanotubes) has been shown to increase their solubility in different solvents and facilitate their processing. It has also been demonstrated that the chemical functionalization affects the CNTs rope size and results in exfoliation into smaller bundles and individual nanotubes. These issues are especially important for the fabrication of high-performance composite materials. On the other hand, CNTs are considered to be the ideal reinforcing fibers due to their exceptional mechanical, electronic and thermal properties, low density and high aspect ratio. However, the incorporation of CNTs into the polymer matrix - is often problematic due to the chemical inertness of CNTs. The covalent functionalization of CNTs is a valuable route towards the development of high-performance composites. It provides homogeneously dispersed CNTs incorporated in the polymer and a strong interfacial bonding between the polymer and CNTs. We have applied chemically functionalized SWNTs-COOH to prepare a number of SWNT-polymer composite materials. CNTs can be divided into two main categories: single-walled carbon nanotubes (SWNTs) and multi-walled carbon nanotubes (MWNTs). The first are formed by a single graphene sheet. The latter are formed by additional graphene sheets wrapped around the SWNT core. As a result of the first report concerning the preparation of a CNTs/polymer composite by Ajayan et al., many efforts have been made to combine CNTs and polymers to produce functional composite materials with superior properties. Apart from possible improvements in the mechanical and electrical properties of polymers, the formation of CNTs/polymer composites has been and still is explored as a promising approach for an effective incorporation of CNTs into practical devices [18] -[26] .

In this study is described electrochemical synthesis of nanocomposite films from conducting polymers and functionalized single-walled carbon nanotubes (FSWCNTs) with different dopants and electrochemical characterization of these nanocomposites by cyclic voltammetry, electrochemical impedance spectroscopy, FT-IR and scanning electron microscopy (SEM), have been presented in this paper.

2. EXPERIMENTAL

The electrochemical polymerizations were carried out using a conventional three electrodes system. A platinum electrode and a saturated calomel electrode (SCE) were used as counter and reference electrode, respectively. The reference electrode was placed in a separate cell and was connected to the electrolytic cell via a salt bridge that ends as a Luggin capillary in the electrolytic cell. This arrangement helps in reducing the ohmic resistance of the electrochemical system. The working electrode was made from a platinum disk with surface area of 0.5 cm^2. In this study were used single-walled carbon nanotubes functionalized with carboxylic acid (FSWCNTs) were provided by CarbonSolutions, Inc. (www.carbonsolution.com, Riverside, CA), and 3,4-ethylenedioxy thiophene (EDOT) (99.5% Fluka) was used as supplied. Which were purchased from Sigma-Aldrich with the following characteristics: 80% - 90% carbon basis, D × L 2 - 10 nm × 0.5 - 2 μm, bundle dimensions. The organic carboxyl groups formed on a nanotube surface are localized at the defects in functionalized single-walled and suitable reactive organic groups from other chemical chains are prone to react in this zone. Bidistilled water was used for all sample preparations. All chemicals were of the highest quality commercially available and were used as received. Cyclic voltammetry and electrochemical impedance spectroscopy were used to investigate the electrochemical properties of the composite films. Electrochemical experiments were carried out with an automated model VoltaLab 40 potentiostat/galvanostat with EIS dynamic controlled by a personal computer. All the following potentials reported in this work are against the SCE (saturated calomel electrode). Scanning electron microscopy (SEM) was used to compare the microstructures of the deposited films.

2.1. Preparation of Modified Electrodes

The Pt electrode was carefully polished with aqueous slurries of fine alumina powder 0.05 micron on a polishing cloth until a mirror finish was obtained. After 20 min sonication, the electrodes were immersed in concentrated H_2SO_4, followed by thorough rinsing with water and ethanol. The prepared electrodes were dried and used for modification immediately. Nanocomposite films of CPs/FSWCNTs were prepared by electrochemical polymerization from a solution containing both the functionalized carbon nanotubes (FSWCNTs) and the corresponding monomer (2,3-dioxythiophene). One kind of FSWCNTs was used in this work namely: single wall carbon nanotubes (SWCNTs with carboxylic acids, see Scheme 1). In the first step, the FSWCNTs aqueous suspension (usually 10 mg/L) was prepared via sonication (1 h) and then the synthesis solution was prepared straightaway by dissolving the monomer and the corresponding supporting electrolyte in the FSWCNTs aqueous suspension. The negatively charged FSWCNTs in solution acted as sole supporting electrolyte and dopant for the (3,4-ethylenedioxy thiophene)-EDOT depositions. For this reason FSWCNTs are enwrapped in polymers during the electropolymerization process in the form of counter ions or dopants. Thus, PEDOT-FSWCNTs composite films were prepared, from an aqueous solution containing 0.025 mol/L of 2,3-dioxythiophene + 0.1 M of $LiClO_4$ + 10 mg/L of FSWCNTs by cyclic voltammetry in the potential scanning range of 0 mV to +1250 mV at a scan rate of 10 mV/s and for a cycles number of 10. The pure poly (3,4-ethylenedioxy thiophene) films were prepared from an aqueous solution of 0.025 mol/L 2,3-dioxythiophene + 0.1 M $LiClO_4$, without adding FSWCNTs, also by cyclic voltammetry in the potential range 0 mV to + 1250 mV at a scan rate of 10 mV/s.

2.2. Characterization of the Modified Electrodes

Electrochemical impedance spectroscopy (EIS) and cyclic voltammetry (CV) were used to investigate the electrochemical properties of the composite films. The electrochemical characterization of the PEDOT-FSWCNTs and PEDOT films was carried out in 0.1 M $LiClO_4$ cycling aqueous solutions for comparison and because the dopant anion of the polymeric films is the same with the anion of the cycling solution. The working electrode potential was cycled in the potential range of 0 mV to +1250 mV with a scan rate of 50 mV/s. The impedance measurements were performed using a VoltaLab 40 potentiostat/galvanostat with EIS dynamic in the frequency range of 100 kHz to 1 mHz with an AC wave of 5 mV (peak-to-peak) overlaid on a DC bias potential and the impedance data were obtained at a rate of 10 points per decade change in frequency. All tests have been performed at 25°C under atmospheric oxygen without agitation.

3. RESULTS AND DISCUSSIONS

The PEDOT film was obtained on the platinum substrate in a synthesis solution of 0.025 mol/L 2,3- dioxythiophene + 0.1 M $LiClO_4$ by cyclic voltammetry in the scanning potential range of 0 to +1250 mV at a scan rate of 10 mV/s and for 10 cycles. It is clearly seen from Figure 1 that, the reaction current increases sharply when the applied potential is larger than 0.7 V, which reveals the formation of the radical cations. The increased current implies that the EDOT radical cations start to electropolymerize onto the platinum electrode. During the first potential scan, the polymerization potential of PEDOT on Pt is around

0.975 V, while the onset of polymerization shifts significantly on following scans to more positive potentials and stabilizes at 1.095 V.

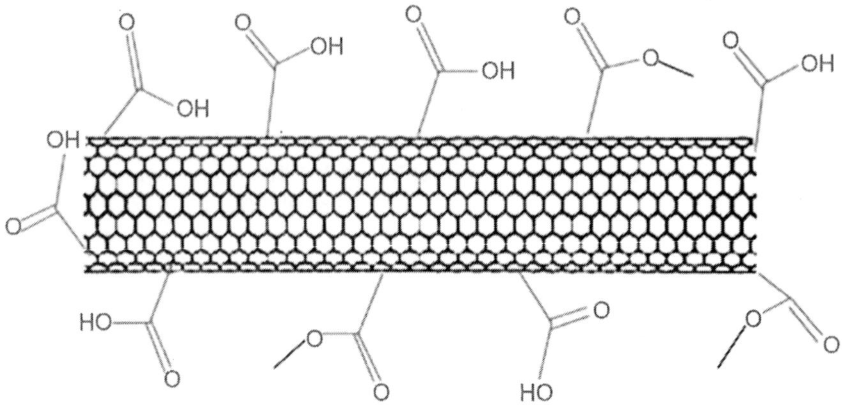

Scheme 1. Chemical structure of single walled carbon nanotube functionalized with carboxylic acids (SWCNTs-COOH).

Composite films of poly (3,4-ethylenedioxythiophene) and functionalized, single-walled carbon nanotubes (PEDOT-FSWCNTs) were fabricated by a simple oxidative electropolymerization method. These films were electrodeposited on a platinum working electrode using a classical system with three electrodes. The PEDOT/ CNT-COOH film was obtained on the platinum substrate in a synthesis solution of 0.025 mol/L 2,3-dioxythiophene +10 mg/L carbon nanotubes polycarboxylic acids.

The obtained cyclovoltammograms are given in Figure 2 and have the same shapes with those obtained for PEDOT/Pt modified electrode cycled in the same conditions, but in this case the anodic peaks are much higher and larger than those for PEDOT film. This fact can be explained taking into account that, FSWCNTs (in our case SWCNTs-COOH) are negatively charged and they can act as doped anions and consequently, the conductivity of PEDOT/FSWCNTs composite film increases.

Hence, the PEDOT/FSWCNTs composites exibit higher currents than PEDOT films, which can be translated into larger capacitance. This is an indication of faster kinetics in the composite, which can be attributed to the higher electronic conductivity of the FSWCNTs network. This means that the redox processes which take place in nanocomposite film are more intense and more complex than those in the pure PEDOT film.

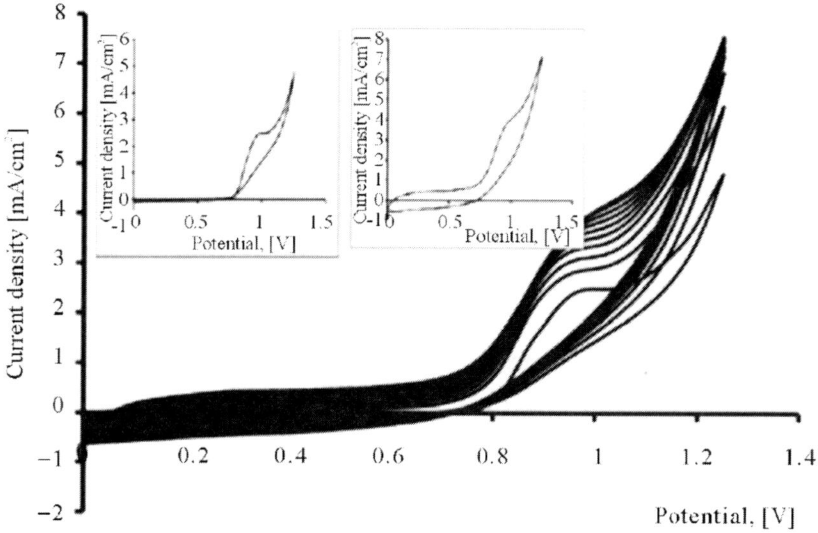

(a)

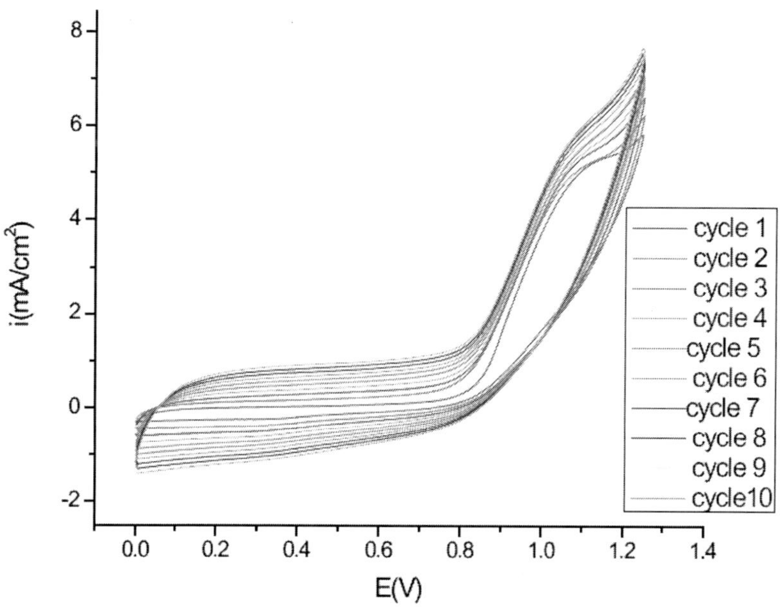

(b)

Figure 1. Polymerization cyclovoltammograms of 2,3-dioxythiophene in 0.1 M $LiClO_4$aqueous solution. Inset 1st Cycle (a) and 10th Cycle (b).

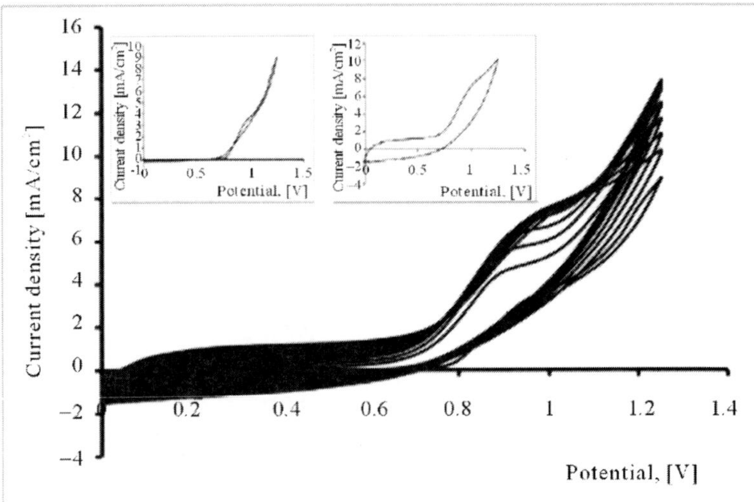

(a)

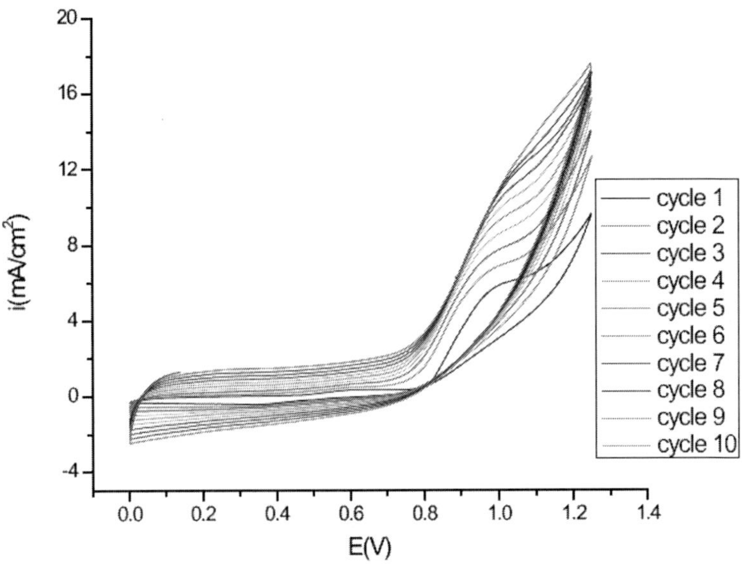

(b)

Figure 2. Polymerization cyclovoltammograms of 2,3-dioxythiophene + CNT-COOH in 0.1 M $LiClO_4$ aqueous. solution. Inset 1st Cycle (a) and 10th Cycle (b).

The PEDOT/CNT-COOH/dopant films were obtained on the platinum substrate in a synthesis solution of 0.025 mol/L 2,3-dioxythiophene + 10 mg/L CNT-COOH + 0.01 mol/L dopant substance in 0.1 M $LiClO_4$ by cyclic voltammetry in the scanning potential range of 0 to +1250 mV at a scan rate of 10 mV/s and for 10 cycles. The dopant substances used were SDS (Sodium dodecyl sulfate) and tiron (1,2-Dihydroxybenzene-3,5-disulfonic acid disodium salt hydrate). In the presence of dopant (SDS and tiron, see Figure 3), it is observed that the current response of doped PEDOT films decreases with addition of the dopant. Moreover, the current increases with continuous cycling in the potential range as seen in Figure 3. During the first potential scan, the polymerization potential of PEDOT/CNT-COOH/dopant on Pt is around 1.05 V, while the onset of polymerization shifts significantly on following scans to more positive potentials and stabilizes at 1.10 V. All PEDOT composite films were uniform, smooth and adherent to the surface of Pt electrode.

The electrochemical behavior of the PEDOT, PEDOT/CNT-COOH films deposited electrochemically from aqueous solution in the absence and presence of dopant (SDS and tiron) was characterized further using cyclic voltammetry. The electrochemical characteristics of obtained PEDOT film was study in the cycling solutions, an aqueous solution of 0.1 M $LiClO_4$. The electrode potential was cycled on the potential range from 0 up to 1250 mV with a sweep rate of 50 mV/s and for a cycles number of 20 (see Figure 4).

The electrochemical characteristics of obtained PEDOT/CNT-COOH film was study in the cycling solutions, an aqueous solution of 0.1 M $LiClO_4$. The electrode potential was cycled on the potential range from 0 up to 1250 mV with a sweep rate of 50 mV/s and for a cycles number of 20 (seeFigure 5). These films can be cycled repeatedly between the conducting (oxidized) and insulating (neutral) state without significant decomposition of the material, which is consistent with the results reported in the literature [17] .

As can be seen, the curves of these polymer films have nearly rectangular shape, which is typical of the pure capacitive behavior of the tested object [21] -[31] . However, the current of film obtained by potentiodynamic method in presence of dopant is nearly 3 times of the film gained in absence of dopant.

The electrochemical characteristics of obtained PEDOT and PEDOT/CNT-COOH/dopant films were study in the cycling solution, an aqueous solution of 0.1 M $LiClO_4$. The electrode potential was cycled on the potential range from 0 up to 1250 mV with a sweep rate of 50 mV/s and for a cycles number of 20 (see Figure 6).

Analyzing in comparison the obtained results it can be observed that, in all the cases the PEDOT/CNT-COOH composite film reveals current much higher than PEDOT film. This fact can be explained thus, the PEDOT chains became neutral and the negative charge of immobile FCNTSF should be balanced by the cations with small size from the supporting electrolyte solution.

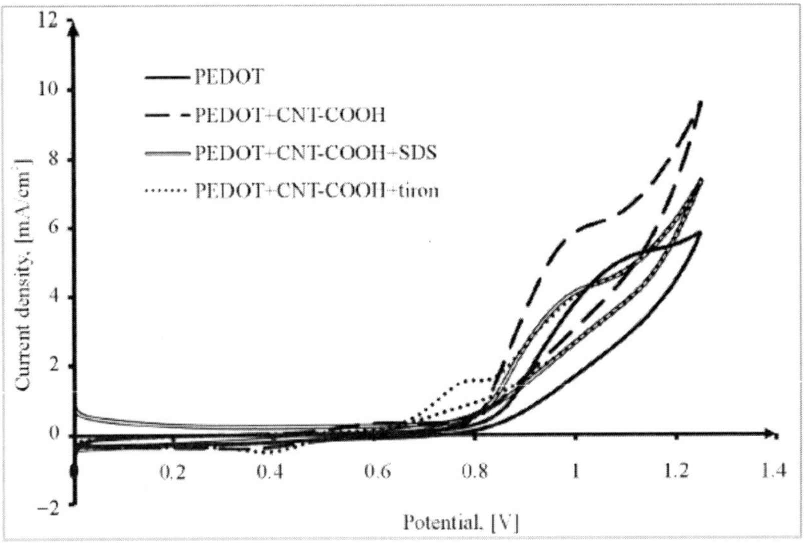

(a)

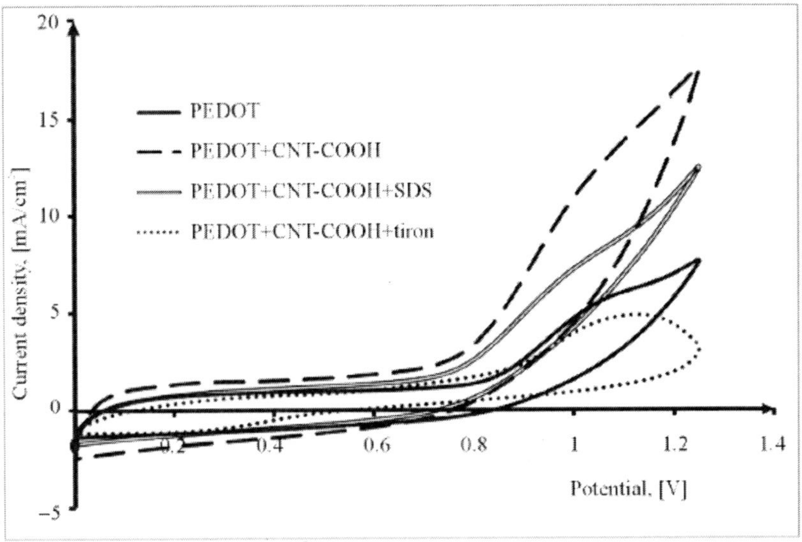

(b)

Figure 3. Comparative polymerization cyclovoltammograms of 2,3-dioxythiophene + CNT-COOH + dopat in 0.1 M $LiClO_4$ aqueous solution, 1st Cycle (a) and 10th Cycle (b).

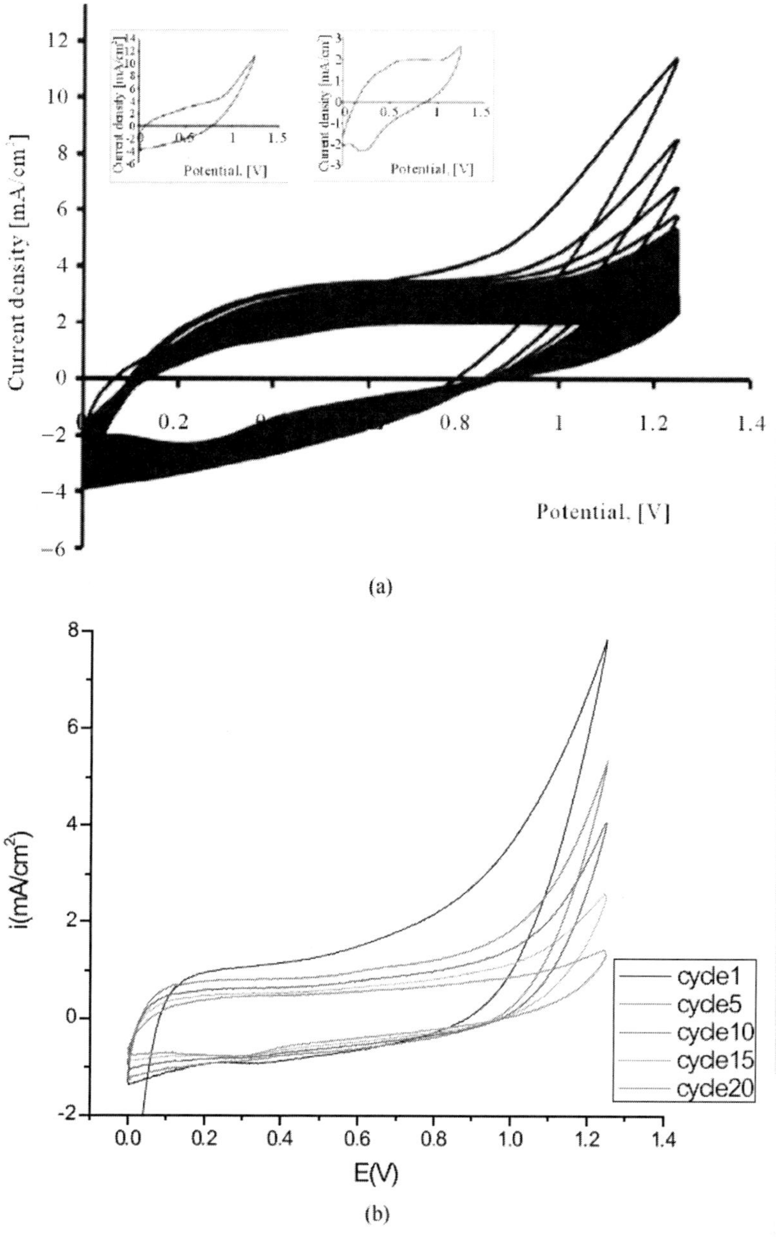

Figure 4. Cyclic voltammograms of PEDOT film in cycling solution (monomer free) of 0.1 M $LiClO_4$. Inset 1st Cycle (a) and 50th Cycle (b).

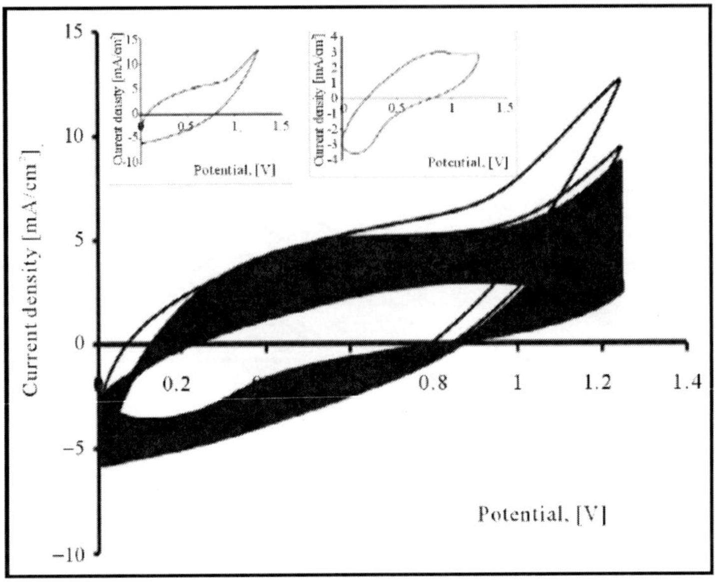

(a)

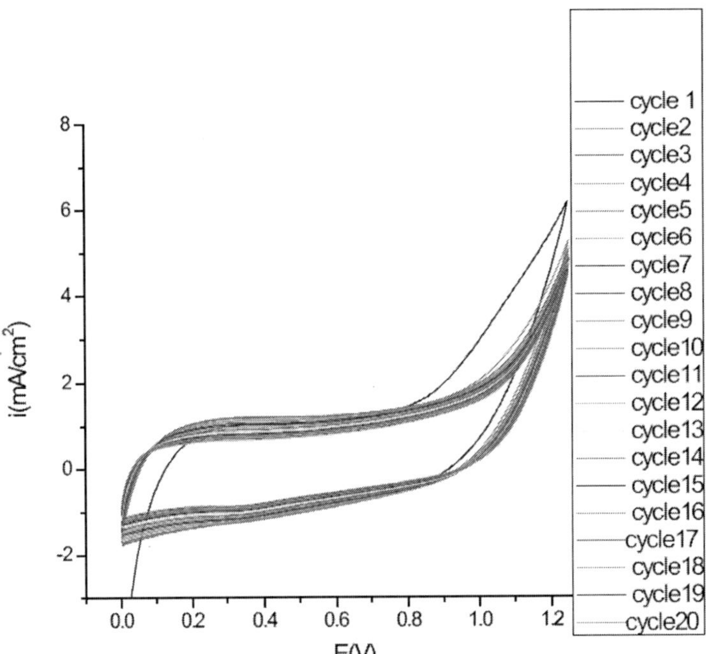

(b)

Figure 5. Cyclic voltammograms of PEDOT/CNT-COOH film in cycling solution (monomer free) of 0.1 M $LiClO_4$. Inset 1st Cycle (a) and 50th Cycle (b).

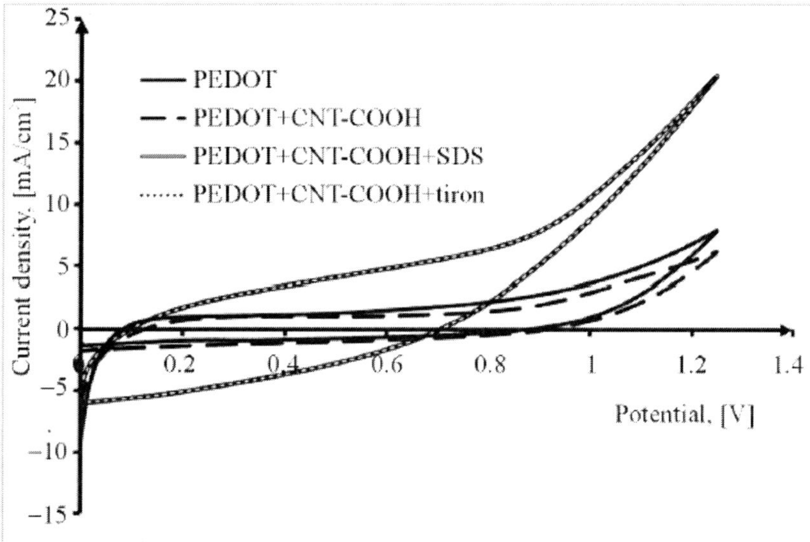

(a)

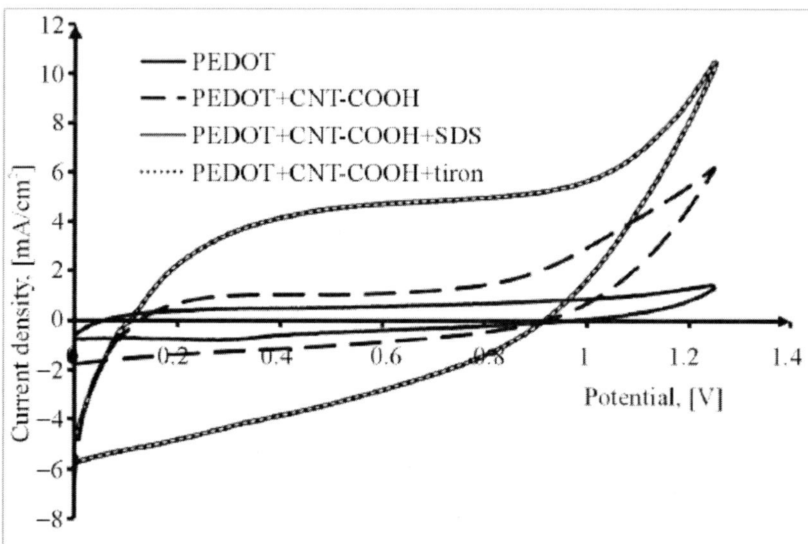

(b)

Figure 6. Comparative cyclic voltammograms of PEDOT, PEDOT/ CNT-COOH with dopants films in cycling solution (monomer free) of 0.1 M $LiClO_4$, 1st Cycle (a) and 20th Cycle (b).

The surface morphology, formation mechanism and electrochemical nature of PEDOT-FSWCNTs films were investigated using scanning electron microscopy (SEM), cyclic voltammetry (CV) and alternating current (AC) impedance spectroscopy (EIS). CV (cyclic voltammetry) and EIS (electrochemical impedance spectroscopy) revealed that the PEDOT-FSWCNTs electrode had higher electrocatalytic activity for the redox reaction and a smaller charge transfer resistance than the PEDOT electrodes.

Further, the composite and pure polymer films were studied by EIS at open circuit potential, in an aqueous solution of 0.1 M $LiClO_4$ and 25°C. The resulting Nyquist plots and Bode plots for PEDOT, PEDOT/SWCNTsCOOH and PEDOT/SWCNTs-COOH/dopant systems are shown in Figure 7 andFigure 8. The impedance plot is composed of a semicircle at high frequencies and a capacitive slope at low and middle frequencies (see Figure 7 and Figure 8). The semicircle appeared at high frequencies is considered to owe to the charge transfer resistance, which originates from the interface structure between the porous electrode surface and the electrolyte [19] . At low frequencies, the impedance plot becomes a near vertical line. The Nyquist plots for both PEDOT, PEDOT/SWCNTs-COOH and PEDOT/SWCNTs-COOH/dopant composite films are featured by a vertical trend at low frequencies, indicating a capacitive behaviour according to the equivalent circuit theory. Bode diagrams point out also the capacitive behaviour in concordance with Nyquist plots (see in comparison Figures 7(a)-(d) and Figures 8(a)-(d)).

The capacitances of the electrode materials were calculated, according to the equation:

$$C = -1/\left(2\pi f Z_{\mathrm{im}}\right).$$

(f = frequency; Z_{im} = imaginary impedance), from the slope of the linear correlation between the imaginary impedance and the reciprocal of the frequency at low frequencies.

From these Figures 7-9 and Table 1, one can observe higher capacitance the one order of magnitude value for PEDOT/FSWCNTs + dopant film in respect with PEDOT pure polymeric films. Higher capacitance of the composite films results obviously from the contribution of the embedded FSWCNTs that provide interconnected pathways for electrons through the FSWCNTs and ions through the pore network or the direct interaction between the delocalised electrons on polymer chains and the FSWCNTs.

The real impedance at low frequencies, where the capacitive behaviour dominates, is an indication of the combined resistance of the electrolyte and the film including both electronic and ionic contributions. The values of the real impedance at 0.01 Hz are given also in Table 1. It can be seen that the PEDOT/FSWCNTs films were significantly lower in resistance than PEDOT films. It can also be seen that PEDOT/FSWCNTs offered much higher overall conductivity compared with the PEDOT film. It has been already mentioned that, in general, the real impedance of an electrode material also decreases as the material's porosity increases due to improved ionic accessibility [20] -[27] [32] -[41] .

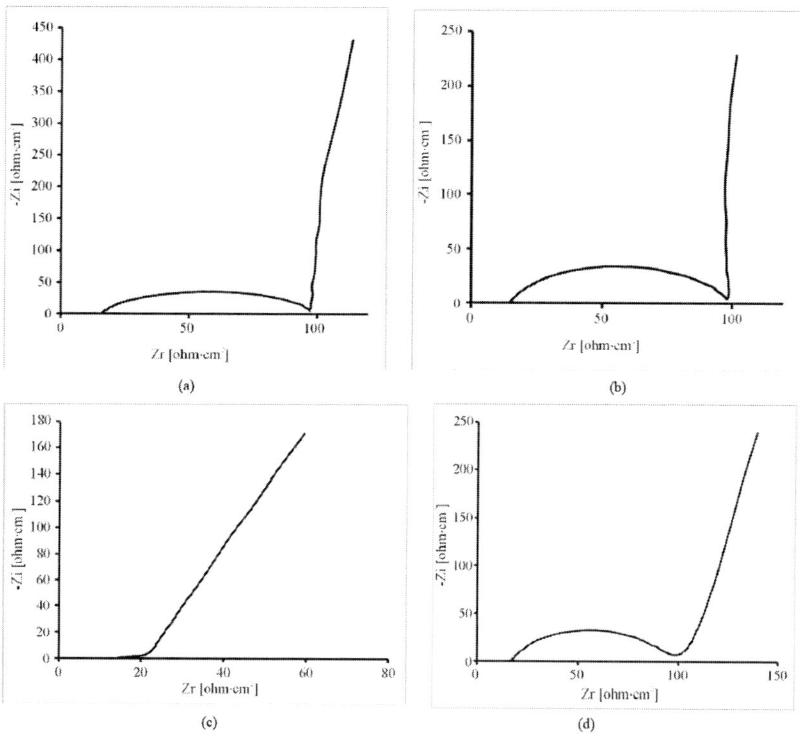

Figure 7. The Nyquist diagrams for modified electrodes (a) PEDOT/Pt, (b) PEDOT/CNsT-COOH/Pt, (c) PEDOT/CNTsCOOH/SDS/Pt and (d) PEDOT/CNTs-COOH/tiron/Pt at open circuit potential in an aqueous solution of 0.1 M $LiClO_4$.

Figure 10 shows the SEM images of the PEDOT film. In Figure 10(a), a nodular accumulating structure is evidenced. The size of the nodules ranged a few hundred nanometers in diameter and they aggregate to form gobbets. In Figure 10(b), the effect of dopant on the morphology of PEDOT film can be observed: lamellar structure with almost vertical orientation to the substrate. It is evident that dopant changes the morphology of PEDOT film into a more porous structure with higher interface area. This is in agreement with the SEM results presented below that illustrated a smaller porosity in the PEDOT film than in the case of the composite film with dopant. The SEM images showed that nanocomposite films were more porous than PEDOT films.

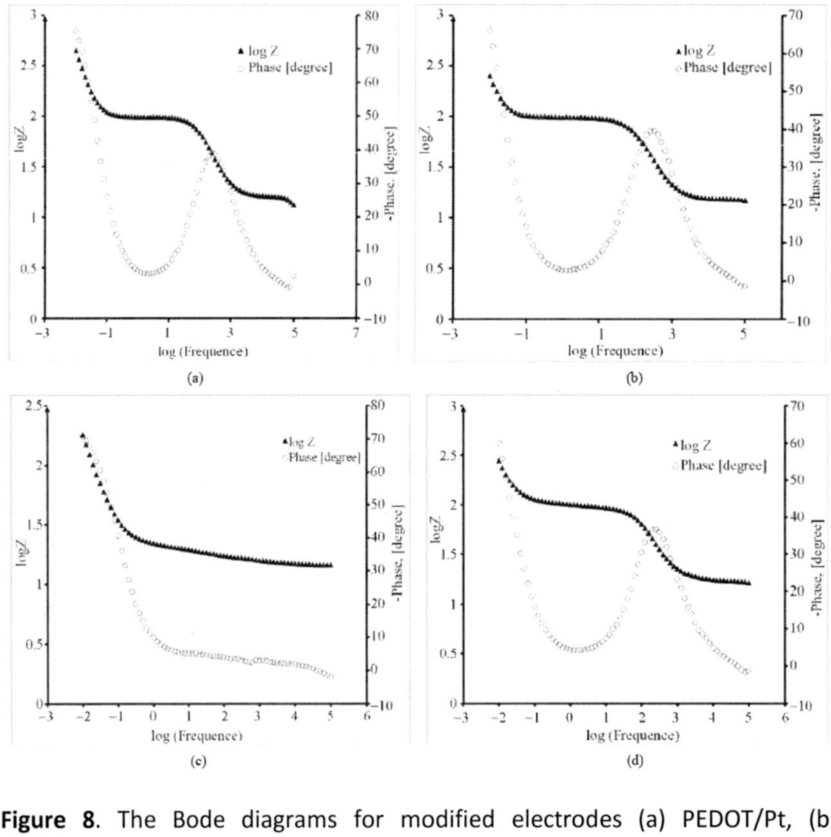

Figure 8. The Bode diagrams for modified electrodes (a) PEDOT/Pt, (b) PEDOT/CNTs-COOH/Pt, (c) PEDOT/CNTsCOOH/SDS/Pt and (d) PEDOT/CNTs-COOH/tiron/Pt at open circuit potential in an aqueous solution of 0.1 M $LiClO_4$.

Fourier transform infrared (FT-IR see Figure 11) spectra were carried with a Bruker optics spectrometer at room temperature. All spectra in this paper were obtained at a resolution 4 cm^{-1}in the region 4000 - 600 cm^{-1}.

Table 1. Real impedance and capacitance values of pure PEDOT film and PEDO/FSWCNTs nanocomposite film with different dopants obtained by co-polymerization using the cyclic voltammetry (CV) at 0.01 Hz.

Polymeric film	Slope values obtained from graph $-Z'' = f(1/2\pi f)$	C $[F/cm^2]$	Z_r (Ωcm^2) at 0.01 Hz
PEDOT	103	0.009	357
PEDOT/CNT-COOH	72	0.014	267
PEDOT/CNT-COOH/SDS	15	0.067	120
PEDOT/CNT-COOH/tiron	11	0.091	139

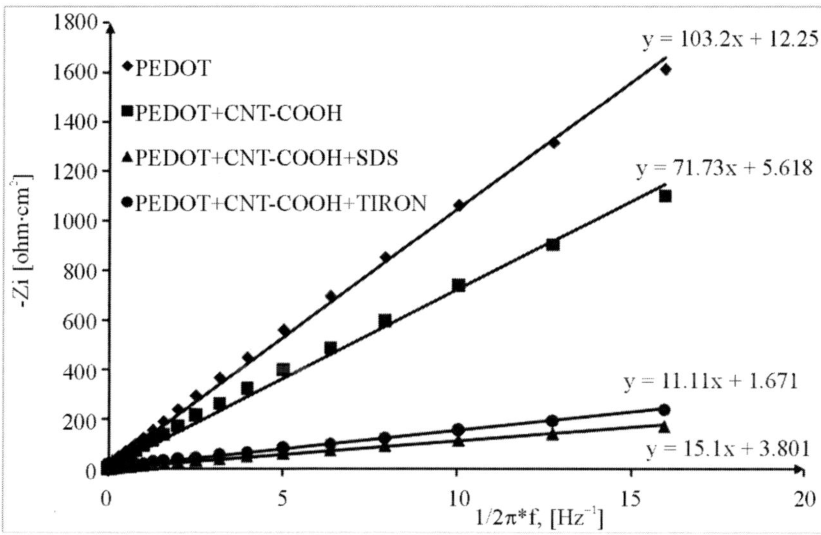

Figure 9. Capacitance evaluation for PEDOT/Pt, PEDOT/CNT/Pt, PEDOT/ CNT-COOH/SDS/Pt and PEDOT/CNT-COOH/ tiron/Pt modified electrode.

FT-IR spectrometer is a powerful instrument that can be used to determine type of bonding for to obtain a new composite. The characteristic bands in the FT-IR spectrum for PEDOT and nanocomposite are the following: a very weak and a medium band at 3000 - 4000 cm^{-1} is assigned to the C-H and OH stretching modes; vibration at 1521, 1476, 1328 and 922 cm^{-1} for PEDOT+ nanocompozit and 1501, 1403, 1181 and 916 cm^{-1} for PEDOT are attributed to the stretching modes of C=C, C-C, and C-S in the thiophene ring [40] [42] -[44] . The bands at 1203 and 1046 cm^{-1} are assigned to the stretching modes of the ethylenedioxy (alkylenedioxy) group and the band around 1046 and 922 cm^{-1} is due to the ethylenedioxy ring deformation mode. The spectrum shows several bands of which the band at 822 cm^{-1} for nanocompozit and 815 cm^{-1} for PEDOT is assigned to the symmetric C-S-C deformation. The band at 1046 cm^{-1} (nanocompozit) and 1412 cm^{-1} (PEDOT) is assigned to the symmetric C-O-C ether bond. A difference that can be observed is the intensity ratio of bands the spectrum (1500 - 600 cm^{-1}) of nanocompozit film than PEDOT band, exhibits a clear difference in intensity [45] . This may suggest that the dopant promotes and stabilizes the structure of the film (nanocomposite).

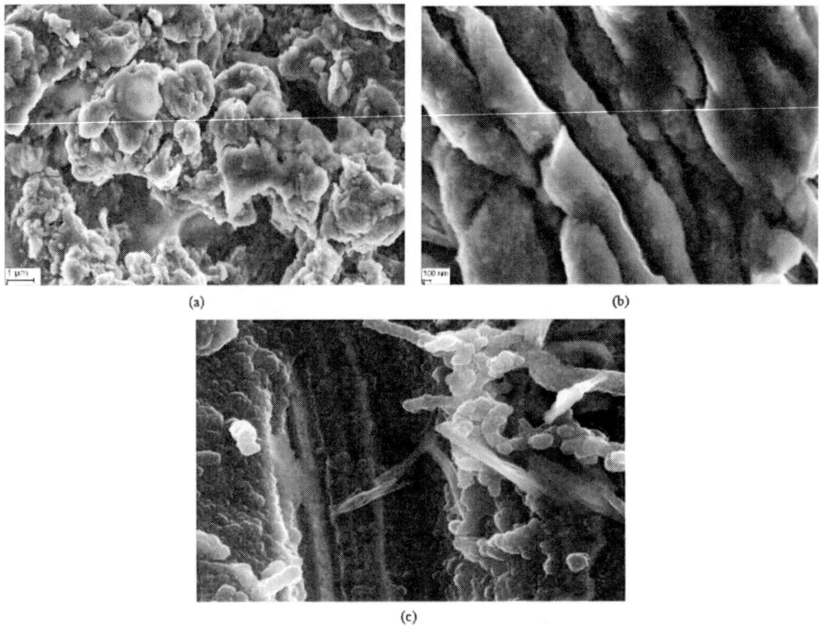

Figure 10. SEM images of the film surface of (a) PEDOT and (b) PEDOT doped film formation by cyclic voltammetry (0 to 1250 mV at a scan rate of 10 mV/s), (c) PEDOT/CNT-COOH.

4. CONCLUSIONS

Electrochemically synthesized composite films of conducting polymers (PEDOT) and FSWCNTs with different dopants have in common a porous structure at nano-meter scales. They have better mechanical integrity and a higher conductivity than the similarly prepared pure polymer. The electrochemical activity of PEDOT/FSWCNTs/Pt modified electrode in 0.1 M $LiClO_4$ cycling solution is much more higher than of PANI/Pt modified electrode in the same cycling solution.

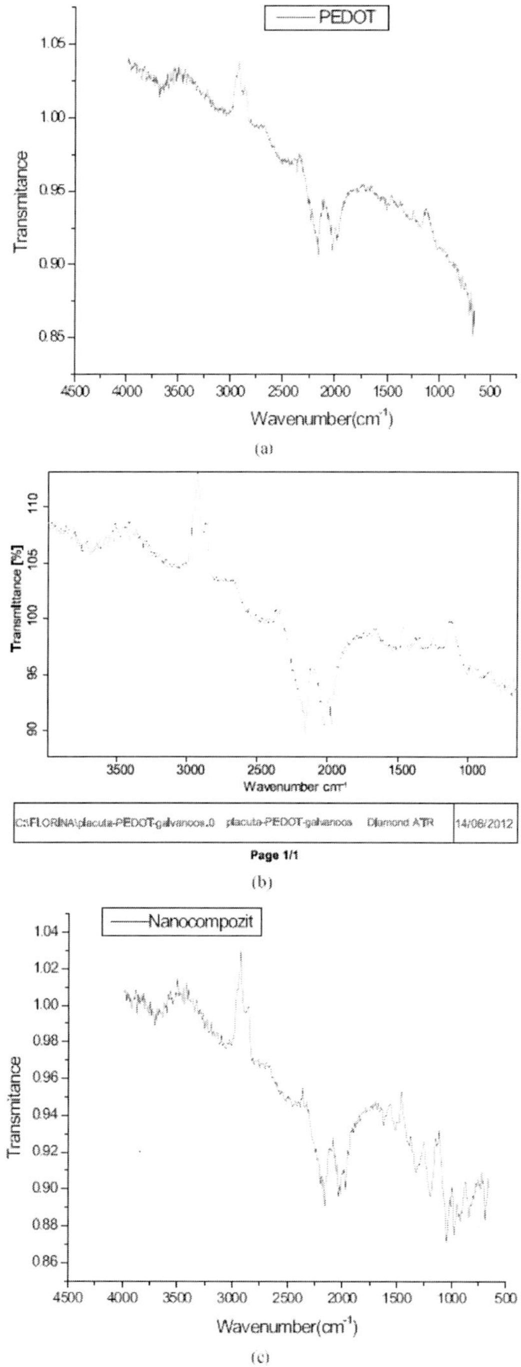

Figure 11. FT-IR spectra of PEDOT and nanocompozit.

The Nyquist plots for both PEDOT and composite films are featured by a vertical trend at low frequencies, indicating a capacitive behaviour according to the equivalent circuit theory. The higher capacitance of the composite films results obviously from the contribution of the embedded FSWCNTs that provide interconnected pathways for electrons through the FSWCNTs and ions through the pore network or the direct interaction between the delocalised electrons on polymer chains and the FSWCNTs.

The PEDOT and FSWCNTsF with dopants films reveal a morphological structure composed of nanofibers, but for the pure polymeric ones they are slightly thicker than those in the PANI/FSWCNTs composite film.

Microstructures of these composites suggests that PEDOT was wrapped around FSWCNTs. The obtained results can be employed to get the desired value of the capacitance by choosing the adequate preparation method and so, by controlling the microstructure of the composites.

ACKNOWLEDGEMENTS

Financial support from PN-II-ID-PCE-2008-2 contract number 596, code ID_716 (The National University Research Council) is gratefully acknowledged.

REFERENCES

1. Ajayan, P.M., Stephan, O., Colliex, C. and Trauth, D. (1994) Aligned Carbon Nanotube Arrays Formed by Cutting a Polymer Resin—Nanotube Composite. Science, 265, 1212-1214.

2. Dai, L. and Mau, A.W.H. (2001) Controlled Synthesis and Modification of Carbon Nanotubes and C_{60}: Carbon Nanostructures for Advanced Polymeric Composite Materials. Advanced Materials, 13, 899-913.

3. Baughman, R.H., Zakhidov, A.A. and Heer, W.A. (2002) Carbon Nanotubes—the Route toward Applications. Science, 297, 787-792.

4. Liao, Y.Z., Zhang, C., Zhang, Y., Strong, V., Tang, J.S., Li, X.-G., Kalantar-Zadeh, K., Hoek, E., Wang, K. and Kaner, R. (2011) Carbon Nanotube/Polyaniline Composite Nanofibers: Facile Synthesis and Chemosensors. Nano Letters, 11, 954-959.

5. Dai, L. (1999) Advanced Syntheses and Microfabrications of Conjugated Polymers, C_{60}-Containing Polymers and Carbon Nanotubes for Optoelectronic Applications. Polymers for Advanced Technologies, 10, 357-420.

6. Chen, G.Z., Shaffer, M.S.P., Coleby, D., Dioxan, G., Fray, D.J. andWindle, A.H. (2000) Carbon Nanotube and Polypyrrole Composites: Coating and Doping. Advanced Materials, 12, 522-526.

7. Saito, Y., Uemura, S. and Hamaguchi, K. (1998) Cathode Ray Tube Lighting Elements with Carbon Nanotube Field Emitters. Japanese Journal of Applied Physics, 37, L346-L348.

8. Liao, Y., Zhang, C., Wang, X., Li, X.G., Ippolito, S.J., Kalantar-Zadeh, K. and Kaner, R.B. (2011) Carrier Mobility of Single-Walled Carbon Nanotube-Reinforced Polyaniline Nanofibers. The Journal of Physical Chemistry C, 115, 16187-16192.

9. Pruně, A., Branzoi, V. and Branzoi, F. (2011) Ordered Arrays of Copper Nanowires Enveloped in Polyaniline Nanotubes. Journal Applied Electrochemistry, 41, 77-81.

10. Shirakwa, H. (2001) The Discovery of Polyacetylene Film: The Dawning of an Era of Conducting Polymers (Nobel Lecture). Angewandte Chemie International Edition, 40, 2574-2580.

11. Groenendaal, B.L., Jonas, F., Freitag, D., Pielartzik, H. and Reynolds, J.R. (2000) Poly(3,4-Ethylenedioxythiophene) and Its Derivatives: Past, Present, and Future. Advanced Materials, 12, 481-494.

12. Crispin, X., Jakobsson, F.L.E., Crispin, A., Grim, P.C.M., Andersson, P., Volodin, A., van Haesendonck, C., Van der Auweraer, M., Salaneck, W.R. and Berggren, M. (2006) The Origin of the High Conductivity of Poly(3,4-Ethylenedioxythiophene)-Poly(Styrenesulfonate) (PEDOT-PSS) Plastic Electrodes. Chemistry of Materials, 18, 4354-4360.

13. Breiby, D.W., Samuelsen, L.B. and Groenendaal, B. (2003) Smectic Structures in Electrochemically Prepared Poly(3,4-Ethylenedioxythiophene) Films. Journal of Polymer Science Part B: Polymer Physics, 41, 945-952.

14. Winther-Jensen, B. and West K. (2006) Stability of Highly Conductive Poly-3,4-Ethylene-Dioxythiophene. Reactive and Functional Polymers, 66, 479-483.

15. Kros, A., Sommerdijk, N.A.J.M. and Nolte R.J.M. (2005) Poly(Pyrrole) versus Poly(3,4-Ethylenedioxythiophene): Implications for Biosensor Applications. Sensors and Actuators B: Chemical, 106, 289-295.

16. Luo, S.C., Mohamed Ali, E., Tansil, N.C., Yu, H.H., Gao, S. and Ying, J.Y. (2008) Poly(3,4-ethylenedioxythiophene) (PEDOT) Nanobiointerfaces: Thin, Ultrasmooth, and Functionalized PEDOT Films with in Vitro and in Vivo Biocompatibility. Langmuir, 24, 8071-8077.

17. Du, X. and Wang, Z. (2003) Effects of Polymerization Potential on the Properties of Electrosynthesized PEDOT Films. Electrochimica Acta, 48, 1713-1717.

18. Li, L., Loveday, D.C., Mudigonda, D.S.K. and Ferraris, J.P. (2002) Effect of Electrolytes on Performance of Electrochemical Capacitors Based on Poly[3-(3,4-difluorophenyl)thiophene]. Journal of the Electrochemical Society, 149, A1201-A1207.

19. Branzoi, V., Prună, A. and Branzoi, F. (2008) Electrosynthesized Doped Polypyrrole Films on Pure Zinc Electrode. Molecular Crystal and Liquid Crystals, 485, 853-861.

20. Branzoi, V., Branzoi, F. and Pilan, L. (2009) Characterization of Electrodeposited Polymeric and Composite Modified Electrodes on Cobalt Based Alloy. Materials Chemistry and Physics, 118, 197-203.

21. Ago, H., Petritch, K., Shaffer, M.S.P., Windle, A.H. and Friend, R.H. (1999) Composites of Carbon Nanotubes and Conjugated Polymers for Photovoltaic Devices. Advanced Materials, 11, 1281-1285.

22. Hughes, M., Chen, G.Z., Shaffer, M.S., Fray, D.J. and Windle, A.H. (2002) Electrochemical Capacitance of a Nanoporous Composite of Carbon Nanotubes and Polypyrrole. Chemistry of Materials, 14, 1610-1613.

23. Downs, C., Nuget, J., Ajayan, P.M., Duquette, D.J. and Santhanam, K.S.V. (1999) Growth from Surface Methodology for the Fabrication of Functional Dual Phase Conducting Polymer Polypyrrole/Polycarbazole/Polythiophene (cp/Polypyr/ Polycbz/Polyth)-Carbon Nanotube (cnt) Composites of Controlled Morphology and Composition—Sidewall Vers. Advanced Materials, 12, 1028.

24. Peng, C., Jin, J. and Chen, G.Z. (2007) A Comparative Study on Electrochemical Co-Deposition and Capacitance of Composite Films of Conducting Polymers and Carbon Nanotubes. Electrochimica Acta, 53, 525-537.

25. Wang, J., Dai, J. and Yarlagadda, T. (2005) Carbon Nanotube-Conducting-Polymer Composite Nanowires. Langmuir, 21, 9-12.

26. Tsai, Y.C., Li, S.C. and Liao, S.W. (2006) Electrodeposition of Polypyrrole-Multiwalled Carbon Nanotube-Glucose Oxidase Nanobiocomposite Film for the Detection of Glucose. Biosensors and Bioelectronics, 22, 495-500.

27. Chen, J., Hamon, M.A., Hu, H., Chen, Y., Rao, A.M., Eklund, P.C. and Haddon, R.C. (1998) Solution Properties of Single-Walled Carbon Nanotubes. Science, 282, 95-98.

28. Chen, Y., Haddon, R.C., Fang, S., Rao, A.M., Eklund, P.C., Lee, W.H., Dickey, E.C., Grulke, E.A., Pendergrass, J.C., Chavan, A., Haley, B.E. and Smalley, R.E. (1998) Chemical Attachment of Organic Functional Groups to SingleWalled Carbon Nanotube Material. Journal of Materials Research, 13, 2423-2431.

29. Holzinger, M., Vostrowsky, O., Hirsch, A., Hennrich, F., Kappes, M., Weiss, R. and Jellen, F. (2001) Sidewall Functionalization of Carbon Nanotubes. Angewandte Chemie International Edition, 40, 4002-4005.

30. Dyke, C.A. and Tour, J.M. (2003) Unbundled and Highly Functionalized Carbon Nanotubes from Aqueous Reactions. Nano Letters, 3, 1215-1218.

31. Saini, R.K., Chiang, I.W., Peng, H., Smalley, R.E., Billups, W.E., Hauge, R.H. and Margrave, J.L. (2003) Covalent Sidewall Functionalization of Single Wall Carbon Nanotubes. Journal of the American Chemical Society, 125, 3617- 3621.

32. Dyke, C.A. and Tour, J.M. (2003) Solvent-Free Functionalization of Carbon Nanotubes. Journal of the American Chemical Society, 125, 1156-1157.

33. Williams, K.A., Veenhuizen, P.T.M., de la Torre, B.G., Eritja, R. and Dekker, C. (2002) Nanotechnology: Carbon Nanotubes with DNA Recognition. Nature, 420, 761.

34. Niyogi, S., Hamon, M.A., Hu, H., Zhao, B., Bhowmik, P., Sen, R., Itkis, M.E. and Haddon, R.C. (2002) Chemistry of Single-Walled Carbon Nanotubes. Accounts of Chemical Research, 35, 1105-1113.

35. Kamaras, K., Itkis, M.E., Hu, H., Zhao, B. and Haddon, R.C. (2003) Covalent Bond Formation to a Carbon Nanotube Metal. Science, 301, 1501.

36. Chen, J., Rao, A.M., Lyuksyutov, S., Itkis, M.E., Hamon, M.A., Hu, H., Cohn, R.W., Eklund, P.W., Colbert, D.T., Smalley, R.E. and Haddon, R.C. (2001)

Dissolution of Full-Length Single-Walled Carbon Nanotubes. Journal of Physical Chemistry B, 105, 2525-2528.

37. Zhao, B., Hu, H. and Haddon, R.C. (2004) Synthesis and Properties of a Water-Soluble Single-Walled Carbon Nanotube-Poly(m-aminobenzene sulfonic acid) Graft Copolymer. Advanced Functional Materials, 14, 71-76.

38. Jurewiez, K., Delpeux, S., Bertagna, V., Beguin, F. and Frackowiak, E. (2001) Supercapacitors from Nanotubes/Polypyrrole Composites. Chemical Physics Letters, 347, 36-40.

39. Gupta, V. and Miura, N. (2006) Polyaniline/Single-Wall Carbon Nanotube (PANI/SWCNT) Composites for High Performance Supercapacitors. Electrochimica Acta, 52, 1721-1726.

40. Xiao, Q. and Zhou, X. (2003) The Study of Multiwalled Carbon Nanotube Deposited with Conducting Polymer for Supercapacitor. Electrochimica Acta, 48, 575-580

41. Zeng, J.X., Wei, W.Z., Wu, L., Liu, X.Y., Liu, K. and Li, Y. (2006) Fabrication of Poly(toluidine blue O)/Carbon Nanotube Composite Nanowires and Its Stable Low-Potential Detection of NADH. Journal of Electroanalytical Chemistry, 595, 152-160.

42. Ferrer-Anglada, N., Kaempgen, M. and Roth, S. (2006) Transparent and Flexible Carbon Nanotube/Polypyrrole and Carbon Nanotube/Polyaniline pH Sensors. Physica Status Solidi (b), 243, 3519-3523.

43. Santhosh, P., Manesh, K.M., Lee, K.P. and Gopalan, A.I. (2006) Enhanced Electrocatalysis for the Reduction of Hydrogen Peroxide at New Multiwall Carbon Nanotube Grafted Polydiphenylamine Modified Electrode. Electroanalysis, 18, 894-903.

44. Lu, G.W., Li, C., Shen, J.Y., Chen, Z.J. and Shi, G.Q. (2007) Preparation of Highly Conductive Gold-Poly(3,4-ethylenedioxythiophene) Nanocables and Their Conversion to Poly(3,4-ethylenedioxythiophene) Nanotubes. Journal of Physical Chemistry C, 111, 5926-5931.

45. Ingan, O. (1997) Handbook of Conducting Polymers. Vol. 2, Wiley, New York, 785.

Chapter 6

Polyvinyl Alcohol/Polyaniline/ZnO Nanocomposite: Synthesis, Characterization and Bactericidal Property

Alireza Samzadeh-Kermani*, Mohsen Mirzaee, Mansour Ghaffari-Moghaddam

Department of Chemistry, Faculty of Science, University of Zabol, Zabol, Iran

ABSTRACT

In this study, a new nanocomposite based on Polyvinylalcohol/Polyaniline/Zinc oxide (PVA/PANi/ ZnO) was successfully synthesized. ZnO powder was used to produce its nanoparticles (NPs) using conventional method. PVA/PANi composite was used to produce ZnO NPs. Polyaniline was synthesized in sol solution of PVA in the presence of ammonium persulphate. Morphology and the size of ZnO NPs in nanocomposites were studied via Scanning Electron Microscopy (SEM), X-ray Diffraction (XRD) analysis and Infrared (IR) spectroscopy. According to XRD results, the average size of prepared ZnO NPs was found to be at the range 22 - 34 nm. SEM images were shown hexagonal shape for ZnO NPs, which was used for the investigation of antibacterial property. Finally, the antibacterial activity of the prepared ZnO NPs and nanocomposites was evaluated against two bacteria of Gram positive Staphylococcus aureus and Gram negative Escherichia coli using paper disc diffusion method. The results were shown that the prepared nanocomposites were effective against two bacteria.

KEYWORDS

Nanostructure, Polymers, Electron Microscopy (SEM), Infrared Spectroscopy, Antibacterial

1. INTRODUCTION

Nanotechnology is fundamentally about preparation of nanoscale materials [1] . NPs are subjected in an increased interest, because they combine the features of polymers with a small quantity of nanoparticles (less than 5% by weight) [2] . Recently, all researches about NPs have gained considerable importance in the areas of biology, medicine, and electronics because of their unique particle size and shape dependent physical, chemical and biological properties [3] [4] . ZnO is a wide band gap (3.37 eV) II-VI compound semiconductor which has received a lot of attentions over the last decade because, it has a wide range of useful properties including electrical [5] , optical [6] , chemical [7] and magnetic [8] . Therefore, ZnO has become extraordinarily important because of its wide applications such as in catalysis [9] , gas sensing [10] , semiconductors [11] , piezoelectric devices [12] , UV-shielding materials [13] , polymer coating [14] , photochemical [15] and antibacterial agents [16] . As ZnO is a multifunctional material, the study on its NPs has been focused on their preparation using various methods including solid state reaction [9] , flame spray pyrolysis [17] , microwave-assisted method [18] , sonochemical route [19] , sol-gel [20] , hydrothermal [21] and mechanochemical methods [22] . ZnO powder inhibits the growth of gram-positive bacteria more strongly than gram-negative bacteria [23]. This research was focused on metal oxides like ZnO as antibacterial agents to prevent bacterial growth. Recently, there have been several reports regarding the antimicrobial activity of ZnO NPs [24] . ZnO displays many interesting properties at nano- scale, and has significant applications as antibacterial and antifungal agents. It also holds importance in biomedical sciences [25] . ZnO has shown a considerable antibacterial activity at pH values at the range 7 - 8 without the presence of light [26] . Polyvinyl alcohol (PVA) is an important water-soluble transparent polymer, and is extensively used in industries due to the excellent chemical and physical properties, non-toxicity, good chemical resistance, good film formation capacity, biodegradability and high crystal modulus. PVA is used here in hydrolyzed form with the degree of 85% hydrolysis [27] . It is a polymer with many technological, pharmaceutical and biomedical applications [28] . It can be used for coating of ZnO [29] , titanium dioxide [30] and Ag [31] . PANi, an intrinsically conducting polymer, has received great attentions because of its wide potential formany technological applications, such as sensors, anticorrosion coatings, electro rheological fluids [32] and good environmental stability [33] . PANi is usually obtained as powder via precipitation polymerization [34] .

2. MATERIALS AND METHODS

2.1. Materials

All reagents in this work were of analytical grade and purchased from Merck (Darmstadt, Germany). Zinc oxide as a precursor for the preparation of ZnO NPs, polyvinyl alcohol as a matrix for stabilization of NPs and pure aniline as a starting material for synthesis of PANi, were used.

2.2. Microorganisms

The antibacterial activity of prepared ZnO NPs and nanocomposites was carried out using two different bacterial strains including gram positive Staphylococcus aureus (Staph.) and gram negative Escherichia coli (E. coli). All bacterial strains were provided from the Faculty of Pharmacy, Zabol University of Medical Sciences, Zabol, Iran.

2.3. Preparation of ZnO Nanoparticles

First, ZnO powder (0.1 - 1 g) was dissolved in 100 ml of 1% acetic acid and the solution was sonicated for 30 minutes in orther to obtain Zn^{2+} cations. Then, 50 ml of 1 M NaOH was added dropwise together with magnetic stirring until the solution attained pH 10. The solution was heated in water bath at (40˚C - 80˚C) for about 3 hours. It was filtered and washed with distilled water three times, and then dried in an oven at 50˚C for 1 hour according to the conventional procedure [35] .

2.4. Preparation of Aniline Hydrochloride

Aniline hydrochloride was prepared by the reaction of pure aniline with concentrated hydrochloric acid via mixing at the ratio of 1:2 [36] . First, 20 ml of hot aniline was added dropwise to 40 ml of hydrochloric acid, and the mixture stirred consequently using magnetic stirrer. The mixture color was changed from colorless to violet, then cooled at room temperature. The mixture was filtered and washed with concentrated hydrochloric acid. Finally, the obtained white sediment was dried at 40°C - 50°C in an electrical oven.

2.5. Preparation of PVA/PANi/ZnO Nanocomposites

PVA/PANi/ZnO nanocomposites were synthesized via an in situ chemical polymerization of aniline hydrochloride into the suspensions containing PVA and different concentrations of ZnO NPs. In a typical synthesis, a suspension was prepared by dissolving 6.5 g PVA in 100 ml 0.1 M HCl. The mixture was heated to attain a transparent solution, and then cooled at room temperature for half an hour. Different amounts of ZnO NPs were added to the mixture and then, 1.3 g aniline hydrochloride added slowly to every mixture and stirred at −5°C for 30 minutes. 10 ml of aqueous solution of ammonium persulphate $[(NH_4)_2S_2O_8]$ 0.1 M was added to the mixture and allowed to complete the reaction for 12 hours under constant stirring at −3°C. Finally, the blue green mixture was prepared, and then dried at 40°C for 48 h.

2.6. Characterization

2.6.1. FT-IR Spectroscopy

The prepared ZnO NPs and nanocomposites were analyzed using a Fourier transform infrared (FT-IR) spectrometer (Bruker Optics Ft Tensor 27, Germany).

2.6.2. X-Ray Diffraction (XRD) Analysis

The structure of PVA/PANi/ZnO nanocomposites was determined by X-ray diffraction using the Bruker-D8 Advanced (Germany) instrument. A wavelength (λ) 0.15418 nm was used for these analyses. The XRD patterns were recorded at the scan speed of 2° min^{-1}.

2.6.3. Field Emission Scanning Electron Microscopy (FESEM)

FESEM was applied to observe the surface morphology of PVA/PANi/ZnO nanocomposites using a Hitachi S4160 instrument. Thin films of the samples were prepared on graphite adhesives, and then the surface of samples was coated by gold powder using sputter hummer instrument.

2.7. Antibacterial Activity

The antibacterial activity of ZnO NPs was investigated against Staphylococcus aureus (Staph.) and Escherichia coli (E. coli) microorganisms by disk diffusion method in accordance with the procedure described by Hwang and Ma [37] . This method is a mean of measuring the effects of an antimicrobial agent on bacteria growth in a culture. Muller-Hinton Agar (MHA) powder was used as a culture medium for bacteria growth. To prepare the culture medium, 19 g of agar was dissolved into 500 ml of distilled water, and then a transparent brown solvent was achieved via boiling the solution. MHA medium (15 ml) was sterilized at 120°C for about 1 hour in autoclave, cooled to room temperature and then poured into sterilized petri dishes (10 - 90 mm). After cooling over 24 hours, the bacteria are swabbed uniformly across the culture plate. Filter-paper disks were placed on the surface of agar. To evaluate the antibacterial activity, 40 μl of each of the samples was dropped moderately on disks' surface using a sampler. If the samples are effective against the bacteria at a certain concentration, then no colonies would grow and the concentration in agar is greater than or equal to the effective concentration. This region is called inhibition zone. The size of inhibition zone measures the efficacy of sample. A more effective sample produces a larger clear area around the disk. All tests were done under laminar flow hood. Finally, all petri dishes containing bacteria and antibacterial reagents were incubated at 37°C for 24 h. At the end of incubation period, the diameters of inhibition zones formed around disks, were determined and presented in mm. The results concerning antibacterial activity were expressed as strong activity (>13 mm), moderate activity (6 - 12 mm), weak activity (5 mm) or no activity (inhibition zone < 5 mm).

3. RESULTS AND DISCUSSION

3.1. FT-IR Spectroscopy

FT-IR spectra of prepared ZnO NPs and nanocomposites were recorded and some partial changes in the absorption peaks of different concentration of nanocomposites C1, C2, C3 and C4 were observed (Figure 1). The broad band near 3429 cm^{-1} represents the hydrogen bonded O-H stretching vibration [38] . The band 2987 cm^{-1} represents asymmetric C^3_{sp}-H stretching frequencies of alkyl groups. The band at 448 - 550 cm^{-1} was pointed out Zn-O stretching frequency [39] . The free OH functional group have a broad peak at 3600 - 3650 cm^{-1}, and this peak goes toward 3100 - 3500 cm^{-1}, if the OH group engaged in formation of hydrogen bond or complex with metal particles, so the broad peak of hydroxyl groups were appeared for nanocomposites in this range. The presence of C-H and CH_2 bonds of Alkanes in the PVA structure, were confirmed with the stretching peaks around 2855 - 3000 cm^{-1} and bending at 1403 cm^{-1}. A single medium peak in 3100 - 3500 cm^{-1} for aromatic C-H and an intense peak in 1100 - 1350 cm^{-1} for C-N bond were shown for PANi. The presence of benzene rings in polyaniline structure was also confirmed with an intense C-H out of plane bending peak around 1475 - 1600 cm^{-1}.

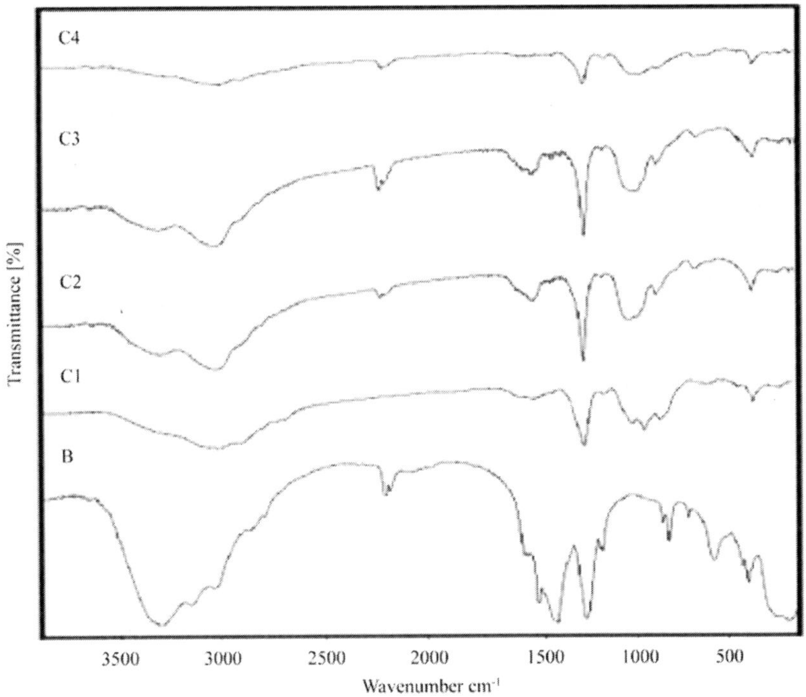

Figure 1. FT-IR spectra of prepared ZnO NPs (B), nanocomposites 1% (C1), 1.5% (C2), 2% (C3) and 2.5% (C4).

3.2. XRD Analysis

The sharp and intense peaks around 2θ values of 31.8, 34.4, 36.2, 47.5, 56.6, 62.8, 67.90 and 68.5 with 100, 002, 101, 102, 110, 103, 112 and 201 diffractions, were related to ZnO crystalline hexagonal structure which formed benign stabilizing complex by polymeric matrix. All the diffraction peaks are in good agreement with those of hexagonal wurtzite structure of prepared ZnO nanoparticles [40] . The XRD patterns were clearly exhibited the presence of ZnO NPs in nanocomposites (Figure 2). The average particle size of every PVA/PANi/ZnO nanocomposite was obtained using Scherrer's Equation:

$$D = \frac{K\lambda}{\beta \cos\theta}$$

In which, D is the crystallite size, k is the shape factor that assumes a value of 0.89 for ZnO, λ is the X-ray wavelength (1.5418 A°), β is the half height width of XRD peak and θ is the diffraction angle [16] . The diameters of ZnO nanoparticles were estimated to be at the range 22.57 - 33.90 nm for all samples.

X-ray diffraction histograms for the size distribution of ZnO NPs were demonstrated that their concentration increment in nanocomposites, would cause a scale down in smaller sizes of ZnO NPs. This is due to the agglomeration of ZnO NPs with increasing their concentrations (Figure 3).

3.3. FESEM Analysis

SEM images of pure and coated ZnO NPs were shown in Figure 4. The images were shown that, uncoated ZnO NPs were aggregated and compressed, while coated NPs had a better distribution and homogeneity. SEM images were also shown that, samples with medium concentraction (like C2 and C3) had a better distribution. This is due to the less agglomeration of NPs in medium concentraction of samples. This fact confirms the results of XRD analysis.

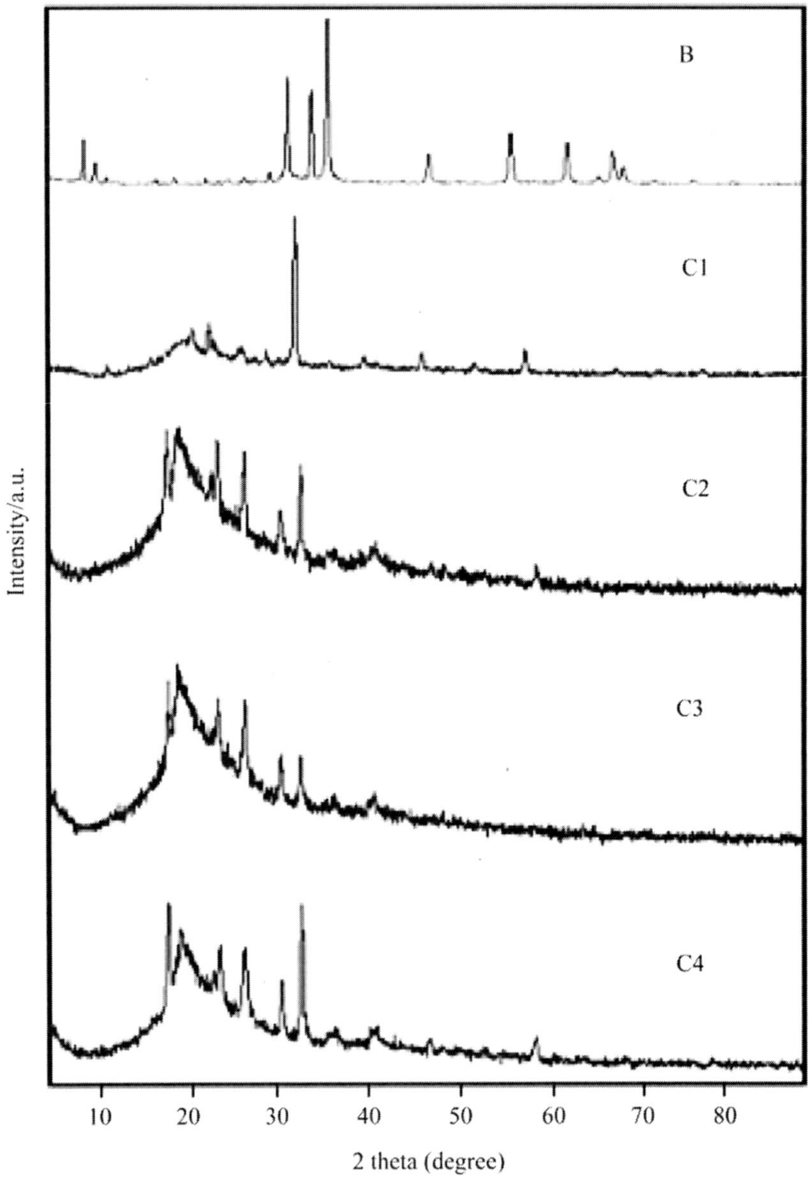

Figure 2. XRD spectra of prepared ZnO NPs (B) and nanocomposites 1% (C1), 1.5% (C2), 2% (C3) and 2.5% (C4).

3.4. Antibacterial Properties of Prepared PVA/PANi/ZnO Nanocomposites

The results of antibacterial activity of ZnO NPs and nanocomposites against bacteria were depicted in Table 1. It was shown that the inhibition zone increased with increasing ZnO NPs' content of nanocomposites, and also observed that the inhibition zone for Staph. was greater than that for E. coli. However, antibacterial activity could be attributed to antibacterial effect of ZnO NPs. It was suggested that ZnO NPs may affect the membrane of bacteria and inhibit their growth [41] . Antimicrobial activity of nanocomposites for low and high concentrations of ZnO NPs was low. This is due to the presence of small amounts of ZnO NPs in low concentration nanocomposite (C1), and agglomeration of ZnO NPs in high concentration nanocomposite (C4). At this circumstance the amount of interaction with cell membrane was decreased which would caused a reduction in the inhibition zone. In this test the antibacterial activity against Staph. was stronger than E. coli (Figure 5). Figure 6 was illustrated the scatter plot of the inhibition zone values for ZnO NPs and various concentrations of nanocomposites 1% (C1), 1.5% (C2), 2% (C3) and 2.5% (C4).

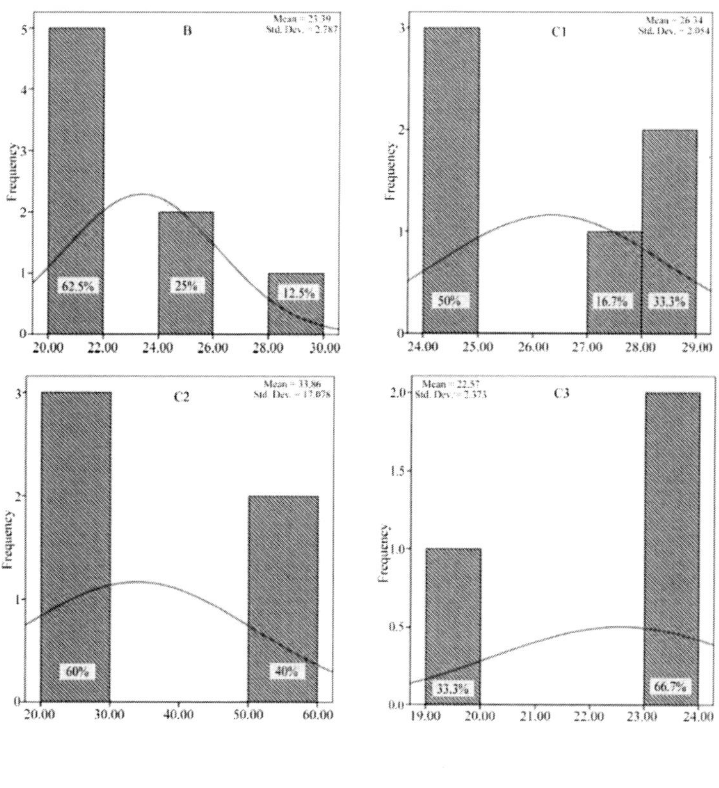

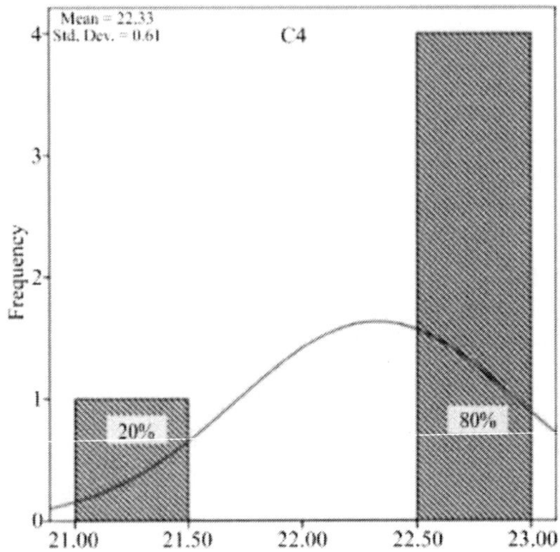

Figure 3. XRD histograms for the size distribution of ZnO nanoparticles (B) and different nanocomposites containing ZnO nanoparticles 1% (C1); 1.5% (C2); 2% (C3); 2.5% (C4).

4. CONCLUSION

ZnO NPs were synthesized using ZnO powder. PVA/PANi/ZnO nanocomposites were prepared in different concentrations of ZnO NPs [1% (C1), 1.5% (C2), 2% (C3) and 2.5% (C4)]. The antibacterial activity of PVA/ PANi/ZnO nanocomposites at different sizes of ZnO NPs was shown a strong antibacterial activity against Gram-positive and Gram-negative bacteria. These results were shown that the antibacterial resistance of ZnO NPs in nanocomposites could be modified according to their size distribution, and decreased with increasing in particle size. Further studies should be focused on the investigation of cytotoxic effects and the bactericidal effects of PVA/PANi/ZnO nanocomposites against different types of bacteria for potential widening of their applications, such as in surgical devices and in drug-delivery. Moreover, the XRD analysis peaks for ZnO NPs were in good agreement with those of hexagonal wurtzite structure. SEM images were shown an appropriate size distribution of ZnO NPs onto the polymer kip, and smaller sizes for ZnO NPs in lower concentration nanocomposites. The size distribution for different concentrations of nanocomposites was at the range 22.57 - 33.90 nm.

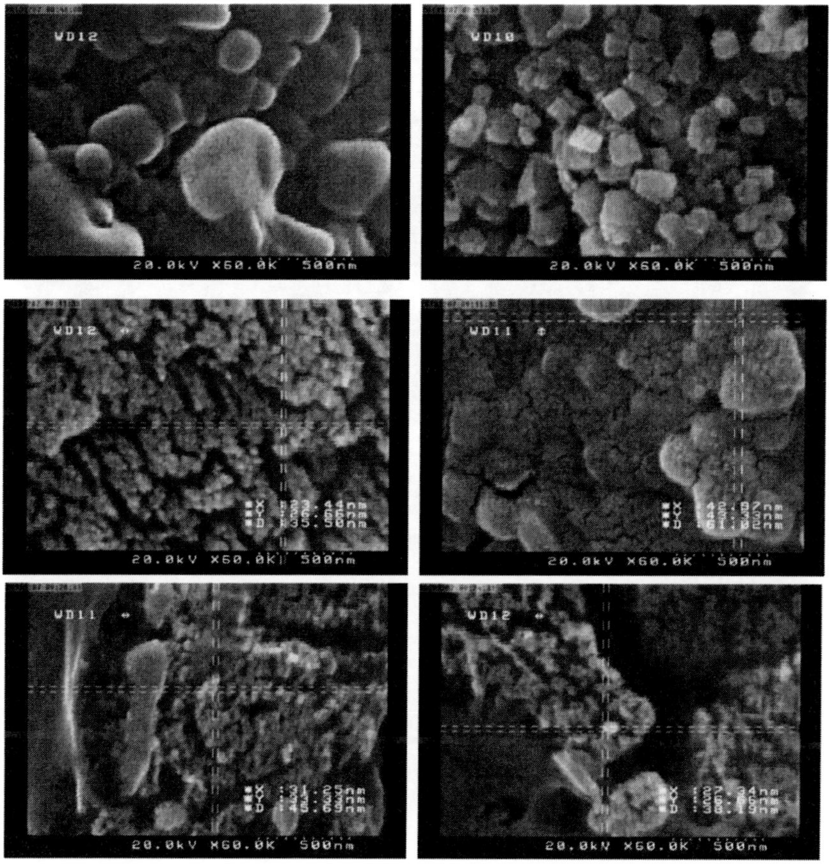

Figure 4. SEM images of PVA/PANi (A), ZnO NPs (B), nanocomposites 1% (C1), 1.5% (C2), 2% (C3) and 2.5% (C4).

Table 1. Average inhibition zone and standard deviation for ZnO NPs and various concentrations of nanocomposites 1% (C1), 1.5% (C2), 2% (C3) and 2.5% (C4).

Sample	Formation diagonal circulars	
	Escherichia coli	*Staphylococcus aureus*
B	21 mm	25 mm
C1	18 mm	20 mm
C2	20 mm	24 mm
C3	22 mm	20 mm
C4	20 mm	19 mm

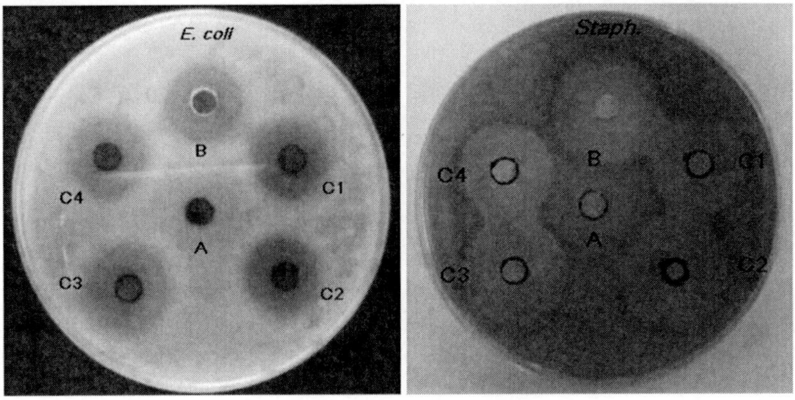

Figure 5. A schematic illustration of created inhibition zones for antibacterial activity of PVA/PANi polymer (A), ZnO NPs (B) and various concentrations of nanocomposites 1% (C1), 1.5% (C2), 2% (C3) and 2.5% (C4).

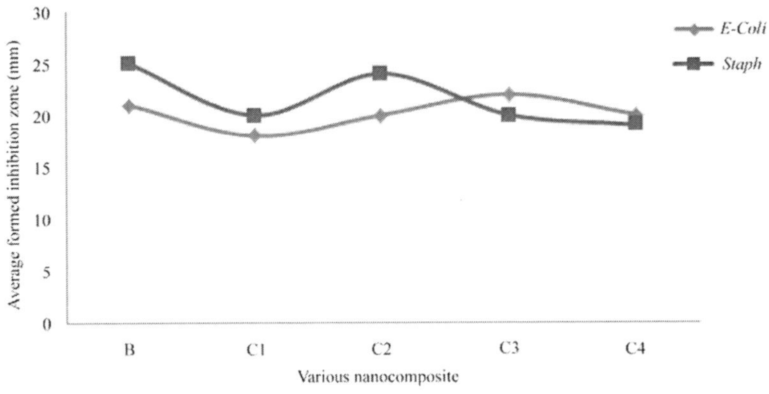

Figure 6. Scatter plot of inhibition zone versus ZnO NPs and various concentrations of nanocomposites 1% (C1), 1.5% (C2), 2% (C3) and 2.5% (C4).

ACKNOWLEDGEMENTS

This research has been accomplished under financial support of the Department of Chemistry, University of Zabol and cordially laboratorial collaboration of the Faculty of Pharmacy, Zabol University of Medical Sciences, which should be appreciated.

REFERENCES

1. Zhou, B. and Balee, R. (2005) Groenendaal, Nanoparticle and Nanostructure Catalysts: Technologies and Markets. Nanotech. Law Business, 2, 222.

2. Peukert, W., Schwarzer, H.C., Gotzinger, M., Gunther, L. and Stenger, F. (2003) Control of Particle Interfaces—The Critical Issue in Nanoparticle Technology. Advanced Powder Technology, 14, 411-426.

3. Sun, C., Lee, J.S. and Zhang, M. (2008) Magnetic Na-noparticles in MR Imaging and Drug Delivery. Advanced Drug Delivery Reviews, 60, 1252-1265.

4. Ko, S.H., Park, I., Pan, H., Grigoropoulos, C.P., Pisano, A.P., Luscombe, C.K. and Fréchet, J.M. (2007) Direct Nano- imprinting of Metal Nanoparticles for Nanoscale Electronics Fabrication. Nano Letters, 7, 1869-1877.

5. Patil, S., Shinde, S. and Rajpure, K. (2013) Physical Properties of Spray Deposited Ni-Doped Zinc Oxide Thin Films. Ceramics International, 39, 3901-3907.

6. Labuayai, S., Promarak, V. and Maensiri, S. (2009) Synthesis and Optical Properties of Nanocrystalline ZnO Powders Prepared by a Direct Thermal Decomposition Route. Applied Physics A, 94, 755-761

7. Suwanboon, S., Amornpitoksuk, P., Bangrak, P. and Randorn, C. (2014) Physical and Chemical Properties of Multifunctional ZnO Nanostructures Prepared by Precipitation and Hydrothermal Methods. Ceramics International, 40, 975-983

8. Yilmaz, S., Nisar, J., Atasoy, Y., McGlynn, E., Ahuja, R., Parlak, M. and Bacaksiz, E. (2013) Defect-Induced Room Temperature Ferromagnetism in B-Doped ZnO. Ceramics International, 39, 4609-4617.

9. Suwanboon, S., Amornpitoksuk, P., Sukolrat, A. and Muensit, N. (2013) Optical and Photocatalytic Properties of La-Doped ZnO Nanoparticles Prepared via Precipitation and Mechanical Milling Method. Ceramics International, 39, 2811-2819.

10. Yu, A., Qian, J., Pan, H., Cui, Y., Xu, M., Tu, L., Chai, Q. and Zhou, X. (2011) Micro-Lotus Constructed by Fe-Doped ZnO Hierarchically Porous Nanosheets: Preparation, Characterization and Gas Sensing Property. Sensors and Actuators B: Chemical, 158, 9-16.

11. Zhang, C. (2007) The Influence of Post-Growth Annealing on Optical and Electrical Properties of P-Type ZnO Films. Materials Science in Semiconductor Processing, 10, 215-221.

12. Seo, M., Jung, Y., Lim, D., Cho, D. and Jeong, Y. (2013) Piezoe-lectric and Field Emitted Properties of Controlled ZnO Nanorods on CNT Yarns. Materials Letters, 92, 177-180.

13. Li, R., Yabe, S., Yamashita, M., Momose, S., Yoshida, S., Yin, S. and Sato, T. (2002) Synthesis and UV-Shielding Properties of ZnO- and CaO-Doped CeO2 via Soft Solution Chemical Process. Solid State Ionics, 151, 235-241.

14. Qin, L., Shing, C., Sawyer, S. and Dutta, P.S. (2011) Enhanced Ultraviolet Sensitivity of Zinc Oxide Nanoparticle Photoconductors by Surface Passivation. Optical Materials, 33, 359-362.

15. Mousa, M., Bayoumy, W. and Khairy, M. (2013) Characterization and Photo-Chemical Applications of Nano-ZnO Prepared by Wet Chemical and Thermal Decomposition Methods. Materials Research Bulletin, 48, 4576-4582.

16. Talebian, N., Amininezhad, S.M. and Doudi, M. (2013) Controllable Synthesis of ZnO Nanoparticles and Their Morphology-Dependent Antibacterial and Optical Properties. Journal of Photochemistry and Photobiology B: Biology, 120, 66-73.

17. Lee, S.D., Nam, S.H., Kim, M.H. and Boo, J.H. (2012) Synthesis and Photocatalytic Property of ZnO Nanoparticles Prepared by Spray-Pyrolysis Method. Physics Procedia, 32, 320-326.

18. Sharma, D., Sharma, S., Kaith, B., Rajput, J. and Kaur, M. (2011) Synthesis of ZnO Nanoparticles Using Surfactant Free In-Air and Microwave Method. Applied Surface Science, 257, 9661-9672.

19. Kazeminezhad, I., Sadollahkhani, A. and Farbod, M. (2013) Synthesis of ZnO Nanoparticles and Flower-Like Nanostructures Using Nonsono- and Sono-Electrooxidation Methods. Materials Letters, 92, 29-32.

20. Ba-Abbad, M.M., Kadhum, A.A.H., Bakar Mohamad, A., Takriff, M.S. and Sopian, K. (2013) The Effect of Process Parameters on the Size of ZnO Nanoparticles Synthesized via the Sol-Gel Technique. Journal of Alloys and Compounds, 550, 63-70.

21. Aneesh, P.M., Vanaja, K.A. and Jayaraj, M.K. (2007) Synthesis of ZnO Nanoparticles by Hydrothermal Method. Proceedings of SPIE, 6639, 66390J-1-66390J-9.

22. Stankovic, A., Veselinovic, L., Skapin, S., Markovic, S. and Uskokovic, D. (2011) Controlled Mechanochemically Assisted Synthesis of ZnO Nanopowders in the Presence of Oxalic Acid. Journal of Materials Science, 46, 3716-3724.

23. Sawai, J., Igarashi, H., Hashimoto, A., Kokugan, T. and Shimizu, M. (1995) Evaluation of Growth Inhibitory Effect of Ceramics Powder Slurry on Bacteria by Conductance Method. Journal of Chemical Engineering of Japan, 28, 288-293.

24. Tankhiwale, R. and Bajpai, S. (2012) Preparation, Characterization and Antibacterial Applications of ZnO-Nanoparticles Coated Polyethylene Films for Food Packaging. Colloids and Surfaces B: Biointerfaces, 90, 16-20.

25. Sharma, D., Rajput, J., Kaith, B., Kaur, M. and Sharma, S. (2010) Synthesis of ZnO Nanoparticles and Study of Their Antibacterial and Antifungal Properties. Thin Solid Films, 519, 1224-1229.

26. Tam, K., Djurisic, A., Chan, C., Xi, Y., Tse, C., Leung, Y., Chan, W., Leung, F. and Au, D. (2008) Antibacterial Activity of ZnO Nanorods Prepared by a Hydrothermal Method. Thin Solid Films, 516, 6167-6174.

27. Wang, H.H., Shyr, T.W. and Hu, M.S. (1999) The Elastic Property of Polyvinyl Alcohol Gel with Boric Acid as a Cross-Linking Agent. Journal of Applied Polymer Science, 74, 3046-3052.

28. Scotchford, C., Cascone, M., Downes, S. and Giusti, P. (1998) Osteoblast Responses to Collagen-PVA Bioartificial Polymers in Vitro: The Effects of Cross-Linking Method and Collagen Content. Biomaterials, 19, 1-11.

29. Chandrakala, H., Ramaraj, B. and Lee, J.H. (2013) Poly-vinyl Alcohol/Carbon Coated Zinc Oxide Nanocomposites: Electrical, Optical, Structural and Morphological Characteristics. Journal of Alloys and Compounds, 580, 392-400.

30. Hebeish, A., Abdelhady, M. and Youssef, A. (2013) TiO2 Nanowire and TiO2 Nanowire Doped Ag-PVP Nanocomposite for Antimicrobial and Self-Cleaning Cotton Textile. Carbohydrate Polymers, 91, 549-559.

31. Shin, J., Kim, Y., Lee, K., Lim, Y.M. and Nho, Y.C. (2008) Significant Effects of Sodium Acetate, an Impurity Present in Poly(vinyl alcohol) Solution on the Radiolytic Formation of Silver Nanoparticle. Radiation Physics and Chemistry, 77, 871-876.

32. Cruz-Silva, R., Ruiz-Flores, C., Arizmendi, L., Romero-Garcia, J., Arias-Marin, E., Moggio, I., Castillon, F. and Farias, M. (2006) Enzymatic Synthesis of Colloidal Poly-aniline Particles. Polymer, 47, 1563-1568.

33. Sarno, D.M., Manohar, S.K. and MacDiarmid, A.G. (2005) Controlled Interconversion of Semiconducting and Metallic Forms of Polyaniline Nanofibers. Synthetic Metals, 148, 237-243.

34. Stejskal, J. and Gilbert, R. (2002) Polyaniline. Preparation of a Conducting Polymer (IUPAC Technical Report). Pure and Applied Chemistry, 74, 857-867.

35. AbdElhady, M. (2012) Preparation and Characterization of Chitosan/Zinc Oxide Nanoparticles for Imparting Antimicrobial and UV Protection to Cotton Fabric. International Journal of Carbohydrate Chemistry, 2012, 1-6.

36. Bhadra, J. and Sarkar, D. (2010) Size Variation of Polyaniline Nanoparticles Dispersed in Polyvinyl Alcohol Matrix. Bulletin of Materials Science, 33, 519-523.

37. Hwang, J.J. and Ma, T.W. (2012) Preparation, Morphology, and Antibacterial Properties of Poly Acrylonitrile/Montmorillonite/Silver Nanocomposites. Materials Chemistry and Physics, 136, 613-623.

38. Maensiri, S., Laokul, P. and Promarak, V. (2006) Synthesis and Optical Properties of Nanocrystalline ZnO Powders by a Simple Method Using Zinc Acetate Dihydrate and Poly(vinyl pyrrolidone). Journal of Crystal Growth, 289, 102-106.

39. Sankara Reddy, B., Venkatramana Reddy, S., Koteeswara Reddy, N. and Pramoda, K. (2013) Synthesis, Structural, Optical Properties and Antibacterial Activity of Co-Doped (Ag, Co) ZnO Nanoparticles. Research Journal of Material Sciences, 1, 11-20.

40. Guo, M., Diao, P. and Cai, S. (2005) Hydrothermal Growth of Well-Aligned ZnO Nanorod Arrays: Dependence of Morphology and Alignment Ordering upon Preparing Conditions. Journal of Solid State Chemistry, 178, 1864-1873.

41. Brayner, R., Ferrari-Iliou, R., Brivois, N., Djediat, S., Benedetti, M.F. and Fiévet, F. (2006) Toxicological Impact Studies Based on Escherichia coli Bacteria in Ultrafine ZnO Nanoparticles Colloidal Medium. Nano Letters, 6, 866-870.

Chapter 7

Development of ricehusk ash reinforced bismaleimide toughened epoxy nanocomposites

K. Kanimozhi[1,2], K. Sethuraman[1], V. Selvaraj[2] and M. Alagar[1*]

[1]Department of Chemical Engineering, Alagappa College of Technology, Anna University, Chennai, India

[2]Department of Chemistry, University College of Engineering Villupuram, (A Constituent College of Anna University, Chennai), Villupuram, India

Recent past decades have witnessed remarkable advances in composites with potential applications in biomedical devices, aerospace, textiles, civil engineering, energy, electronic engineering, and household products. Thermoset polymer composites have further enhanced and broadened the area of applications of composites. In the present work epoxy-BMI toughened-silica hybrid (RHA/DGEBA-BMI) was prepared using bismaleimide as toughener, bisphenol-A as matrix and a silica precursor derived from rice husk ash as reinforcement with glycidoxypropyltrimethoxysilane as coupling agent. Differential scanning calorimetry, electron microscopy, thermogravimetric analysis, and goniometry were used to characterize RHA/DGEBA-BMI composites developed in the present work. Tensile, impact and flexural strength, tensile and flexural modulus, hardness, dielectric properties were also studied and discussed. The hybrid nanocomposites possess the higher values of the glass transition temperature (Tg) and mechanical properties than those of neat epoxy matrix.

KEYWORDS

epoxy, bismaleimide, rice husk ash, 3-glycidoxypropoyltrimethoxysilane (GPTMS), AFM, TEM, mechanical and thermal properties

1. INTRODUCTION

The present work involves the development of an organic and inorganic hybrid in combination with a recycling waste material (rice husk ash) to enhance the properties of organic polymer. The wide spread uses and properties of epoxy resin have been extensively studied by our research group for the past two decades (Dinakaran et al., 2003; Rajasekaran and Alagar, 2007; Premkumar et al., 2008; Ramesh et al., 2010; Selvaganapathi et al., 2010; Chandramohan and Alagar, 2011; Vengatesan et al., 2011; Kanimozhi et al., 2013; Prabunathan et al., 2013). In the present study an attempt has been made to form a covalent bond between organic polymers and inorganic components through coupling agent to enhance the compatibility of components involved (Farhadyar et al., 2005; Bagherzadeh and Mahdavi, 2007; Xinghong et al., 2007; Qingming et al., 2008; Balamurugan and Kannan, 2010; Sea-Fue et al., 2010).

Bismaleimide resins (BMI) are thermosetting polyimides that can be polymerized via multiple carbon–carbon bond formations without generating volatiles. The thermal curing properties of BMI and their incorporation with epoxy matrices lead to possess superior thermal and flame-retardant properties in comparison with those of conventional epoxy resins. BMI-modified epoxy resin matrices have been shown to have high crosslinking ability and glass transition temperature, high thermal stability and char yield, excellent fire resistance, specific strength and specific modulus and lower water absorption (Gu et al., 1996; Chanda and Rahabi, 1997; Dinakaran et al., 2003; Dinakaran and Alagar, 2004; Qilang et al., 2010). Rice husk ash possesses hard surface, high silica content (80–90%), insoluble in water, possess high chemical stability and mechanical strength, abrasive in nature, inherent resistance behavior, small bulk density, non-toxicity, low cost, and stable to bacteria (Krishnarao et al., 1998; Hanafi et al., 2001; Qiang et al., 2008; Li et al., 2011; Yue et al., 2011). Due to the high content of silica moiety in the rice husk ash it is expected to function similar to OMMT-clay with epoxy polymer and hence it has been chosen as the inorganic reinforcement in the present work. (Khalf and Ward, 2009; Kumagai and Sasaki, 2009;Bhagiyalakshmi et al., 2010; Adam et al., 2012; Chand et al., 2012). The present investigation was focused on the development of new epoxy based hybrid matrices reinforced chemically with functionalized rice husk ash and toughened by N,N′ bismaleimido 4,4′ diaminodiphenylmethane cured using DDM. The resulting hybrid organic-inorganic composites were characterized by different analytical methods and the data resulted are reported and discussed.

1.1 Experimental and Materials

GPTMS, ethanol, diaminodiphenylmethane (DDM), were obtained from SRL (India) and were used as received. The matrix resin used in the present study was diglycidyl ethers of bisphenol—A (DGEBA epoxy resin) (LY556) was received from Ciba-Geigy Ltd. Rice husk ash was synthesized as per the reported procedure (Bhagiyalakshmi et al., 2010).

1.2 Preparation of Rice Husk Ash (RHA)

Acid treatment is one of the most useful routes used to remove the lignin, wax and oils covering the external surface of the fiber cell wall of natural fibers. Rice husk ash is a solid obtained after burning of rice husk and was washed with distilled water, dried in an oven at about 60°C for 2 h. Then bleached with conc. HCl to remove the dirt and other contaminants present in it and subsequently washed with water till the pH become neutral, then dried in oven at 60°C for 4 h. Then it was heated at 500°C for 5 h in a muffle furnace, to obtain rice husk ash.

1.3 Surface Functionalization of Rice Husk Ash (GRHA)

3-glycidoxypropyltrimethoxysilane (GPTMS) was used as coupling agent to functionalize the rice husk ash. Four milliliter ml of GPTMS was mixed with 95% absolute ethanol and 5% deionized water and the resulting solution was sonicated for 15 min. The pH of the solvent was initially adjusted to 4.5 using acetic acid and subsequently sonicated for 1 h in order to get complete hydrolysis of GPTMS. Then 10 g of rice husk ash was added and the resulting mixture was sonicated for 2 h. Then the mixture was refluxed for 24 h at 80°C and centrifuged with addition of water followed by ethanol and hexane. The rice husk ash thus functionalized was further dried in hot air oven at 100°C in order to remove the moisture (Figure 1).

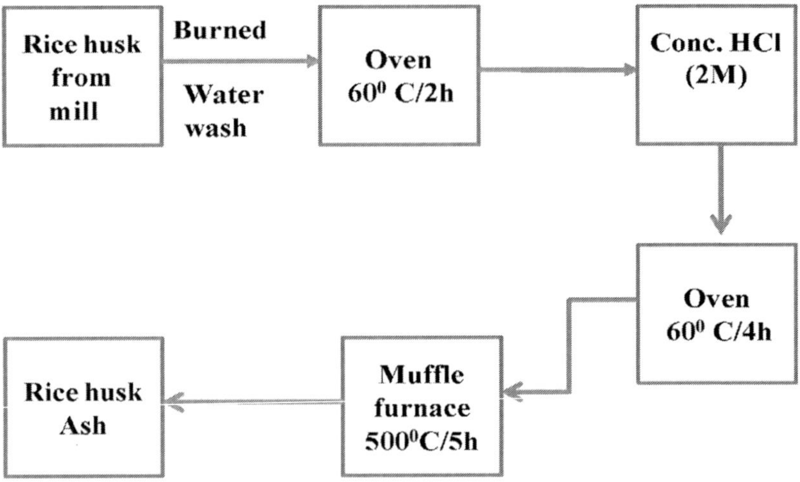

Figure 1. Synthesis of rice husk ash (RHA) from rice hull.

1.4 Synthesis of N,N′ Bismaleimido 4,4′ Diaminodiphenyl Methane (BMI)

To a 1-l three-necked round-bottom flask fitted with paddle stirrer, reflux condenser and nitrogen inlet, were added, 600 ml acetone, 1.0 mole (98.1 g) maleic anhydride and 0.5 mole of the diaminodiphenylmethane. Rapid formation of precipitate of the bismaleiamic acid occurred on mixing the reactants together, and the mixture was allowed to stand for 30 min to complete the reaction (Ashok kumar et al., 2002). To the above, 1.0 g of nickel acetate and 25 ml of triethylamine were added and the entire mixture was heated slowly to reflux. Then by means of pressure equalizing funnel 117.9 ml acetic anhydride was added to the refluxing reaction mixture and heating was continued for an additional 3 h. The reaction mixture was diluted with 500 ml water and chilled to crystallize the bismaleimide (Yield 92%).

1.5 Preparation of RHA/DGEBA-BMI Nanocomposites

The 5 wt.% of BMI and 95 wt.% DGEBA epoxy resin were mixed with the desired amount of glycidyl functionalized rice husk ash (0.5, 1.0, 1.5 wt.%) and mechanically stirred at 50°C for 24 h. A stoichiometric amount of DDM, corresponding to epoxy equivalents was also added. The resulting product was poured into a pre-heated mold. The mold was pre heated at 120°C for an hour, to remove the moisture and trapped air. The samples were cured successively at 120°C for 2 h, post-cured at 180°C for 3 h, and finally removed from the mold and characterized. The preparation of the RHA/DGEBA-BMI nanocomposites is illustrated in Scheme 1 and in Figure 2.

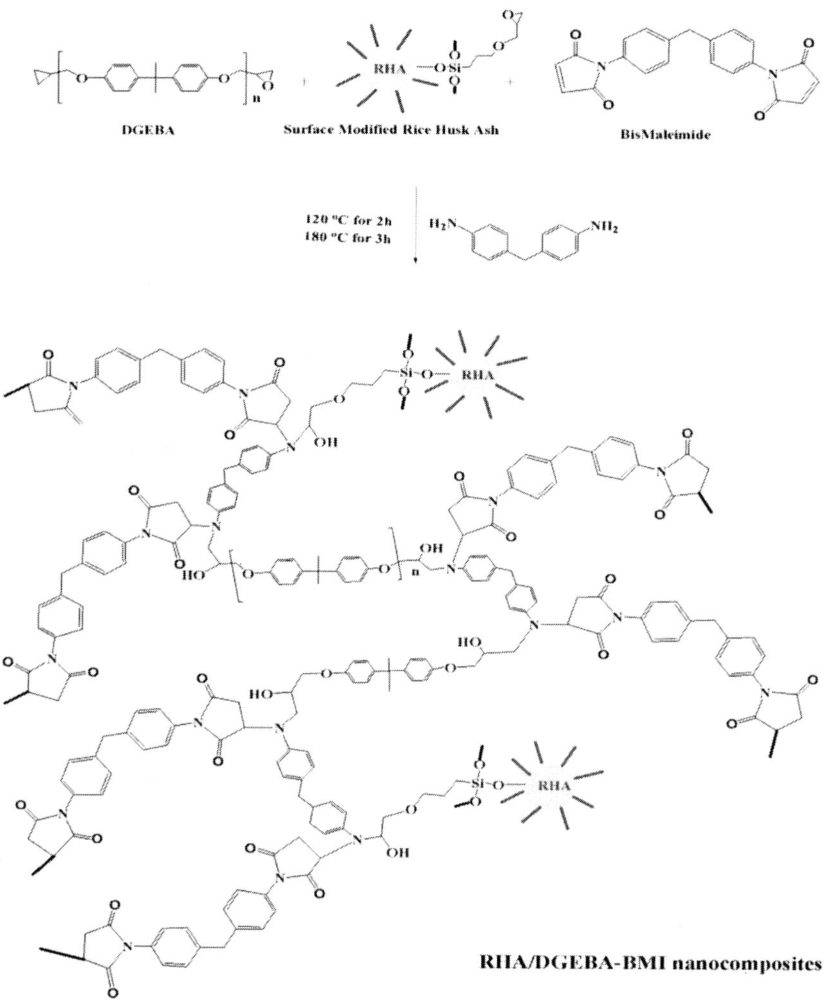

Scheme 1. Schematic representation of RHA/DGEBA-BMI nanocomposites.

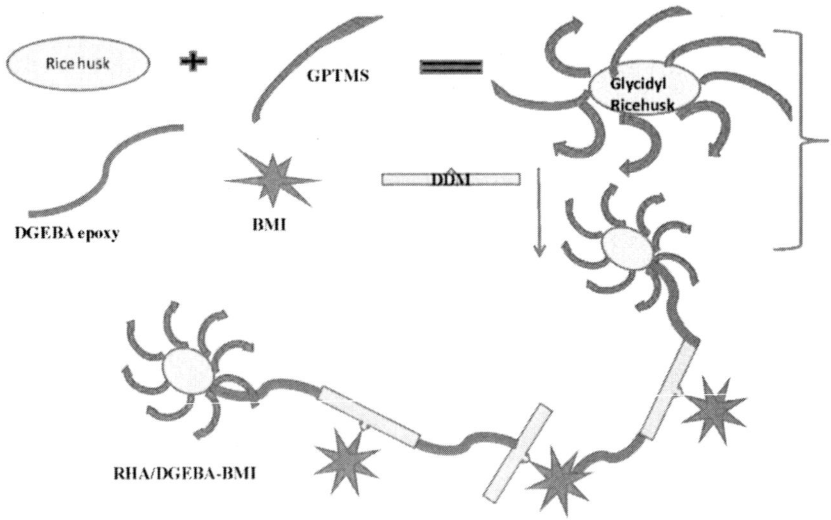

Figure 2. Preparation of surface functionalized rice husk ash reinforced (GRHA-EP) nanocomposites (Graphical abstract).

1.6 Instrumentation

Fourier transform infrared (FT-IR) spectra for the samples were recorded on a Perkin Elmer 6X FT-IR spectrometer. The glass transition temperature (Tg) of the samples was determined, using DSC 200 PC differential scanning calorimeter (DSC) (Netzsch Gerateban GmbH). Thermogravimetric analysis (TGA) was carried out, using the DSTA 409 PC analyzer (Netzsch Gerateban GmbH). The tensile (stress–strain) properties were determined, using INSTRON (Model 6025 UK) as per ASTM D 3039. The flexural properties were measured by the INSTRON (Model 6025 UK) as per ASTM D 790. The un-notched Izod impact strength of each sample was studied as per ASTM D 256. The water absorption behavior of the samples was tested as per ASTM D 570. The percentage of water absorbed by the specimen was calculated, using the following equation:

$$\%\text{Water absorption} = (w_2 - w_1) \times 100/w_1 \tag{1}$$

where w_1 is the initial weight of the sample and w_2 is the weight of the sample after immersion in distilled water for 48 h at 30°C. The dielectric studies of the neat epoxy and RHA/DGEBA-BMI nanocomposites were determined with the help of an impedance analyser. Contact angle measurements were carried out using 210 a Rame-hart Inc. goniometer (Succasunna, NJ, USA) with 5 μl of deionized water and diiodomethane (DIM).

X-ray diffraction patterns were recorded at room temperature, by monitoring the diffraction angle 2θ from 10 to 70° as the standard, on a Rich Seifert (Model 3000) X-ray powder diffractometer. The surface morphology of the fractured surface of the samples was examined, using a scanning electron microscope (SEM; JEOL JSM Model 6360). A JEOL JEM-3010 analytical transmission electron microscope, operating at 80 kV with a measured point-to-point resolution of 0.23 nm, was used to characterize the phase morphology of the developed nanocomposites. TEM samples were prepared by dissolving the powdered composite samples in ethanol mounted on carbon-coated Cu TEM grids and dried for 1 h at 70°C to form a film of <100 nm.

2. RESULTS AND DISCUSSION

2.1 FT-IR Spectra of RHA/DGEBA-BMI Nanocomposites

Figure 3 showed the FT-IR spectra of neat epoxy and bismaleimide reinforced epoxy system, where the two set of epoxies showed their similar behavior with identical peaks. Figure 4 showed the FT-IR spectra of neat rice husk ash (RHA) and glycidyl functionalized rice husk (GRHA). The FT-IR observed for neat rice husk ash (RHA) shows the peak for OH at 3434 cm^{-1}. The absence of peak at 2338 cm^{-1} is due to the chain lengthening with GPTMS and the presence of peak at 965 cm^{-1} in GRHA confirms the glycidyl functionality has been introduced into RHA. The presence of broad peak at 3468 cm^{-1} shows the stretching vibration of O-H bond. The TGA thermogram from Figure 5 also confirms the formation of glycidyl functionality in GRHA, whereas the rice husk ash possess 78 wt.% of char yield at 800°C and that of GRHA with only 65 wt.% of char yield.

The reaction between the glycidyl functionalized RHA, BMI and epoxy was confirmed by the FT-IR spectra and showed in Figure 6. The peak at 965 cm^{-1} in GRHA indicates the glycidyl functionality in the rice husk ash, where as its absence in the composites confirms the complete curing reaction takes place between the matrix and the reinforcements. The cleavage of oxirane ring and the formation of –OH linkage were also confirmed the occurrence of polymeric network structure. The peak at 1505 cm^{-1} indicates the -Si-O-Si- linkages by the silica group in the rice husk ash. The peaks at 1033 and 2948 cm^{-1} indicates the symmetric and asymmetric CH_2 stretching respectively. The OH group confirms the band appearance at 3393 cm^{-1}. This clearly confirms that the ring opening polymerization takes place between the epoxy, BMI, DDM and the functionalized rice husk ash during curing process.

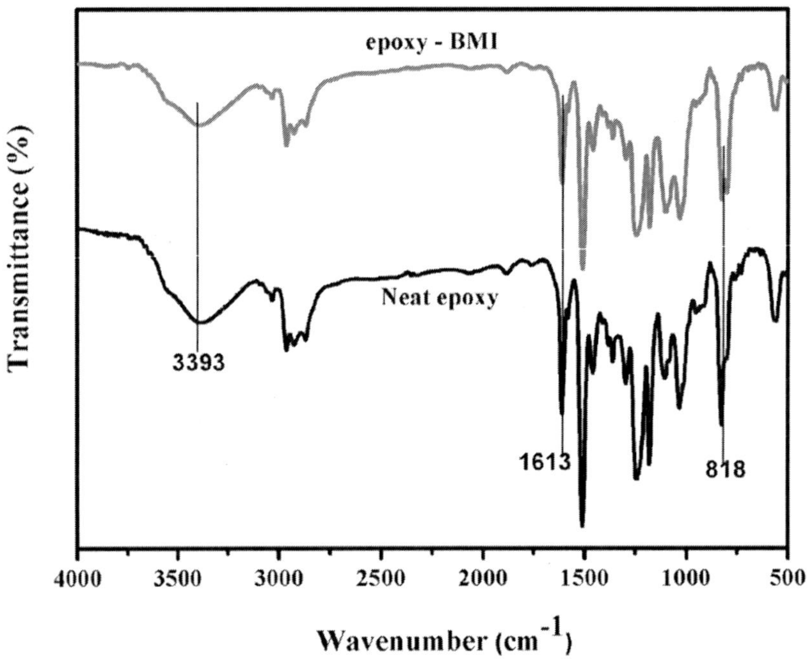

Figure 3. FT-IR spectra of neat epoxy and epoxy-BMI.

2.2 Effect of Thermal Properties of RHA/DGEBA-BMI Nanocomposites

The value of glass transition temperature (Tg) of the neat epoxy matrix and the RHA reinforced bismaleimide toughened epoxy composites were obtained from DSC and the data are presented in Table 1and showed in Figure 7. In the present work RHA/DGEBA-BMI exhibited significantly higher values of Tg than that of the neat epoxy matrix. The Tg value of neat epoxy is 162°C, whereas for varying weight percentages reinforced (0.5, 1.0, and 1.5 wt.%) RHA/DGEBA-BMI nanocomposites possess 163.1, 169, and 171.9°C respectively. The higher values of Tg may be explained by the introduction of silica network by the rice husk ash and the homopolymerization of BMI through Michael addition reaction via hydrogen bonding interaction, which restricts the mobility of epoxy network. An entangled three-dimensional rigid structure was also formed due to the chemical interaction of GPTMS with the rice husk ash, BMI and epoxy (Nagendiran et al., 2008).

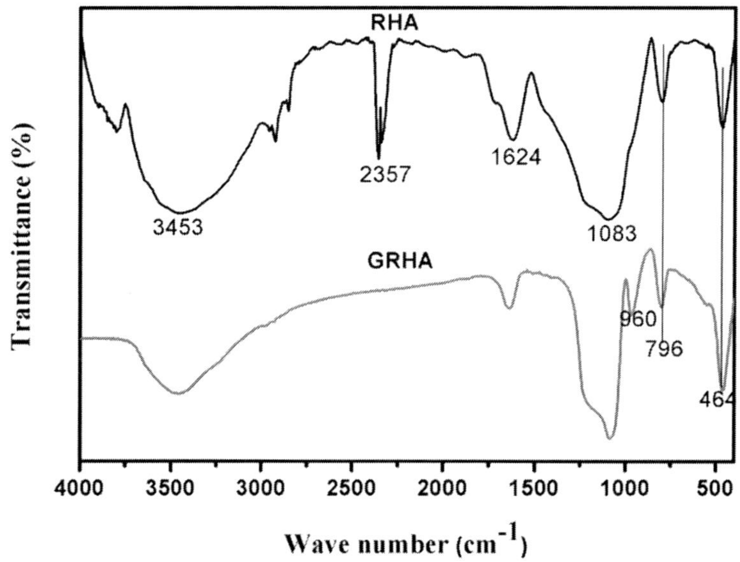

Figure 4. FT-IR spectra of neat rice hush ash (RHA) and glycidyl functionalized rice husk ash.

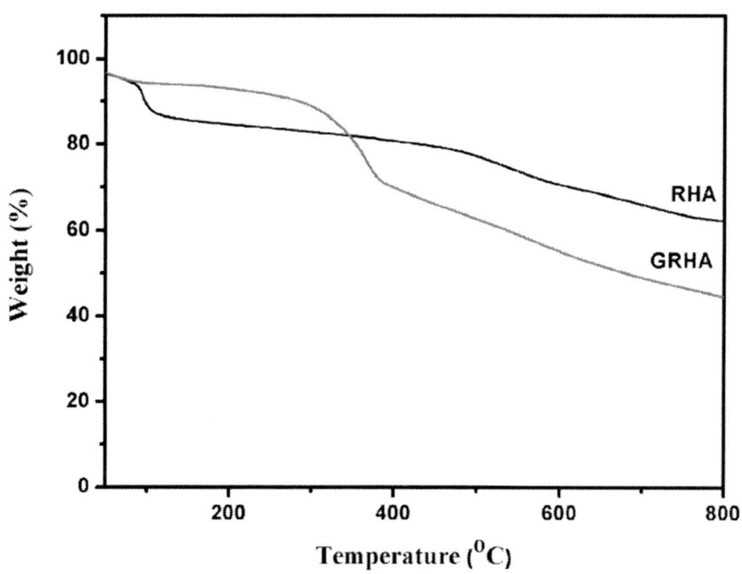

Figure 5. TGA of RHA and glycidyl functionalized rice husk ash.

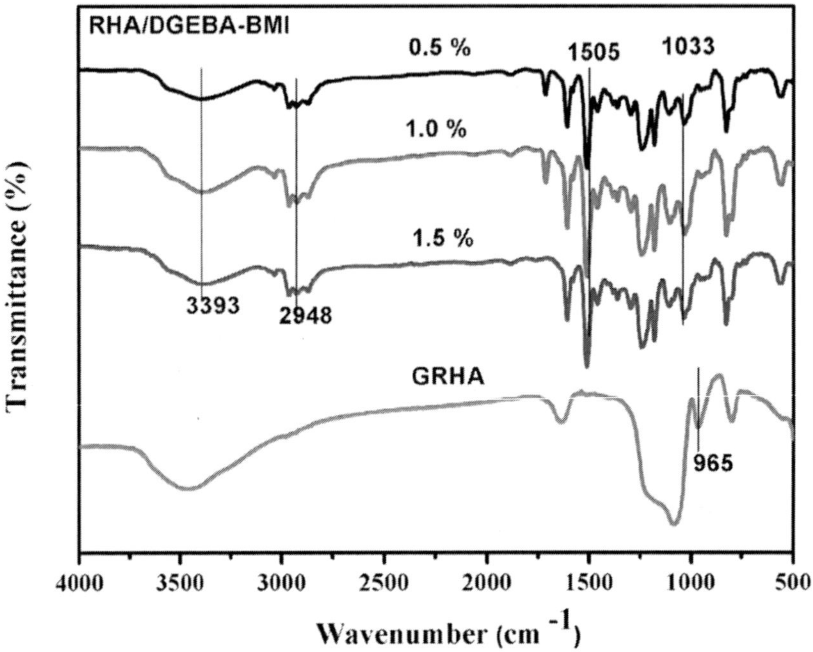

Figure 6. FT-IR spectra of GRHA and RHA/DGEBA-BMI composites.

The thermal degradation behavior of RHA/DGEBA-BMI composites are analyzed using TGA and the data are presented in Figure 8 and Table 1. The 40% weight loss for neat epoxy, BMI-EP, 0.5, 1.0, and 1.5 wt% RHA/DGEBA-BMI nanocomposites were occurred at the temperature of 393, 376.5, 387.8, 401.6, and 410.1°C respectively. Similarly it was noticed that the 60% wt. loss of the composites occurred at the temperature of 420.0, 404.2, 424.0, 476.5, and 507.5°C respectievely. The introduction of BMI into the epoxy first degrades at lower temperature whereas the surface functionalized rice husk ash reinforcement into the matrix gradually increased the thermal stability of the composites (Table 1). This may be due to the efficient interaction of the rice husk ash with BMI and the organic polymeric backbone makes the composites in a state of three dimensional cross linked network structures. As the varying weight % of glycidyl functionalized RHA and 5.0% wt. of BMI content introduced into the epoxy matrix, the system prolongs the degradation temperature of the composites due to the restriction of chain mobility and hinders the movement of molecules (Lin et al., 2007). For example, under nitrogen atmosphere, the char yield of 0.5, 1.0, and 1.5 wt.% rice husk ash reinforced bismaleimide toughened epoxy systems are 14.9, 16.7, and 17.8% respectively (Nagendiran et al., 2010) (Figure 8). The degradation of imide containing functional groups in a relatively high temperature region played an important role in enhancing flame retardation.

Table 1. Thermal properties of neat epoxy and RHA/DGEBA-BMI nanocomposites.

Sample	Percentage composition	Tg (°C)	Degradation Temperature (°C)		Char yield at 800°C (%)	Water absorption (%)	Dielectric constant (1 MHz)
			40%	60%			
Neat epoxy resin	100/0/0	162.0	393	420.0	0	0.123	4.50
BMI-EP	00/95/5	167.0	376.5	404.2	13.0	0.088	3.96
RHA/DGEBA-BMI	0.5/95/5	163.1	387.8	424.0	14.9	0.012	2.65
RHA/DGEBA-BMI	1.0/95/5	169.0	401.6	476.5	16.7	0.008	2.43
RHA/DGEBA-BMI	1.5/95/5	171.9	410.1	507.4	17.8	0.003	1.88

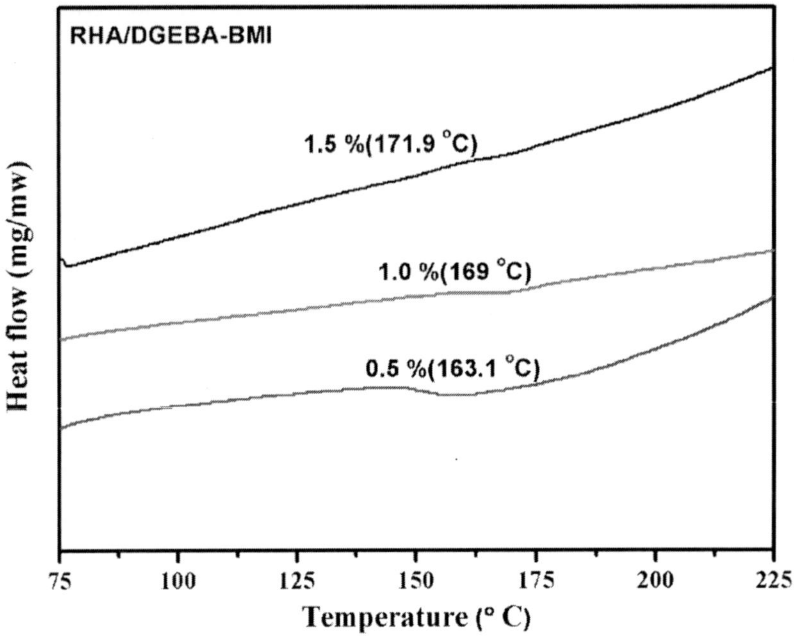

Figure 7. DSC thermogram of RHA-DGEBA/BMI nanocomposites.

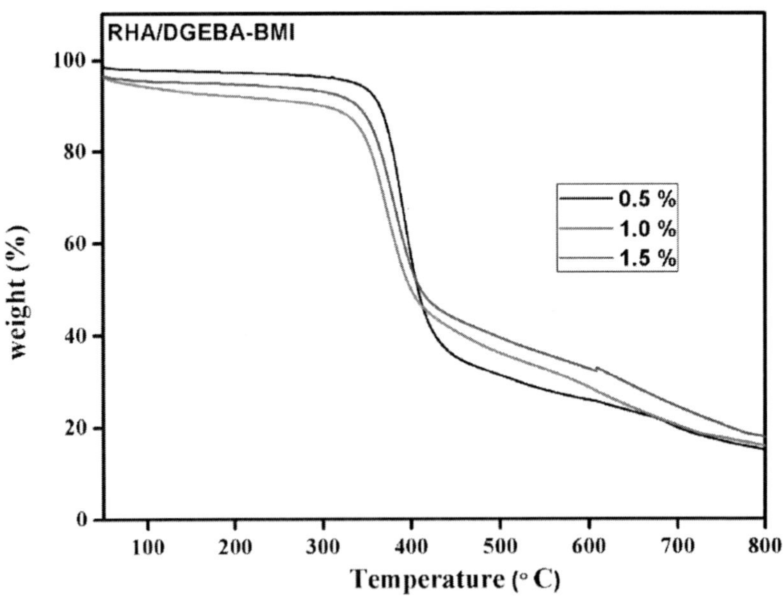

Figure 8. Thermogram of RHA/DGEBA-BMI nanocomposites.

2.3 Mechanical Behavior of RHA/DGEBA-BMI Nanocomposites

Table 2 presents the mechanical properties of neat epoxy, BMI-EP and the RHA/DGEBA-BMI nanocomposites. An incorporation of varying weight percentages (0.5, 1.0, and 1.5 wt.%) of RHA into 5.0 wt.% BMI and epoxy matrix increases the values of tensile and flexural strength. The maximum load bearing capacity of a material per unit area when it is stretched is calculated from tensile values. For neat epoxy the value of tensile strength is 61.3 MPa and for BMI-EP, 0.5, 1.0, and 1.5 wt.% reinforced nanocomposites there is an increment in the values of tensile strength to 76.8, 84.5, 99.8, and 115.9 MPa respectively. The chemically active silica present in the rice husk ash forms a strong covalent linkage with BMI and epoxy results an inter-crosslinked network structure in the reinforced composites primarily improved the value of tensile strength. The curative DDM also open its NH-OH linkage and forms an entanglement molecular structure. Flexural strength values of neat epoxy, BMI-EP and RHA/DGEBA-BMI composites are 106.0, 119.5, 125.0 147.2, and 163.1 MPa respectively. The values of flexural strength for RHA/DGEBA-BMI composites are due to the homogenous dispersion of RHA reinforcement and the BMI toughener into the epoxy matrix which supported the stress transferred from polymer matrix. It was also clearly observed from the SEM analysis that no void growth formation or crack formation occurred in the resulting reinforced composites. Moreover, the addition of rice husk ash and the BMI content into the polymer matrix enhanced the values of tensile and flexural modulus significantly. The increase in modulus value is showing the stiffness of RHA/DGEBA-BMI reinforced composites. This improvement is attributed to the relatively lower strain rates of the composites. An observation similar to this was also noticed in the case of hardness behavior. Due to the increment of the loading, the values of hardness enhanced from 84 to 214.5. Further it was also observed that the reinforced composites have resistant to penetration, scratching and indentation of other foreign agents (Zulfiqar et al., 2007).

The amount of energy absorbed per unit area is calculated by the impact strength. The values of impact strength for neat epoxy, BMI-EP, RHA/DGEBA-BMI nanocomposites are presented in Table 2. The reinforcement RHA, toughener BMI into the epoxy matrix improved the impact behavior according to their percentage content. The oxirane ring in the DGEBA molecule and –OH surface of the reinforcement induces to form hydrogen bond with the maleimide linkage which tightly holds one another and reason out for the increased values of impact strength. The primary long-chain rigid structure and possible secondary hydrogen bond formation is expected to form a chain-entangled network structure (Chanda and Rahabi, 1997). The values of impact strength of toughened and reinforced composites enhanced to a significant extent when compared to that of neat epoxy due to the flexibility and plasticizing effect imparted by the silica moiety in the rice husk ash (Hameed et al., 2007).

Table 2. Mechanical properties of neat epoxy and RHA/DGEBA-BMI nanocomposites.

Sample	Percentage composition	Tensile strength (MPa)	Tensile modulus (GPa)	Flexural strength (MPa)	Flexural modulus (GPa)	Impact strength (Jm^{-1})	Hardness HV1.2
Neat epoxy	100/0/0	61.3	2.71	106.0	2.37	101.7	84
BMI-EP	00/95/5	76.8	3.57	119.5	2.69	110.0	121.6
RHA/DGEBA-BMI	0.5/95/5	84.5	3.84	125.0	2.78	113.1	130.0
RHA/DGEBA-BMI	1.0/95/5	99.8	4.23	147.2	3.26	119.6	166.7
RHA/DGEBA-BMI	1.5/95/5	115.9	4.78	163.1	3.54	125.8	214.5

2.4 Influence of Water Absorption Dielectric Behavior and Contact Angle of RHA/DGEBA-BMI Nanocomposites

The water absorption behavior and dielectric constant of RHA/DGEBA-BMI nanocomposites was studied and presented in Table 2. The experiments showed volume change without any clustering or microvoiding which clearly confirms the water repellent character of the nanocomposites. All the surface-modified rice husk ash reinforced BMI toughened epoxy resin systems showed good resistance to moisture absorption was purely due to the rigid aromatic hydrophobic structure. The decrease in percentage water uptake of the RHA-BMI incorporated systems was due to the inherent hydrophobic nature of the RHA networks. Rice husk ash possesses its original nature that it does not absorbed water and it is also biologically non-degradable in nature. Neat epoxy matrix possesses the values of dielectric constant of 4.50 at 1 MHz whereas that of BMI-EP and 0.5, 1.0, and 1.5 wt.%. RHA/DGEBA-BMI composites are 3.96, 2.65, 2.43, and 1.88 respectively. Decreasing trend of dielectric constant indicates that the increasing incorporation of RHA and BMI which in turn reduce the dipole–dipole interaction. This behavior reduces the polarizability and increases the free volume of the resulting hybrid nanocomposites which obviously reduced the values of dielectric constant. Low water absorption and dielectric constant are very important property for a material used for insulation applications, especially those requiring stable high performance (Devaraju et al., 2013).

The variation of contact angle against blend composition is given in Figure 9 and in Table 3. It can be observed that the reinforced composites possess higher values of contact angle when compared to that of neat epoxy matrix. The values of contact angle of RHA/DGEBA-BMI composites are increased from 82.7 to 89.3 and from 52.7 to 60.4 for water and DIM respectively. The lower affinity of the developed composites indicates the improvement in the hydrophobic behavior contributed by the reinforcement and toughening agent to organic matrix. This may be due to the non-polar nature of the toughened rice husk ash and the BMI which may ultimately increases the roughness of the polymer surface. The acid treatment and the glycidyl functionalization of the rice husk ash are important modification that disrupts the molecular motion of the polymer and induced the crosslinked network structure, thereby increasing the surface roughness. The roughness morphology of polymer surface was also supported by the SEM micrograph (Figure 8). The additional free energy at the surface of the composites is known as surface free energy. High-energy absorption received by the activated curing agents of –NH bonds strike the DGEBA epoxy which interacted with the toughened BMI component and to the surface of rice husk reinforcement thereby forming a network structured composites. Consequently, the developed composites reduced the surface energy and enhanced the substrate surface roughness and make the material surface become super hydrophobic (Yue et al., 2010).

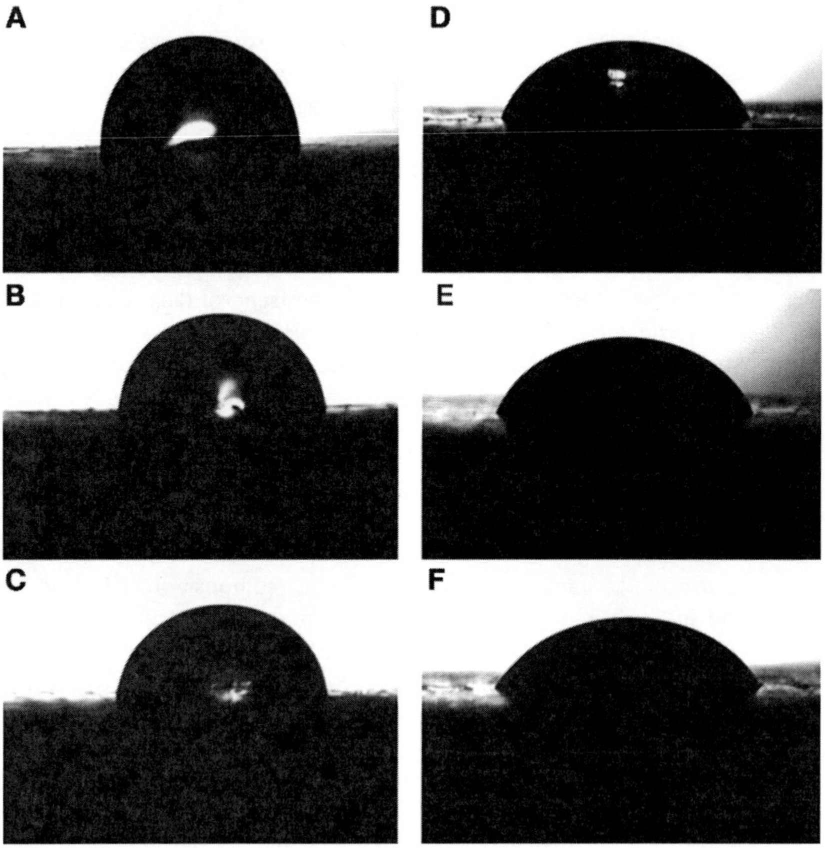

Figure 9. Contact angle of (A-0.5%, B-1.0%, C-1.5% RHA-DGEBA/BMI) for diiodomethane and water (D-0.5%, E-1.0%, F-1.5% RHA-GEBA/BMI).

Table 3. Contact angle and surface energy of RHA/DGEBA-BMI nanocomposites.

Properties	Contact angle		Surface energy		
	Water	Diiodomethane	γ^d	γ^p	Γ
0.5% RHA/DGEBA-BMI	82.7	52.7	13.3	10.9	46.8
1.0% RHA/DGEBA-BMI	88.7	57.8	28.3	5.1	33.5
1.5% RHA/DGEBA-BMI	89.3	60.4	27.8	19.1	24.2

2.5 Morphological Features of RHA/DGEBA-BMI Nanocomposites

The morphology of neat epoxy matrix and RHA/DGEBA-BMI composites were analyzed using SEM micrograph and presented in Figure 10. The neat epoxy matrix (Figure 10A) shows a smooth glassy fractured surface in different places. The brittle behavior was overcome by the addition of glycidyl functionalized rice husk ash and the BMI as toughening agent. There is no observance of crack formation or pulled out flaws in the SEM image (Figures 10B–D), which clearly provides an idea about the exfoliation of reinforcement combined with toughening agent into the matrix. At higher weight percentages also no chance for agglomeration of particles observed in the epoxy matrix which confirms the homogenous and molecular level dispersion of rice husk ash into epoxy matrix. This contributes to the improvement of thermal and mechanical properties (Devaraju et al., 2011).

XRD analysis is used to investigate the 0.5 and 1.5% RHA/DGEBA-BMI composites and the diffractograms of the samples are presented in Figure 11. From Figure 11, it is evident that there is no diffraction peak observed for epoxy composites. The absence of the diffraction peak in the case of reinforced composites is due to the complete exfoliation of the RHA-BMI into the epoxy network structure. The observance of peak at $2\theta = 20°$ showed the amorphous nature and the uniform level dispersion of RHA-BMI concurrently confirms the efficient and effective compatibility between the RHA-BMI and the epoxy matrix. The RHA-BMI dispersion is homogenously observed in the form of individual layers within the polymer matrix and leads to form exfoliated composites which attributes to the improvement of properties of the resulting nanocomposites.

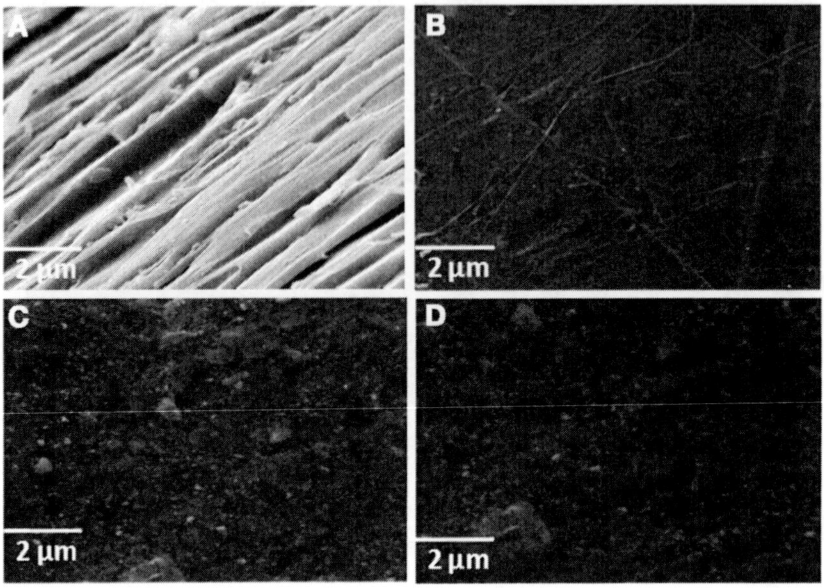

Figure 10. SEM micrograph of (A) neat epoxy, (B) 0.5, (C) 1.0, (D) 1.5 wt.% RHA-DGEBA/BMI nanocomposites.

The TEM micrograph for 1.0 wt.% RHA/DGEBA-BMI nanocomposites is shown in Figure 12. The nanozized distribution of RHA and BMI content can be viewed from the TEM micrograph where the smaller particles correspond to RHA and the bigger particles related to BMI. It was also ascertained that the dark portion of the pictures are due to the homogenous dispersion of rice husk ash and BMI into the epoxy matrix without any crack. No accumulation mode was observed in the particle number size distribution, the particles on the left top corner were clearly visible which indicates the presence of rice husk ash reinforcement and bottom corner represents BMI respectively.

The surface topology of 1.0% wt. RHA/DGEBA-BMI in 2-D and 3-D was characterized using the AFM technique and are shown in Figure 13. The surface of the hybrid thin film is quite smooth and has few visible defects, indicating that the developed composites have a molecular level dispersion of organic and inorganic networks, silica particles and BMI in RHA/DGEBA-BMI are distributed uniformly in the organic phase. The phase images also indicate that the BMI and reinforcement domains were uniformly dispersed throughout the matrix. This is further supported by SEM analysis that the hybrid exhibits excellent structural uniformity, with no cracks or flaws. The surface morphology of the hybrid materials is of great importance for industrial applications. The compatability between the epoxy, BMI and inorganic rice husk ash had a great significant in thermal and mechanical properties.

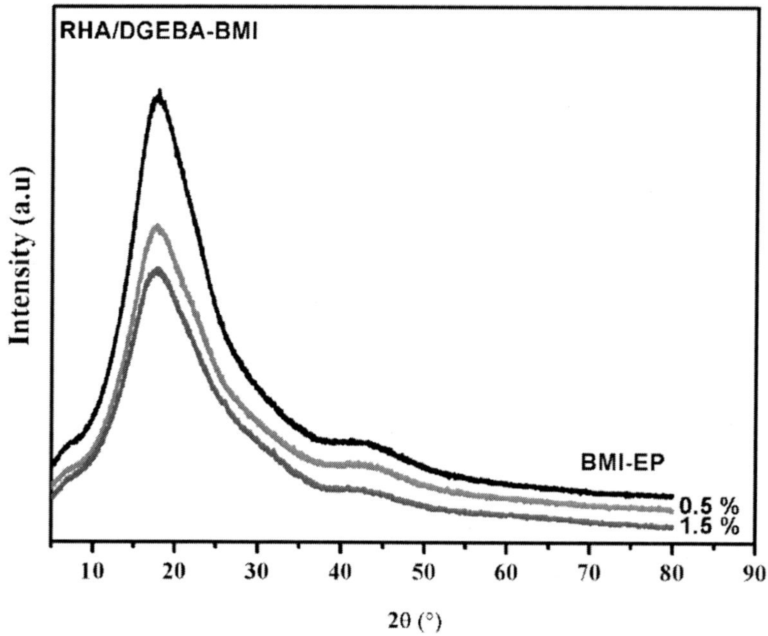

Figure 11. XRD of BMI-EP, 0.5 and 1.5 wt.% RHA/DGEBA-BMI nanocomposites.

3. CONCLUSION

Varying weight percentages of surface functionalized rice husk ash reinforced BMI toughened epoxy nanocomposites (0.5, 1.0, and 1.5 wt.% RHA/DGEBA-BMI) were developed and their mechanical, thermal, dielectric and surface morphology were studied. The developed composites possess improved thermo-mechanical properties due to excellent covalent interaction between them. RHA/DGEBA-BMI exhibited with higher values of glass transition, delay in degradation of the composites, higher char yield and excellent flame retardant character showed that the composite materials can withstand at higher temperature than that of neat epoxy matrix. The homogenous level of dispersion and exfoliation of the composites were confirmed by SEM, TEM, AFM, and XRD analysis. The hydrophobic nature of the composite samples is well understood from the increasing value of contact angle and the surface roughness of the composites experienced by the RHA and the BMI. The enhanced properties of the developed composites make it to use in the form of coatings, adhesives and matrices which can be utilized for the fabrication of advanced composite components in the place of conventional epoxy for high performance applications.

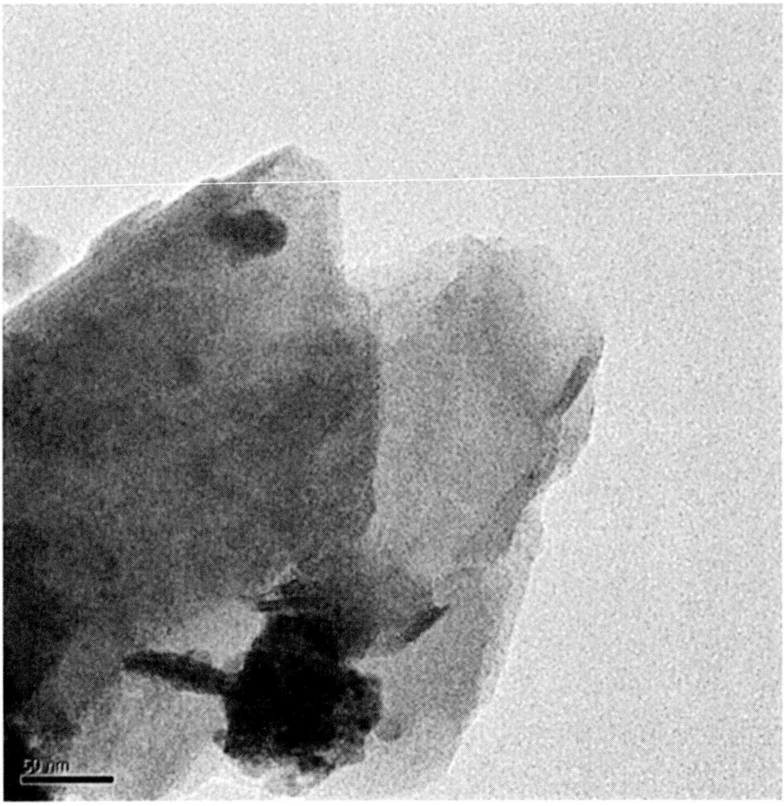

Figure 12. TEM micrograph 1.0% RHA-DGEBA/BMI nanocomposites.

CONFLICT OF INTEREST STATEMENT

The authors declare that the research was conducted in the absence of any commercial or financial relationships that could be construed as a potential conflict of interest.

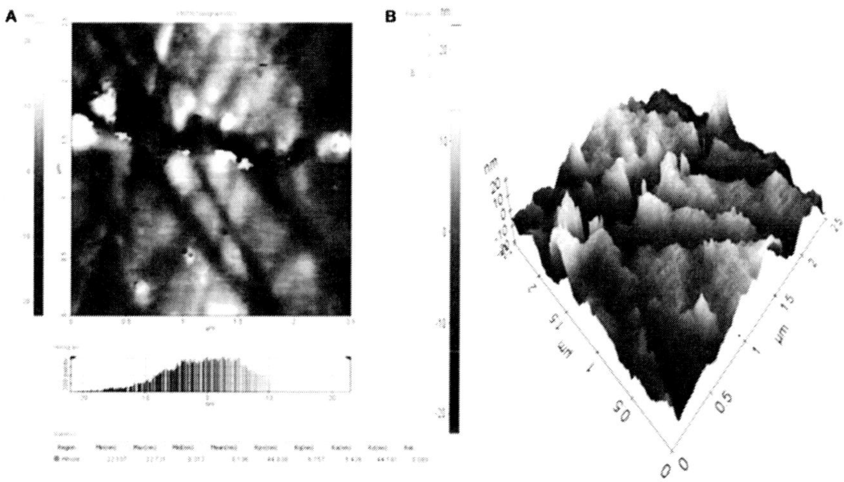

Figure 13. AFM microphotographs of 1.0% RHA-DGEGBA/BMI nanocomposites (A) 2D image (B) 3D image..

ACKNOWLEDGMENTS

The authors thank the Department of Mechanical Engineering and the Department of Ceramic Technology, Anna University, Chennai for providing characterization facility. The authors also thank Dr. M. R. Vengatesan and Dr. S. Devaraju for providing characterization facility and their moral support.

REFERENCES

1. Adam, F., Farook, A., Nelson, J., Iqbal, M., and Anwar, M. (2012). The utilization of rice husk silica as a catalyst: review and recent progress. *Catal. Today* 190, 2–14.

2. Ashok kumar, A., Alagar, M., and Rao, R. M. V. G. K. (2002). Synthesis and characterization of siliconized epoxy-1,3-bis(maleimido) benzene intercross linked matrix materials. *Polymer* 43, 693–702.

3. Bagherzadeh, M. R., and Mahdavi, F. (2007). Preparation of epoxy–clay nanocomposite and investigation on its anti-corrosive behavior in epoxy coating. *Prog. Org. Coatings* 60, 117–120.

4. Balamurugan, R., and Kannan, P. (2010). 1,3,4-Oxadiazole epoxy resin-based liquid crystalline thermosets and their cure kinetics. *J. Mater. Sci.* 45, 1321–1327.

5. Bhagiyalakshmi, M., Yun, L. J., Anuradha, R., and Jang, H. T. (2010). Utilization of rice husk ash as silica source for the synthesis of mesoporous silicas and their application to CO_2 adsorption through TREN/TEPA grafting. *J. Hazard. Mater.* 175, 928–938.

6. Chand, N., Fahimb, M., Sharmaa, P., and Bapat, M. N. (2012). Influence of foaming agent on wear and mechanical properties of surface modified rice husk filled polyvinylchloride. *Wear* 278, 83–86.

7. Chanda, R., and Rahabi, L. (1997). Recent advanced in bismaleimides and epoxy-imide/bismaleimide formulations and composites. *JMS Rev. Marcomol. Chem. Phys. C* 37, 61–96.

8. Chandramohan, A., and Alagar, M. (2011). Synthesis and characterization of 1, 1-bis (3-methyl-4-epoxyphenyl) cyclohexane-toughened DGEBA and TGDDM organo clay hybrid nanocomposites. *High Perform. Polym.* 23, 197–211.

9. Devaraju, S., Vengatesan, M. R., and Alagar, M. (2011). Studies on thermal and dielectric properties of ether linked cyclohexyl diamine (ELCD)-based polyimide POSS nanocomposites (POSS-PI). *High Perform. Polym.* 23, 99–111.

10. Devaraju, S., Vengatesan, M. R., Selvi, M., Ashok Kumar, A., Hamerton, I., and Alagar, M. (2013). Low surface free energy cyanate ester-silica hybrid (CE-SiO_2) nanomaterials for low k dielectric applications. *RSC Adv.* 3, 12915–12921.

11. Dinakaran, K., and Alagar, M. (2004). Mechanical properties of bismaleimide (N,N'-bismaleimido-4,4′-diphenyl methane)—vinylester oligomer (VEO) modified unsaturated polyester intercrosslinked matrices for advanced composites. *Int. J. Polym. Mater.* 53, 11–19.

12. Dinakaran, K., Alagar, M., and Suresh Kumar, R. (2003). Preparation and characterization of bismaleimide/1,3-dicyanatobenzene modified epoxy intercrosslinked matrices. *Eur. Polym. J.* 39, 2225–2233.

13. Farhadyar, N., Rahimi, A., and Langroudi, A. (2005). EIranian preparation and characterization of aromatic amine cured epoxy-silica hybridinorganic-organic coating via *in situ* sol-gel. *Process. Polym. J.* 14, 155–162.

14. Gu, A. J., Liang, G. Z., and Lan, L. W. (1996). A high-performance bismaleimide resin with good processing characteristics. *J. Appl. Polym. Sci.* 62, 799–803.

15. Hameed, N., Thomas, S. P., Abraham, R., and Thomas, S. (2007). Morphology and contact angle studies of poly(styrene-co-acrylonitrile) modified epoxy resin blends and their glass fibre reinforced composites. *eXPRESS Polym. Lett.* 1, 345–355.

16. Hanafi, I., Nizam, J. M., and Abdul Khalil, H. P. S. (2001). The effect of a compatibilize on the mechanical properties and mass swell of white rice husk ash filled natural rubber/linear low density polyethylene blends. *Polym. Test.* 20, 125–133.

17. Kanimozhi, K., Devaraju, S., Vengatesan, M. R., Selvaraj, V., and Alagar, M. (2013). Studies on synthesis and characterization of surface modified mullite fibre reinforced epoxy nanocomposites. *High Perform. Polym.* 25, 658–667.

18. Khalf, A. I., and Ward, A. A. (2009). Use of rice husks as potential filler in styrene butadiene rubber/linear low density polyethylene blends in the presence of maleic anhydride. *Mater. Des.* 31, 2414–2421.

19. Krishnarao, R. V., Mahajan, Y. R., and Kumar, T. J. (1998). Conversion of raw rice husks to sic by pyrolysis in nitrogen atmosphere, journal of the european ceramic society. *J. Eur. Ceram. Soc.* 18, 147–152.

20. Kumagai, S., and Sasaki, J. (2009). Carbon/silica composite fabricated from rice husk by means of binderles shot-pressing. *Bioresour. Technol.* 100, 3308–3315.

21. Li, Y., Ding, X., Guo, Y., Rong, C., Wang, L., Qu, Y., et al. (2011). A new method of comprehensive utilization of rice husk. *J. Hazard. Mater.* 186, 2151–2156.

22. Lin, C. H., Feng, C. C., and Hwang, T. Y. (2007). Preparation, thermal properties, morphology, and microstructure of phosphoruscontaining epoxy/SiO_2 and polyimide/SiO_2 nanocomposites. *Eur. Polym. J.* 43, 725–742.

23. Nagendiran, S., Alagar, M., and Hamerton, I. (2010). Octasilsesquioxane reinforced DGEBA and TGDDM epoxy nanocomposites: characterization of thermal, dielectric and morphological properties. *Acta Mater.* 58, 3345–3356.

24. Nagendiran, S., KarikalChozhan, C., Alagar, M., and Hamerton, I. (2008). Inorganic/organic hybrid nanocomposites involving ommt clay and cyanate ester–siloxane-modified epoxy resin: thermal, dielectric and morphological properties. *High Perform. Polym*. 20, 323–347.

25. Prabunathan, P., Sethuraman, K., and Alagar, M. (2013). Development of bio-based F-SBA-15 reinforced epoxy nanocomposites for low-*k*dielectric applications. *High Perform. Polym*. 25, 283–289.

26. Premkumar, S., Karikal Chozhan, C., and Alagar, M. (2008). Studies on thermal, mechanical and morphological behaviour of caprolactam blocked methylenediphenyl diisocyanate toughened bismaleimide modified epoxy matrices. *Eur. Polym. J*. 44, 2599–2607.

27. Qiang, Z., Jian, T., Richard, C. M. Y., Albert, C. K. M., Robert, L. K. Y., and Cunjiang, S. (2008). Biodegradation behaviour of polycaprolactone / ricehusk composites in simulated soil medium. *Polym. Degrad. Stab*. 93, 1571–1576.

28. Qilang, L., Ruigang, Z., and Penghui, T. (2010). Material characterisation preparation and characterization of BMIresin/graphite oxide nanocomposites. *Polym. Test*. 29, 537–543.

29. Qingming, J., Shaoyun, S., Yaming, W., Lili, G., and Jinbo, L. (2008). Tribological performance and thermal behaviour of epoxy resin nanocomposites containing polyurethane and organoclay. *Polym. Adv. Technol*. 19, 859–864.

30. Rajasekaran, R., and Alagar, M. (2007). Mechanical properties of bismaleimides modified polysulfone epoxy matrices. *Int. J. Polym. Mater*. 56, 911–927.

31. Ramesh, S., Alagar, M., and Sivasamy, A. (2010). Synthesis and characterization of 3, 3-bis (3-maleimidophenyl) phenyl phosphine oxide (BMI)/1, 3-cyanatobenzene epoxy inter-crosslinked matrices for engineering applications. *High Perform. Polym*. 22, 635–648.

32. Sea-Fue, W., Yuh-Ruey, W., Kuo-Chung, C., and Shih-Hsin, C. (2010). Physical and electrical properties of polyimide/ceramic hybrid films prepared via non-hydrolytic sol–gel process. *J. Mater. Sci. Mater. Electron*. 21, 104–110.

33. Selvaganapathi, A., Alagar, M., and Gnanasundaram, P. (2010). Preparation of siloxane-modified tetraglycidyl epoxy-clay nanocomposites—its thermal and morphological properties. *J. Compos. Mater*. 44, 543–557.

34. Silvia, L. F., Milena, S. L., Alberto, G. V. C., Ricardo, R. S., and Eduardo, R (2010). Chemical, morphological, and mechanical analysis of rice husk/post-consumer polyethylene composites. *Composites A* 41, 154–160.

35. Vengatesan, M. R., Devaraju, S., and Alagar, M. (2011). Studies on thermal, mechanical and morphological properties of organoclay filled azomethine modified epoxy nanocomposites. *High Perform. Polym*. 23, 3–10.

36. Xinghong, Z., Shang, C., Tao, C., Xue, K. S., Fei, L., and Qi, G. (2007). Synthesis of a soluble azomethine-containing bisphenol and the properties of its modified epoxy thermosets. *J. Appl. Polym. Sci*. 106, 1632–1639.

37. Yue, C., Yanchao, Z., Zichen, W., Ying, L., Lili, W., Lili, D., et al. (2011). Application studies of activated carbon derived from rice husks produced by chemical-thermal process—a review, advances in colloidal and interface. *Science* 163, 39–52.

38. Zulfiqar, S., Ahmad, Z., and Sarwar, M. (2007). Soluble aromatic polyamide bearing ether linkages: synthesis and characterization.*Polym. Sci.* 285, 1749–1754.

Chapter 8

Synthesis and Characterization of New Biodegradable Chitosan/Polyvinyl Alcohol/Cellulose Nanocomposite

Alireza Samzadeh-Kermani*, Negin Esfandiary

Department of Chemistry, Faculty of Science, University of Zabol, Zabol, Iran

ABSTRACT

In this report, a new nanocomposite based on chitosan/polyvinyl alcohol/nanocrystalline cellulose (Cts/PVA/NCC) was synthesized. The morphology and particle size of NCC and nanocomposites were studied by scanning electron microscopy (SEM), X-ray diffraction (XRD) analysis and Fourier transform infrared (FT-IR) spectroscopy. According to XRD results, the size of NCC was found to be at the range of 15 - 17 nm. SEM images showed the rod-like shape of NCC whiskers. Finally, biodegradation and swelling studies were performed on Cts/PVA/NCC nanocomposites.

KEYWORDS

Nanocomposite, Nanocrystalline Cellulose, Chitosan, Polyvinyl Alcohol, Biodegradation

1. INTRODUCTION

Nanostructures are produced from nanofillers which are dispersed in a polymer matrix. Incorporation of small amounts of nanometer-sized fillers can yield composites with enhanced properties earnestly required for many industrial and technological applications [1] . Naturally biopolymers, like polysaccharides can be made into either coatings or stand-alone films [2] . Many studies on biopolymer films for food packaging applications have been reported [3] . The production of NCC has gained increasing attention in recent years. Cellulose, one of the world's most abundant, natural and renewable biopolymer resources, is widely present in various forms of biomass, such as plants. In cellulosic plant fibers, cellulose is present in an amorphous form, but is associated with crystalline phases through both intermolecular and intramolecular hydrogen bondings in which cellulose does not melt before thermal degradation [4] . Cellulose is organized in fibrils, which are aligned parallel to each other, surrounded by a matrix of lignin and hemicellulose. The properties of cellulose including mechanical state, low density and biodegradability [5] depend on the type of cellulose which is present. Cellulose macromolecule consists of β-1, 4-D-linked glucose chains with molecular formula $(C_6H_{10}O_5)_n$ through an acetal oxygen which covalently bonds C-1 of one glucose ring to C-4 of the other [6] . Owing to the abundance of hydroxyl groups existent on the surface of cellulose nanocrystals, reactive NCC can be modified with different chemical groups to accomplish the expected surface modification like esterification and silylation or polymer grafting, which could successfully functionalize NCC and facilitate its dispersion into different polymer matrices [7] . Therefore, NCC is considered as one of the ideal nano-reinforcing agents for polymer matrices (including water soluble and water-insoluble polymer systems) and has been used into many polymer matrices to produce reinforced nanocomposites [8] . In addition, low density, low energy consumption, inherent renewability, biodegradability and biocompatibility are also good advantages of environmentally-friendly NCC [9] . Because of good dispersion of NCC in water [10] , fabrication and application of hydrogels including NCC without modification have many advantages versus other nanofillers such as polymer and metal nanoparticles [11] . The improved interface between nanofillers and polymer matrix is beneficial to the properties of polymer-based nanocomposites [12] . In nanocomposites, hydrogen bonding between NCC and polymer matrix plays an important role in determining polymer-NCC interaction. Strong hydrogen bonding between PVA and NCC was confirmed by observing the band gap between 3550 and 3200 cm^{-1} in Fourier transform infrared spectra (FTIR). In addition to hydrogen bonding, covalent bonding also provides an enhancement in polymer-nanofiller interface to achieve optimal nanobiocomposite properties [13] . NCC was made in spherical state [14] [15] , or rod-like highly crystalline nanocrystals [6] [16] -[18] , which was obtained by acid hydrolysis of cellulosic fibers and microfibrillated cellulose

(MFC) resulting from disintegration of cellulose fibers under high shearing and impact forces [19] . NCC, which is obtained via acid hydrolysis of cellulose, typically has a wide range in size. With increasing consumption consciousness about petroleum-based polymer materials, the development of fully biodegradable, eco- friendly and bio-based nanocomposites has gained more and more attention both in industrial and academic fields [20] . The so-called "green" composites are derived from different natural resources such as plants or animals [21] . PVA is a material with technological potential which has a wide range of applications due to its unique physical and chemical properties [22] . It is a nontoxic, highly crystalline matter and has good film- forming and high hydrophilic properties. PVA, cellulose and Cts are polar polymers; thus a nonocomposite of PVA, Cts and NCC is likely to produce a material including excellent mechanical properties. Embedding NCC into PVA matrix changes the physicochemical properties of PVA and improves polymer structure in both molecular and morphological levels [23] . We have chosen Cts because it is one of the strongest biopolymers containing reactive amino groups. Moreover, this biopolymer has several important properties such as biocompatibility and anti-bacterial properties, which made it a suitable material for biomedical applications such as drug delivery, tissue engineering, wound healing and various antimicrobial strategies [24] . Also, Cts films have been used for food packaging, mainly due to their non-toxicity and biodegradability [25] . In this study we have prepared NCC from Whatman No. 1 filter paper via acid hydrolysis.

2. EXPERIMENTS

2.1. Materials

The materials used in this study, consist of Whatman No. 1 filter paper as a precourser for NCC, acetic acid (analytical purity), Sulfuric acid (98%) for acidic hydrolysis, PVA and low-molecular-weight Chitosan as matrices, which were obtained from Merk (Darmstadt, Germany) Company Inc.

2.2. Preparation of Nanocrystalline Cellulose

The NCC was obtained via acidic hydrolysis using traditional method. A Whatman No. 1 filter paper (98% α-cellulose, 80% crystallinity) was chopped and transferred into 70 ml of deionized water and blended by a (10 Speed Osterizer) Blender to turn a pulp. The pulp was filtered by a Buchner Funnel under vaccume filtration. Then 40 ml Sulfuric acid (98%) was added dropwise to the solution under vigorous stirring at 45°C until the slurry turned to a milky mixture for about 2 hours. After dilution with 200 ml of cold deionized water, the mixture was subjected to five washing/centrifugation cycles using a centrifuge apparatus (14,000 g, 15 min, Eppendorf, Germany). When pH was reached to about 4 - 5, the fine cellulose particles were dispersed into aqueous supernatant. The residue containing NCC, was then filtered.

2.3. Preparation of Cts/PVA/NCC Nanocomposite

Cts/PVA/NCC was prepared using traditional method. 1 g Cts was dissolved in 100 ml acetic acid (0.1 M) and stirred for 6 hours at room temperature. Then the solution was filtered in order to remove undissolved materials. PVA was dissolved in 20 ml of hot distilled water under constant stirring and then added to Cts solution followed by stirring for 2 hours. Various weight ratios of NCC (0, 5, 10 and 15% w/w) were added directly into Cts/PVA solution and then stirred for 24 hours. The mixtures were casted into petri dishes and placed in an air-convection oven at 60°C for 12 hours. The average thickness of every prepared film was about 1 mm.

3. CHARACTERIZATION METHODS

3.1. Fourier-Transform Infrared (FTIR) Spectroscopy

FT-IR spectroscopy was used to confirm that a reaction was occurred between PVA, Cts and NCC. The IR spectra of PVA/Cts/NCC films were recorded using a Bruker Optics Ft Tensor 27, Germany Instrument. FTIR spectra of cellulosic samples were recorded in the range of 400 - 4000 cm^{-1} by the resolution of 4 cm^{-1}. The samples were grounded into powder, mixed with KBr and then converted into thin pellets.

3.2. X-Ray Diffraction (XRD) Analysis

An X-ray diffractometer (X Pert Prompd, Phillips, Netherlands) was used to investigate the crystallinity of nanocellulose. Milled sample powder was analyzed using a Cu-K_α (λ = 1.54 angstrom) analyser. The angle of incidence was varied from 4° to 80° by the step of 0.02° and at the rate of 3°/min at 40 kV and 30 mA. The Segal method was used to calculate the sample crystallinity, using the height of the 2 0 0 peak (I_{200}, 2θ = 22.93°) and the lowest height between the 2 0 0 and 1 1 0 peaks (I_{AM}, 2θ = 18°). I_{200} represents both crystalline and amorphous regions while I_{AM} represents amorphous material only.

3.3. Scanning Electron Microscopy (SEM) Analysis

Scanning electron microscopy (SEM) photographs of the samples were captured using (Hitachi S4160, Germany) electron microscope. The samples were coated with gold using sputtering technique. Scanning electron microscopy (SEM) was used to investigate the morphology of different types of films.

3.4. Gel Swelling Property

First the films were dried at 40°C for 12 hours in an incubator and then weighed. Dried films were then immersed in distilled water for 1, 2, 4 and 8 hours. Wet weight of the films was measured after removing them from water and obliterating the surface adsorbed water. Then films were weighed immediately [26] . Gel swelling property of films was calculated by the following equation:

$$S = \left[\frac{w_s - w_d}{w_d}\right] \times 100$$

in which, S is the percentage of water adsorption of films at equilibrium state, w_s and w_d are the weights (in grams) of samples in swollen and dried states, respectively.

3.5. Biodegradation Property

Biodegradation property of different PVA/Cts/NCC films was determined by exposing the samples to compost mud. In this study, the degradation of films was evaluated by measurement of their weight loss, which refers to erosion of the molecules from solid phase to aqueous phase. The dissolved components were easily degraded by micro organisms in a natural environment.

4. RESULTS AND DISCUSSION

4.1. FTIR Spectroscopy Analysis

FT-IR spectra of NCC and the nanocomposites were shown in Figure 1. The spectrum of NCC was exhibited a broad band in the region of 3500 - 3200 cm^{-1}, which indicated the inter hydrogen bonded O-H stretching vibration of the OH groups in cellulose molecule. The spectra of all other samples were shown characteristic C-H stretching vibration around 2900 cm^{-1} for C-H stretching and the peak in the region of about 1641 cm^{-1} which was attributed to glycosidic C-O bonds. Besides, the vibrational peak at 1365 cm^{-1} in the samples B, C, D and E was related to the bending vibration of C-H bonds in aromatic ring. The band around 1423 cm^{-1} was attributed to the aliphatic C-H stretching of PVA. Furthermore, the peaks around 1430 cm^{-1} were associated with the C-H in plane bending of NCC.

4.2. X-Ray Diffraction Studies

The crystallinity of NCC was calculated to be 81.5% from Segal Method. This value was in the range of published values 86.7% for commercial microfibrillated cellulose (MFC). Pure Cts was shown two peaks at 2θ of 9.37°, 19.56° and a broad peak appearing at 2θ values in the range 19° - 28° was generally pertinent to the polymeric PANi chains. The peaks at 2θ values of 7.12° for MMT and 6.09° for MMT/Cts were also observed. As shown in Figure 2, all the four diffractograms display two well-defined peaks around 2θ of 12.5° (for 1 1 0 plane) and 2θ of 22.5° (for 2 0 0 plane) characteristic of cellulose [4] . Proper control of the hydrolysis conditions were resulted in selective degradation and preferential removal of amorphous cellulosic fraction. However, the average crystallite size was obtained from X-ray diffraction data using Scherrer's formula:

$$D = \frac{K\lambda}{\beta Cos\theta}$$

where k = 0.94, λ = 0.154056 nm and β is the full width at half maximum (FWHM) in radians. The prepared NCC had rectangular shape with average dimensions of 15.7 nm. The above experimental observations were indicated that NCC belonged to semi-crystalline polymer, which were contained crystalline and amorphous regions. Accordingly, the above results were demonstrated that the hydrolysis was taken place in the amorphous region.

This increase of crystallinity after acidic treatment has been reported by several authors [27] [28] .

4.3. Scanning Electron Microscopy (SEM) Analysis

The morphology of Cts/PVA/NCC nanocomposites was assigned using SEM. Scanning electron micrographs of the nanocomposite films with different contents of NCC were shown in Figure 3. The random orientation and good dispersion of NCC in matrices was observed. These results were indicated an excellent compatibility between the three components of every nanocomposite. According to the SEM results, it was obvious that the whiskers of NCC in nanocomposites were rod like. The morphology of NCC was appeared unchanged after blending with Cts/PVA matrix.

4.4. Gel Swelling Property Analysis

Figure 4 was illustrated the effect of the concentration of NCC on swelling percentage and water uptake of Cts/ PVA films. The presence of NCC in the films was reduced the swelling percentage of Cts/PVA films. Water uptake of nanocomposites was depended on the nature of matrix and filler. The decreased water uptake could be ascribed to the high crystallization of NCC which was less hydrophilic than Cts/PVA composite and the formation of strong filler-matrix interactions [26] [29] [30] . NCC was acted as an interpenetrating network within the matrix and prevented the swelling of the Cts/PVA films when taken into the water [31] .

4.5. Biodegradation Analysis

The weight loss in biodegradation of Cts/PVA and Cts/PVA/NCC films was shown in Table 1. Accordingly, it was observed that the degradation occurred in a faster rate in the presence of NCC in the Cts/PVA matrix. Consequently, the nanocomposites could be easily degraded in natural condition.

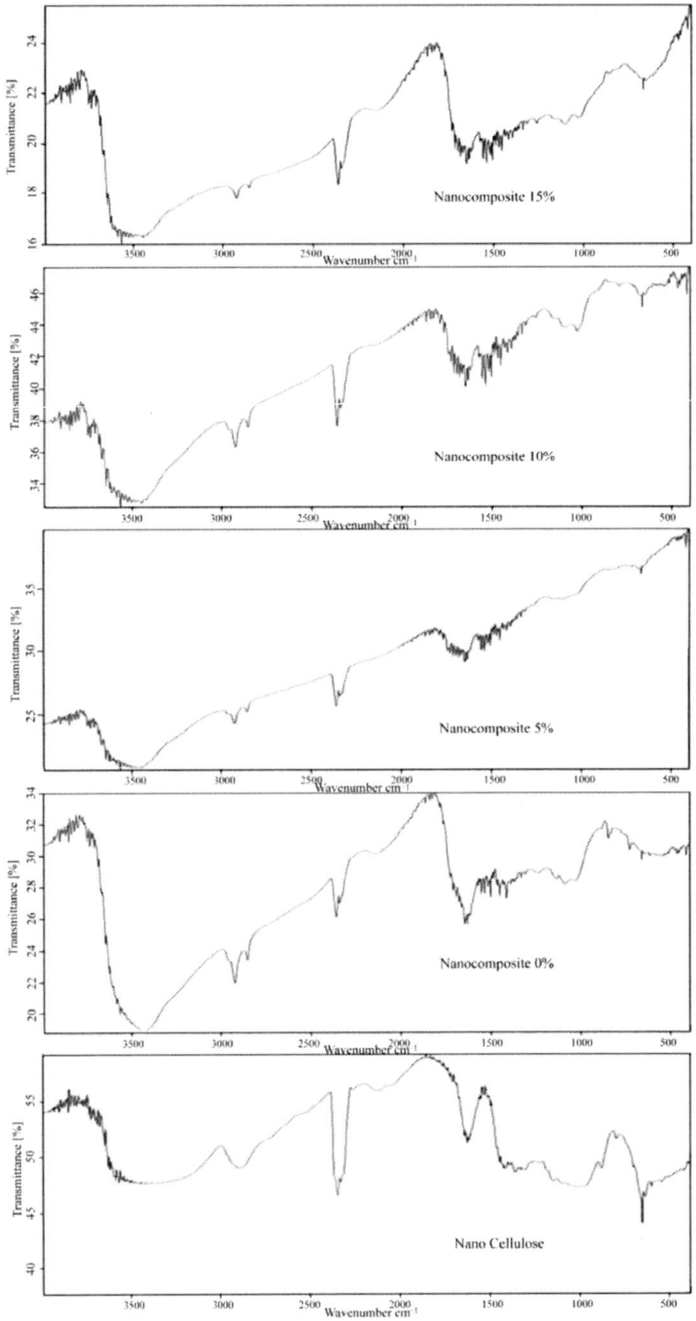

Figure 1. FT-IR spectra of NCC and nanocomposites with different NCC concentrations.

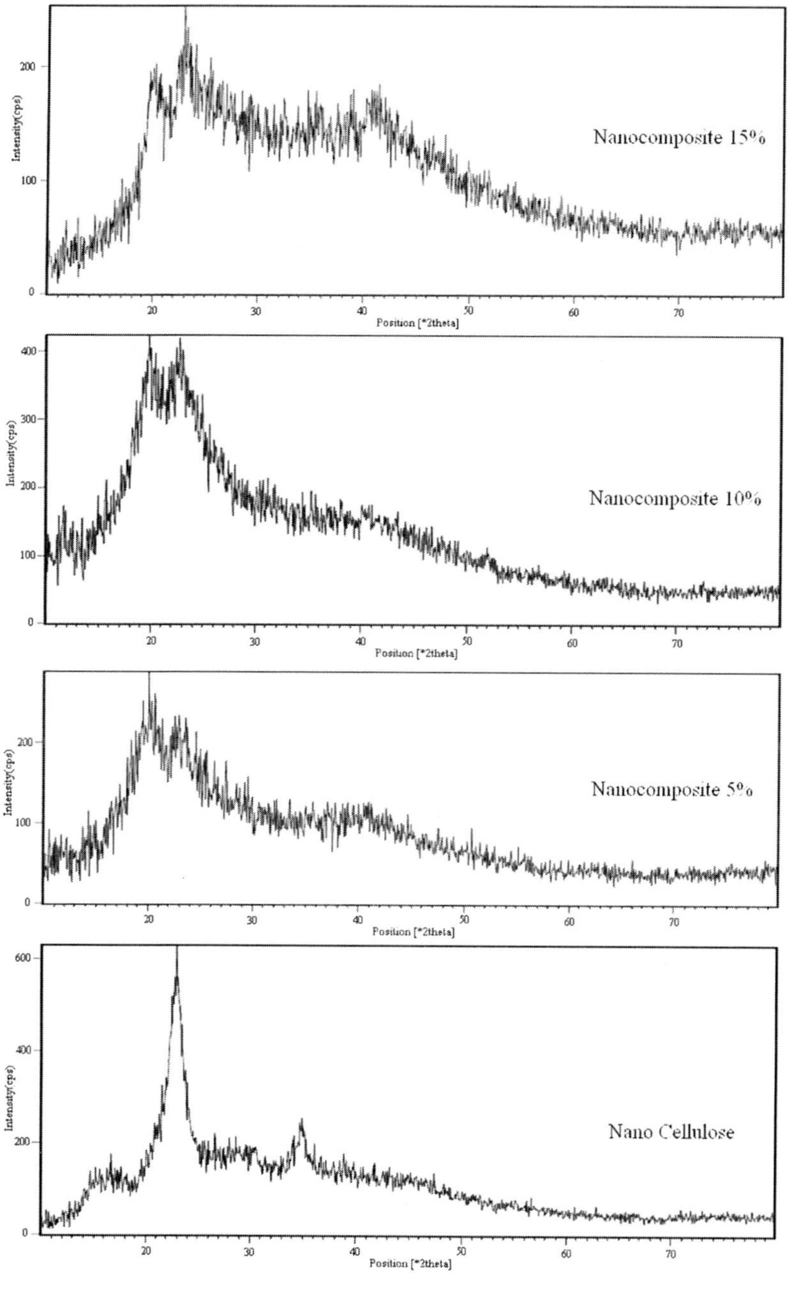

Figure 2. XRD patterns of NCC and nanocomposites with different NCC.

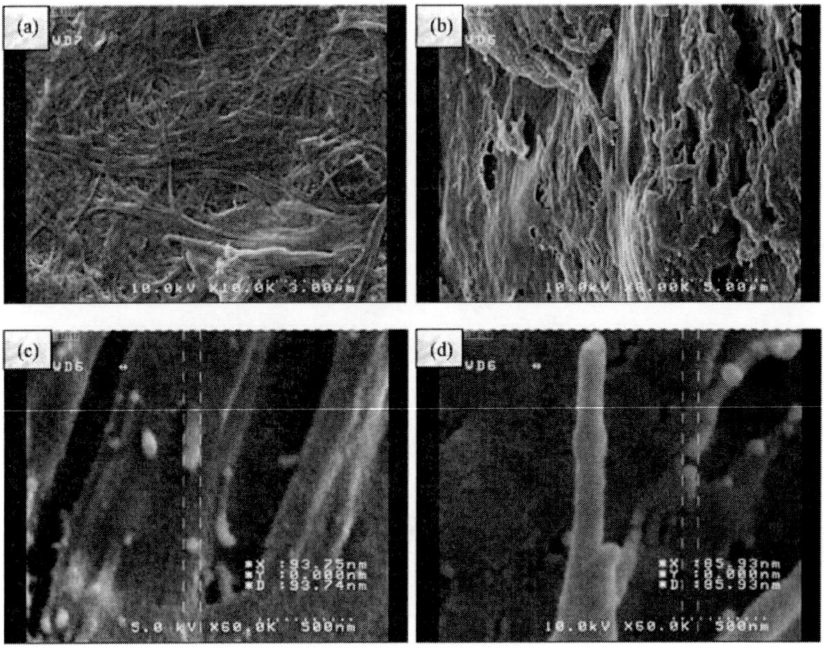

Figure 3. FESEM pictures of (a) NCC, (b) nanocomposite 5%, (c) nanocomposite10% and (d) nanocomposite 15%.

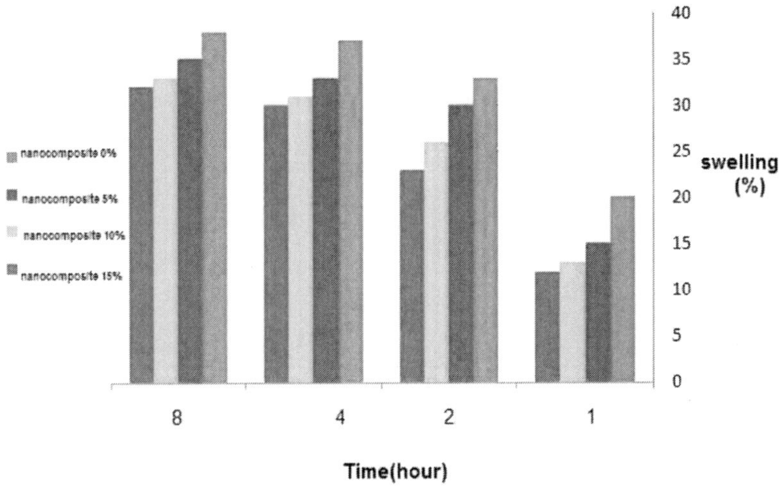

Figure 4. The effect of NCC content in the gel swelling property of Cts/PVA films.

Table 1. Biodegradation for the Cts/PVA and Cts/PVA/NCC films.

Films	Weight loss (g)			
	1st day	2nd day	4th day	5th day
Nanocomposite 0%	0.16	0.12	0.11	0.09
Nanocomposite 5%	0.21	0.23	0.13	0.07
Nanocomposite 10%	0.32	0.30	0.15	0.03
Nanocomposite 15%	0.40	0.33	0.19	0.02

5. CONCLUSION

In this research, NCC was synthesized from Whatmann No. 1 filter paper. Cts/PVA/NCC nanocomposites were prepared in different concentrations of NCC [5% (B), 10% (C) and 15% (D)]. Further studies should be focused on the investigation of other potential applications of NCC and its nanocomposites in combination with other polymers. NCCs which were isolated by sulfuric acid hydrolysis, were rod-like whiskers with an average size of about 15.7 nm observed by XRD analysis, and the crystallinity of NCC was about 81.5%. SEM images of NCC showed the dimensions of about 85 nm wide. NCC improved the barrier properties of Cts/PVA composite films via reduction of swelling property. Although, a comparison with related literature results showed that the nanocomposites which were investigated in this work were amongst the best materials reported so far, this approach appeared really promising for the design of a wide range of new related nanocomposites. Surface morphology of nanocomposite films revealed a homogeneous dispersion of NCC into the Cts/PVA matrix. The prepared Cts/ PVA/NCC films were biodegradable.

ACKNOWLEDGEMENTS

This research has been accomplished under financial support of the Department of Chemistry, Faculty of Science, and cordially laboratorial collaboration of our colleagues at the Department of Agronomy, Faculty of Agriculture, University of Zabol, which should be appreciated.

REFERENCES

1. Whitesides, G.M. (2005) Nanoscience, Nanotechnology, and Chemistry. Nanoscience and Chemistry, 1, 172-179.

2. Kester, J.J. and Fennema, O.R. (1986) Edible Films and Coatings: A Review. Food Technology, 40, 47-59.

3. Gennadios, A. (2002) Protein-Based Films and Coatings. CRC Press, Boca Raton, 367-392.

4. Klemm, D., Heublein, B., Fink, H.P. and Bohn, A. (2005) Cellulose: Fascinating Biopolymer and Sustainable Raw Material. Angewandte Chemie International Edition, 44, 3358-3393.

5. Zimmerman, T., Pohler, E. and Schwaller, P. (2005) Mechanical and Morphological Properties of Cellulose Fibril Reinforced Nanocomposites. Advanced Engineering Materials, 12, 1156-1161.

6. Azizi-Samir, M.A.S., Alloin, F. and Dufresne, A. (2005) Review of Recent Research into Cellulosic Whiskers, Their Properties and Their Application in Nanocomposite Field. Biomacromolecules, 6, 612-626.

7. Habibi, Y., Lucia, L.A. and Rojas, O.J. (2010) Cellulose Nanocrystals: Chemistry, Self-Assembly, and Applications. Chemical Reviews, 110, 3479-3500.

8. Cao, X.D., Habibi, Y., Magalhaes, W.L.E., Rojas, O.J. and Lucia, L.A. (2011) Cellulose Nanocrystals-Based Nanocomposites: Fruits of a Novel Biomass Research and Teaching Platform. Current Science, 100, 1172-1176.

9. Siro, I. and Plackett, D. (2010) Microfibrillated Cellulose and New Nanocomposite Materials: A Review. Cellulose, 17, 459-494.

10. Liang, S.M., Zhang, L.N., Li, Y.F. and Xu, J. (2007) Fabrication and Properties of Cellulose Hydrated Membrane with Unique Structure. Macromolecular Chemistry and Physics, 208, 594-602.

11. Wu, Y.T., Zhou, Z., Fan, Q.Q., Chen, L. and Zhu, M.F. (2009) Facile In-Situ Fabrication of Novel Organic Nanoparticle Hydrogels with Excellent Mechanical Properties. Journal of Materials Chemistry, 19, 7340-7346.

12. Zhang, Y., Yu, J.R., Zhou, C.J., Chen, L. and Hu, Z.M. (2010) Preparation, Morphology, and Adhesive and Mechanical Properties of Ultrahigh-Molecular-Weight Polyethylene/SiO2 Nanocomposite Fibers. Polymer Composites, 31, 684-690.

13. Zhou, C.J., Wang, S.F., Zhang, Y., Zhuang, Q.X. and Han, Z.W. (2008) In Situ Preparation and Continuous Fiber Spinning of Poly(p-

phenylenebenzobisoxazole) Composites with Oligo-Hydroxyamide-Functionalized Multi-Walled Carbon Nanotubes. Polymer, 49, 2520-2530.

14. Pu, Y.Q., Zhang, J.G., Elder, T., Deng, Y., Gatenholm, P. and Ragauskas, A. (2007) Investigation into Nanocellulosics versus Acacia Reinforced Acrylic Films. Composites Part B: Engineering, 38, 360-366.

15. Zhang, J., Jiang, N., Dang, Z., Elder, T.J. and Ragauska, A.J. (2008) Oxidation and Sulfonation of Cellulosics. Cellulose, 15, 489-496.

16. Lima, M.M.D. and Borsali, R. (2004) Rod-Like Cellulose Microcrystals: Structure, Properties and Applications. Macromolecular Rapid Communications, 25, 771-787.

17. Dufresne, A. (2006) Comparing the Mechanical Properties of High Performances Polymer Nanocomposites from Biological Sources. Journal of Nanoscience and Nanotechnology, 6, 322-330.

18. Dufresne, A. (2008) Polysaccharide Nanocrystal Reinforced Nanocomposites. Canadian Journal of Chemistry, 86, 484-494.

19. Teixeira, E.M., Pasquini, D., Curvelo, A.A.S., Corradini, E., Belgacem, M.N. and Dufresne, A. (2009) Cassava Bagasse Cellulose Nanofibrils Reinforced Thermoplastic Cassava Starch. Carbohydrate Polymers, 78, 422-431.

20. Kim, J.T. and Netravali, A.N. (2010) Mechanical, Thermal, and Interfacial Properties of Green Composites with Ramie Fiber and Soy Resins. Journal of Agricultural and Food Chemistry, 58, 5400-5407.

21. Khalil, H.P.S.A., Bhat, A.H. and Yusra, A.F.I. (2012) Green Composites from Sustainable Cellulose Nanofibrils: A Review. Carbohydrate Polymers, 87, 963-979.

22. Masuhiro, T., Giuliano, F. and John, S.C. (1994) Structure and Compatibility of Poly(vinyl alcohol)-Silk Fibroin (PVA/SA) Blend Films. Journal of Polymer Science B: Polymer Physics, 32, 243-248.

23. Siddaramaiah, R.T., Ravi, P., Kumar, K.P. and Jagadeesh, K.S. (1999) Effect of Fillers on the Properties of LDPE Films. Journal of Polymer Materials, 16, 233-237

24. Dash, M., Chiellini, F., Ottenbrite, R.M. and Chiellini, E. (2011) Chitosan-A Versatile Semi-Synthetic Polymer in Biomedical Applications. Progress in Polymer Science, 36, 981-1014.

25. De Azeredo, H.M.C. (2009) Nanocomposites for Food Packaging Applications. Food Research International, 42, 1240-1253.

26. Jin, J., Song, M. and Hourston, D.J. (2004) Novel Chitosan-Based Films Cross-Linked by Genipin with Improved Physical Properties. Biomacromolecules, 5, 162-168.

27. Azizi-Samir, M.A.S., Alloin, F., Paillet, M. and Dufresne, A. (2004) Tangling Effect in Fibrillated Cellulose Reinforced Nanocomposites. Macromolecules, 37, 4313-4316.

28. Tang, L.G., Hon, D.N.S., Pan, S.H., Zhu, Y.Q., Wang, Z. and Wang, Z.Z. (1996) Evaluation of Microcrystalline Cellulose. I. Changes in Ultrastructural Characteristics during Preliminary Acid Hydrolysis. Journal of Applied Polymer Science, 59, 483-488.

29. Dufresne, A., Dupeyre, D. and Vignon, M.R. (2000) Cellulose Microfibrils from Potato Tuber Cells: Processing and Characterization of Starch-Cellulose Microfibril Composites. Journal of Applied Polymer Science, 76, 2080-2092.

30. Li, Q., Zhou, J. and Zhang, L. (2009) Structure and Properties of the Nanocomposite Films of Chitosan Reinforced with Cellulose Whiskers. Journal of Polymer Science Part B: Polymer Physics, 47, 1069-1077.

31. Svagan, A.J., Hedenqvist, M.S. and Berglund, L. (2009) Reduced Water Vapour Sorption in Cellulose Nanocomposites with Starch Matrix. Composites Science and Technology, 69, 500-506.

Chapter 9

Reinforcement Effects of CNTsfor Polymer-Based Nanocomposites

Yuan Li[1], Yaolu Liu and Ning Hu

[1] Chiba University, Japan

1. INTRODUCTION

With large aspect ratio, high strength and stiffness, carbon nanotubes (CNTs) have been widely used as the reinforcement for polymer-based nanocomposites with prospective application in aerospace engineering, automotive industry, etc. Generally, there are two typical representatives: three-phase hybrid CFRP laminates (CNT/Polymer/CF) and two-phase CNT-reinforced polymer (CNT/Polymer) nanocomposites.

For three-phase hybrid CFRP laminates which use CNTs to improve the interlaminar mechanical properties of CFRP laminates, the fabrication methods can be categorized into the following three ways:

- *Matrix reinforcement*, which uses the two-phase CNT/Polymer nanocomposites as the matrix for CFRP plies before lay-up [Yokozeki et al., 2007; Karapappas et al., 2009; Inam et al., 2010];

- *Filler reinforcement*, which employs the CNT-grafted carbon fibres as reinforcement filler for CFRP plies before lay-up [Thostenson et al., 2002 ; Veedu et al., 2006; He et al., 2007; Kepple et al., 2008];

- *Interface reinforcement*, which adds CNTs into the interface between CFRP plies during lay-up [Garcia et al., 2008; Arai et al., 2008].

Although great progress has been made for the first two ways, less work is reported on the third one.

Therefore, in the present work, a newly simple fabrication method, i.e., Powder method *was developed to disperse the CNTs at the interface between CFRP plies. The improvement of interlaminar mechanical properties was investigated by DCB (double cantilever bending) tests and FEM (finite element method) analysis.*

For two-phase CNT/Polymer nanocomposites, numerous experimental investigations have been carried out on the reinforcement effect of CNTs addition in thermosetting ones, e.g. epoxy [Schadler et al., 1998; Gojny et al., 2005; Ci & Bai, 2006], polyimide [Ogasawara et al., 2004; Jiang et al., 2005], and phenolic [Tai et al., 2004], as well as thermoplastic ones, e.g. polypropylene (PP) [Bhattacharyya et al., 2003; Chang et al., 2005], polystyrene (PS) [Thostenson & Chou, 2002 ; Chang et al., 2006], polymethylmethacrylate (PMMA) [Cooper et al., 2002 ; Lee et al., 2006] and polyether ether kentone (PEEK) [Deng et al., 2007; Bangarusampath et al., 2009]. However, the reported improvement of mechanical properties of these nanocomposites is far more less than the expectation because of poor dispersion, difficult alignment of CNTs, and weak interface between CNT and polymer matrix, which are involved in geometrical properties of CNTs themselves, polymer properties, fabrication method of CNT/Polymer nanocomposites.

To overcome these obstacles, various efforts, such as ultrasonication [Safadi et al., 2002; Ding et al., 2003], surface treatment [Curran et al. 1999; Vigolo et al., 2000; Gojny et al., 2003], shear mixing [Andrews et al., 2002], bi/tri-axial rolling [Rosca & Hoa, 2009; Gojny et al., 2004], extrusion [Cooper et al., 2002 ; Wong et al., 2003], and their combination, have been firstly made to effectively achieve good dispersion of CNTs in polymer. Moreover, several techniques have been proposed to control CNTs alignment using shear [Ajayan et al., 1994], elongation [Jin et al., 1998], melt processing [Haggenmueller et al., 2000], as well as magnetic field [Kimura et al., 2003] or electrical spinning [Sen et al., 2004]. Furthermore, great progress has been made on the investigation of the interfacial properties. For example, direct CNT pull-out experiments were made in telescope electron microscopy (TEM) [Qian et al., 2000; Deng, 2008], or by atomic force microscopy (AFM) [Barber et al., 2003, 2004], Raman spectroscopy [Schadler et al., 1998], scanning probe microscope (SPM) [Cooper et al., 2002]. However, difficult manipulation in nano-scale and measurement on force/displacement makes large data scattering inevitable, which makes numerical simulation a powerful alternative approach. Continuum mechanical, molecular mechanics (MM), conventional or ab initio molecular dynamics (MD), or their combinations, have also been attempted to investigate the interfacial properties. For instance, some continuum mechanics based models (e.g. cohesive zone model [Jiang et al., 2006], shear lag model [Xiao & Zhang, 2004; Gao & Li, 2005; Tsai & Lu, 2009], and pull-out model [Lau, 2003; Natsuki et al., 2007]) were developed to investigate the interfacial properties between CNT and polymer matrix. Compared with Natsuki's suggestion [Natsuki et al., 2007] of that maximum interfacial shear stress (ISS) occurs at the pull-out end of CNT, Gao [Gao & Li, 2005] predicted that ISS approaches maximum at the two ends of CNT, but keeps zero at the middle part. On the other hand, most of MM/MD simulations [Lordi & Yao, 2000; Liao & Li, 2001;Frankland et al., 2002; Gou et al., 2004; Zheng et al., 2009; Al-Ostaz et al., 2008; Chowdhury & Okabe, 2007] assumed constant ISS during the whole pull-out process with uniform distribution along the contact surface between CNT and polymer. Obviously, the above totally different distribution forms of ISS leads to great confusion in understanding the inherent interfacial characteristics of CNT/Polymer nanocomposites. Moreover, all of the previous MM/MD simulations are limited to a specified SWCNT with a fixed length and diameter. The influences of CNTs' dimension on the pull-out behaviour have never been investigated.

Therefore, in the present work, the other focus is put on the clarification of interfacial properties between CNT and polymer matrix, in which the effects of CNTs' dimension (i.e. nanotube length, diameter and wall number) were explored for the first time. Moreover, the obtained interfacial properties were combined with the theoretical model of continuum mechanics to develop a sequential multi-scale model for predicting the overall mechanical properties of CNT/Polymer nanocomposites, which were verified using tensile tests and SENB (single-edge notched bending) tests.

Here, two commercial CNTs products, i.e. MWCNT-7 (Nano Carbon Technologies Co., Ltd. Japan) and VGCF® (Showa Denko K.K., Japan) as shown in Fig. 1, were employed as reinforcement fillers at the interface of conventional CFRP laminates to fabricate three-phase hybrid CFRP laminates, and in epoxy to fabricate two-phase CNT/Polymer nanocomposites, respectively. The average diameter of VGCF® is around two times higher than that of MWCNT-7. Moreover, wall thickness of VGCF® is larger, while the central wall is much smaller than that of MWCNT-7. The corresponding geometrical and mechanical properties of these two CNTs are listed in Table 1.

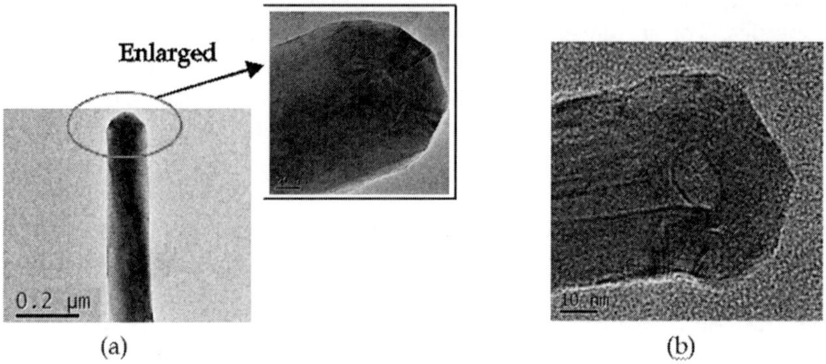

Figure 1. TEM pictures of two commercial CNTs, a) VGCF®; b) MWCNT-7

Table 1. Properties of CFRP prepregs and two commercial CNTs

CFRP prepregs		Commercial CNTs	VGCF®	MWCNT-7
Matrix	#2500	Radius r_f (nm)	75	32.5
PAN-based CF	T700S	Length l_f (μm)	15 (10~20)	7.5 (5~10)
Young's modulus (GPa)		Density ρ_f (kg/m³)	2.0	2.1
Fibre direction	116.8	Young's modulus E_f (GPa)	516.5 (273~760)	850 [Demczyk et al., 2002]
Transverse direction	8.83	Tensile strength σ_{ultf} (MPa)	3100 (2700~3500)	1.5×10⁵ [Demczyk et al., 2002]

2. THREE-PHASE HYBRID CFRP LAMINATES

Powder method was developed to disperse CNTs directly at the interface between CFRP prepregs to fabricate three-phase hybrid CFRP laminates. The improvement on the interlaminar mechanical properties was confirmed by DCB tests and FEM analysis.

2.1. Dcb Experiments

2.1.1. Materials and Specimen Fabrication

The CFRP prepregs (T700S/#2500 Toray, Co. Ltd., Japan) were used, where their physical and mechanical properties are also given in Table 1.

Here, CNTs, as the reinforcement at interface, was dispersed at the mid-plane of unidirectional $[0º/0º]_{14}$ CFRP laminates during the hand lay-up process, where a simple fabrication method with low cost, i.e. powder method, was employed. The detailed process is described in Fig. 2 as below.

a. Initially, 14 pieces of CFRP plies were stacked together to form two pieces of $[0º]_7$ CFRP unidirectional sublaminates, respectively;

b. CNTs powder was spread by half on the surface of lower sublaminate using a sifter with mesh size about 70μm. The zigzagged spreading path makes the distribution of CNTs at the interface as consistent as possible;

c. The left half of CNTs powder was spread after placing a 25 μm thick polyamide film (Kapton, Toray Co. Ltd., Japan) to make an initial crack;

d. The upper sublaminate was piled up;

e. Finally, it was put into an autoclave for 3 hours at the temperature of 130º C to cure.

The obtained laminates are typed in Table 2 according to CNTs' area density ρ_A at the interface. Obviously, the measured thicknesses of formed CNTs interlayer t_I increases linearly with area density ρ_A. For reference, base CFRP laminates were also fabricated.

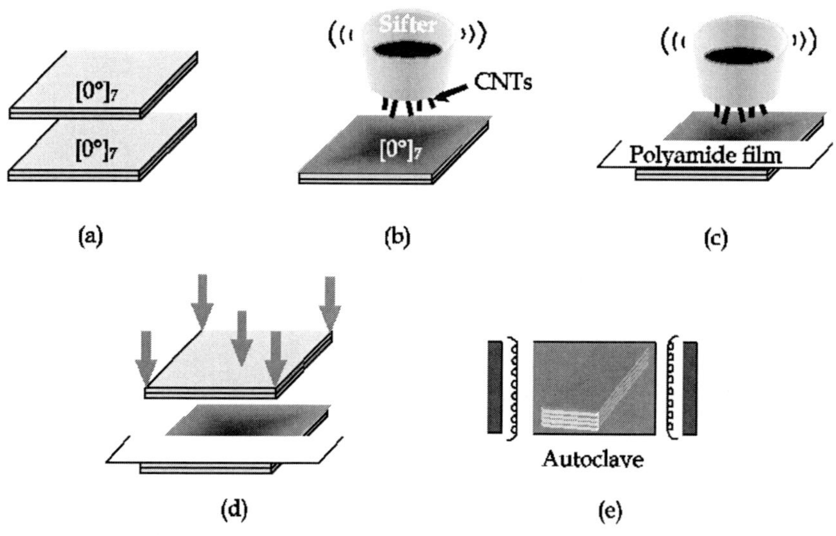

Figure 2. Powder method

Table 2. Fabricated hybrid CFRP laminates

Reinforcement	ρ_A (g/m²)	Notation	t_I (μm)
VGCF®	10	CFRP/VGCF(10)	70
	20	CFRP/VGCF(20)	140
	30	CFRP/VGCF(30)	210
MWCNT-7	5	CFRP/MWCNT(5)	66
	10	CFRP/MWCNT(10)	97
	20	CFRP/MWCNT(20)	148

2.1.2. Dcb Test Procedure

To evaluate Mode-I interlaminar fracture toughness, DCB tests were carried out using a universal material testing machine (AG-100kNE, Shimazu Co. Ltd, Japan) at 20º C according to Japanese Industrial Standards (JIS K7086). Five specimens for each type were cut from the fabricated laminates, where marked lines were painted on side surface for crack length measurement. The specimens were approximately 20mm wide and 120mm long, with the initial crack of 34mm. The thickness of hybrid CFRP laminates was thicker than that of base CFRP laminates with 3.14mm due to the formed CNTs interlayer. The crosshead speed was 0.5mm/min. Tests were terminated when the increment of crack length Δa reaches 70mm.

2.1.3. Mode-I Interlaminar Fracture Toughness

From the obtained load-COD (crack opening displacement) curves of the above various types of hybrid CFRP laminates, the average critical load at crack growth, i.e. peak load P_c, are plotted in Fig. 3awhich shows the obvious increase. The highest P_c occurs at CFRP/VGCF(20) and CFRP/MWCNT(10), which are about 41% and 17% higher than that of base CFRP laminates, respectively. Obviously, VGCF® performs better than MWCNT-7.

Since P_c is dominated by Mode-I interlaminar fracture toughness G_{IC} and interlaminar tensile strength N, it is anticipated that G_{IC} and N also increase. The average G_{IC} and fracture resistance G_{IR} are demonstrated in Fig. 3b and Fig. 3c. Note that G_{IR} is the average value when the crack length varies from 20mm to 60mm in the obtained R-curves. In Fig. 3b, there are 96% and 58% increase of G_{IC} for CFRP/VGCF(20) and CFRP/VGCF(10), respectively, which can be considered as the optimal addition. In Fig. 3c, CFRP/VGCF(20) has the highest G_{IR}, which is about 25% higher than that of base CFRP laminates. However, the effect of MWCNT on G_{IR} is unpromising. The largest value occurs at CFRP/MWCNT(5) which increases only 6%. Moreover, there is even minor negative influence for some cases on G_{IR}.

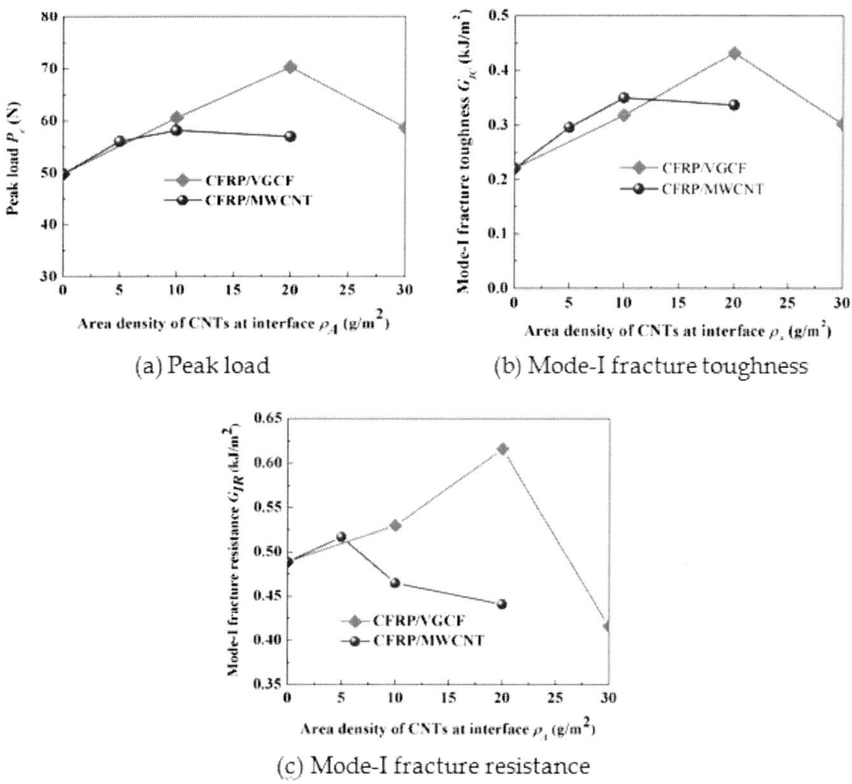

Figure 3. Reinforcement effects of two commercial CNTs in three-phase hybrid CFRP laminates

The reinforcement mechanism can be explained from crack propagation process observed by an optical microscopy. At the initial stage, crack initiates from crack tip at the CNTs interlayer for all cases (Fig. 4a), which explains the increase of G_{IC}. Moreover, for CFRP/VGCF(10,20) and CFRP/MWCNT(5), crack extends forward in a zigzag pattern CNTs interlayer (Fig. 4a) creating much more fracture surfaces and consuming much more energy, which leads to the improved G_{IR}. However, for CFRP/VGCF(30) and CFRP/MWCNT(10,20), crack transits from toughened CNTs interlayer toward CFRP plies and propagates forward (Fig. 4c), which results in the decreased G_{IR} compared with that of base CFRP laminates.

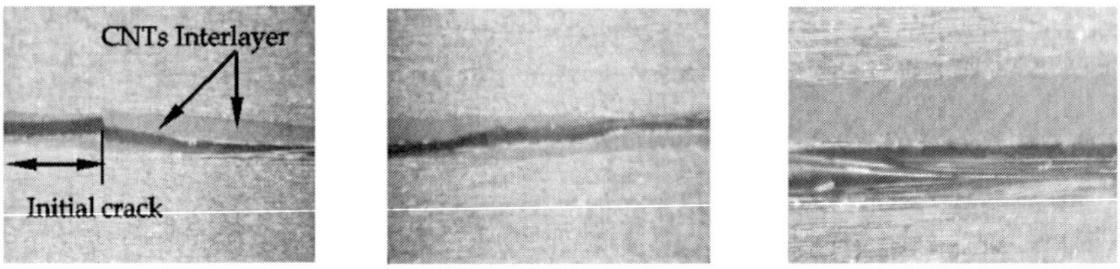

(a) CFRP/VGCF(20) (Δa=0mm) (b) CFRP/VGCF(20) (Δa=10mm) (c) CFRP/VGCF(30) (Δa=10mm)

Figure 4. Crack propagation process for CFRP/VGCF laminates,

Moreover, fracture surfaces of DCB specimens were also investigated using scanning electron microscopy (SEM). For CFRP/VGCF(10) and CFRP/MWCNT(5), naked carbon fiber (Fig. 5a)indicates insufficient addition of CNTs at interface. On the other hand, for CFRP/VGCF(30) and CFRP/MWCNT(20), there are obvious defects (Fig. 5c) caused by the poor CNTs dispersion. It explains that CFRP/VGCF(20) and CFRP/MWCNT(10) with good dispersion of CNTs (Fig. 5b)provide best reinforcement effect, respectively.

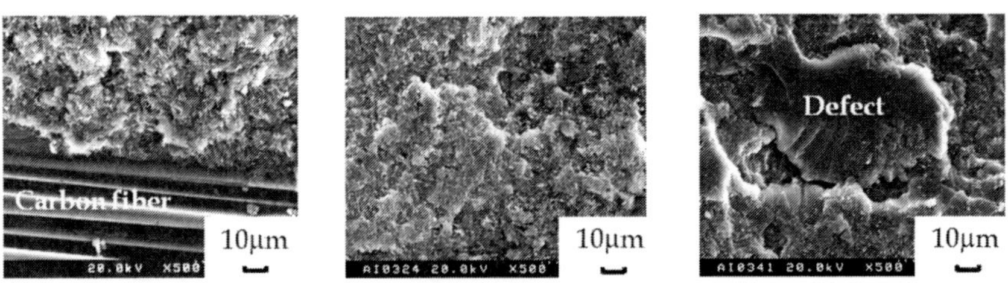

(a) CFRP/VGCF(10) (b) CFRP/VGCF(20) (c) CFRP/VGCF(30)

Figure 5. Fracture surfaces of CFRP/VGCF laminates

2.2. Numerical Simulation Using Fem

Instead of experiment, such as butt-joint test under tensile load, FEM simulation on delamination propagation was employed to predict approximately Mode-I interlaminar tensile strength N.

Here, a brick element of 8 nodes [Cao et al., 2002] and a cohesive element [Camanho et al., 2002] were used to model CFRP sublaminates and CNTs interlayer, respectively. The corresponding material properties are listed in Table 3, where the latter was obtained by matching the numerical load-COD curves to experimental ones (see Fig.6). Here, experimental G_{IC} was directly adopted. Interlaminar tensile strength N and initial stiffness K were determined by matching peak load P_c and the initial slope of the numerical load-COD curves before the peak load, respectively. In Table 3, although N increases with CNTs addition, which are about 75% and 50% higher than that of base CFRP laminates, respectively, there is no any difference among them for different CNTs addition. This can be attributed to that peak load is mainly dominated by G_{IC} while N influences it slightly.

In conclusion, the improvement of interlaminar tensile strength N was confirmed by FEM simulation, and the correctness of experimental G_{IC} was verified from good consistence between numerical and experimental results (Fig.6).

Table 3. Material properties for brick element and cohesive element in FEM simulation

Brick element		Cohesive element	N (MPa)	K (N/mm³)	G_{IC} (kJ/m²)
E_{11} (GPa)	120	CFRP	20	3.5×10^{-3}	0.221
E_{22}=E_{33} (GPa)	8.8	CFRP/VGCF(10)	35	3.5×10^{-3}	0.318
G_{12}=G_{13} (GPa)	6.0	CFRP/VGCF(20)	35	3.5×10^{-3}	0.432
G_{23} (GPa)	3.7	CFRP/VGCF(30)	35	3.5×10^{-3}	0.302
$\upsilon_{12}=\upsilon_{13}$	0.25	CFFRP/MWCNT(5)	30	3.5×10^{-3}	0.296
υ_{23}	0.45	CFFRP/MWCNT(10)	30	3.5×10^{-3}	0.350
		CFFRP/MWCNT(20)	30	3.5×10^{-3}	0.337

3. TWO-PHASE CNT/POLYMER NANOCOMPOSITES

Pull-out simulations based on MM were carried out to investigate the interfacial properties between CNT and polymer matrix, which were further incorporated into the theoretical model of continuum mechanics to develop a sequential multi-scale model for predicting the overall mechanical properties of two-phase CNT/Polymer nanocomposites. The present method is expected to be applied in CNT-reinforced nanocomposites with various matrices (e.g. metal, ceramics, etc.), which provides useful information for material design.

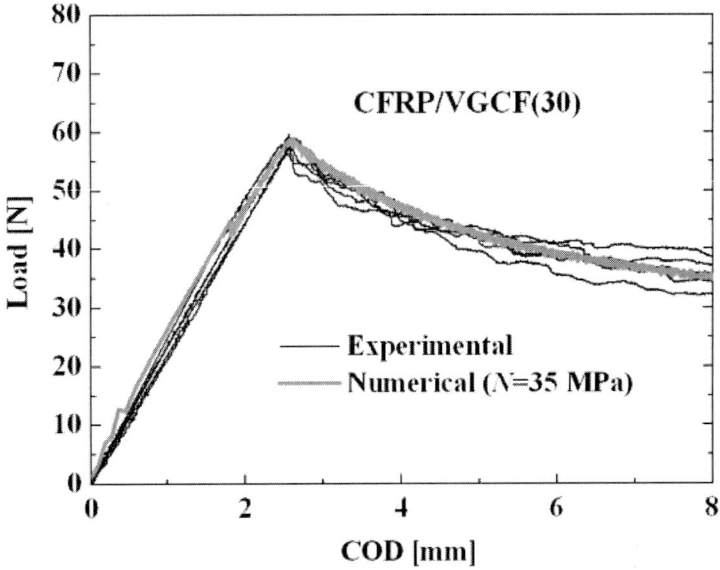

Figure 6. Comparison of experimental and numerical load-COD curves for CFRP/VGCF(30)

3.1. Pull-Out Simulation on Interfacial Properties Between Cnt And Polymer Matrix

To investigate the interfacial properties between CNT and polymer matrix in two-phase CNT/Polymer nanocomposites only due to van der Waals (vdW) interactions, a series of CNT pull-out simulations from polyethylene (PE) matrix were carried out. Here, the condensed phase optimization molecular potentials for atomistic simulation studies (COMPASS) force field [Sun, 1998] was employed to calculate the systematic potential energy, which has been broadly testified for CNT itself and CNT-reinforced nanocomposites because of providing reliable results as compared with more accurate tight-binding or ab initio density functional theory based methods. The PE was chosen as matrix because its structural simplicity can effectively reduce the computational cost. Moreover, PE as a representative polymer matrix can give a general picture of the possible interfacial behaviour of various CNT-reinforced polymer nanocomposites. The simulation cell was composed of a fragment of CNT totally embedded inside PE matrix. Taking SWCNT(5,5)/PE nanocomposites as a benchmark, the processes of model building and CNT pull-out are described in detail in the following.

3.1.1. Model Building

The unit cell of simulation system, which was of periodic boundary conditions in y-z plane, was initialized by randomly generating 36 PE chains with initial density of 1.2g/cm^3 surrounding an open-ended SWCNT(5,5). Each PE chain had 20 repeating units of $-CH_2$. The length and diameter of the SWCNT(5,5) fragment were l=4.92nm and D=0.68nm, respectively. Note that the unsaturated boundary effect was avoided by adding hydrogen atoms at both ends of the SWCNT. The hydrogen atom had charge of +0.1268e and the connected carbon atom had charge of -0.1268e, which made the SWCNT neutral. The corresponding computational cell constructed was in the range of 2.69nm$\subseteq$2.69nm$\subseteq$4.9nm in which the volume fraction of SWCNT was V_f=9.0vol.%. Note that the size of computational cell at x axis (i.e. the axial direction of CNT) had to be set large enough to eliminate the interaction among polymer itself [Marietta-Tondin, 2006]. As shown in Fig. 7a, the vacuum layer was set for the CNT pull-out without extending the cell. The size of vacuum layer was about the sum of the cut-off distance of vdW interaction (0.95nm) and nanotube length.

The equilibrated structure of SWCNT/PE nanocomposites in Fig. 7 was obtained as follows:

1 While holding SWCNT as a rigid, by virtue of MD simulations, the model was first put into a constant-temperature, constant-volume (NVT) ensemble for 50ps and then a constant-temperature, constant-pressure (NPT) ensemble for another 50ps with temperature of T=298K, pressure of P=10atm, and time step of Δt=1fs after the initial minimization. The purpose of this step is to slowly compress the structure of the PE to generate an initial amorphous matrix with correct density and low residual stress.

2 The nanocomposite system was further put into NVT ensemble and equilibrated for 50ps at the same time step of Δt=1fs with releasing all rigid constraints on SWCNT. This step is to create a zero initial stress state.

3 Finally, the nanocomposite system was minimized again using MM to obtain the equilibrated configuration in vacuum.

The equilibrated separation distance h between SWCNT and PE matrix due to vdW interaction was about 0.23nm in the present work as shown in Fig. 7b, which is very close to the value of 0.18nm obtained by Han et al. [Han & Elliott, 2007] for SWCNT/PMMA nanocomposites. The difference can be attributed to the different types of polymer.

3.1.2. Pull-Out Process

The pull-out simulations of SWCNT from PE matrix were carried out by applying displacement-controlled load on the atoms at the right end of CNT. The displacement increment along the axial (x-axis) direction of CNT was Δx=0.2nm. Snap shots of the atom configurations for SWCNT/PE nanocomposites system during the pull-out process are shown in Fig. 8. Note that the deformation of PE matrix during the pull-out process was neglected by fixing the matrix to reduce the computational cost, since it was confirmed numerically that the influence of the enforced conditions on polymer was very small.

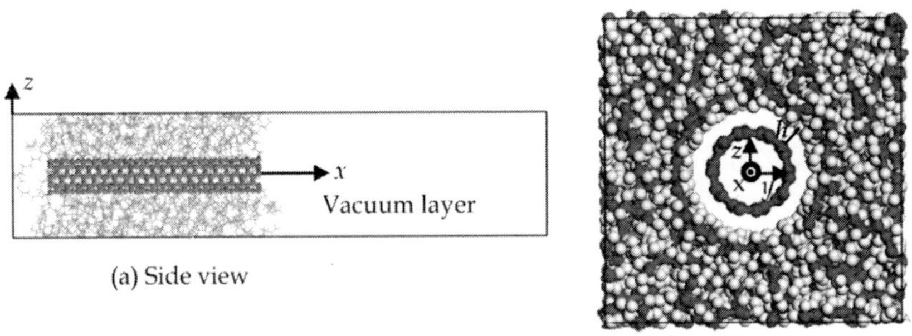

(a) Side view

(b) Enlarged cross section

Figure 7. (CNT: l=4.92nm, D=0.68nm, C atoms: purple, H atoms: grey)Equilibrated structure of simulation cell for SWCNT(5,5)/PE nanocomposites

After each pull-out step, the molecular structure was relaxed to obtain the minimum systematic potential energy E by MM. The potential energies of the SWCNT/PE nanocomposites were monitored and recorded during the whole pull-out process, which will be discussed in the following.

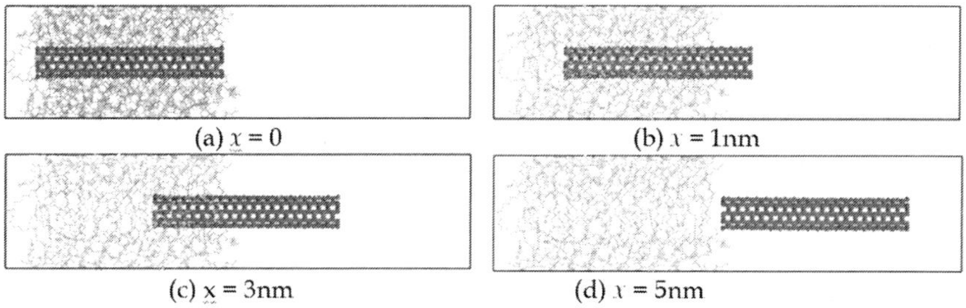

Figure 8. Snap shots of CNT pull-out from PE matrix

3.1.3. Variation Of Potential Energy During Pull-Out

In view of that the work done by the pull-out force equals energy increment of nanocomposite system at each pull-out step, the trend of potential energy variation, and the energy increment at each pull-out step becomes very important for analyzing the corresponding pull-out force, and the ISS between CNT and polymer matrix.

The obtained systematic potential energy E during the pull-out is shown in Fig. 9a, which increases gradually accompanied with the CNT pull-out. This trend is just identical to all of the previous simulation results of CNT/Polymer nanocomposites [Liao & Li, 2001; Frankland et al., 2002; Gou et al., 2004; Zheng et al., 2009].

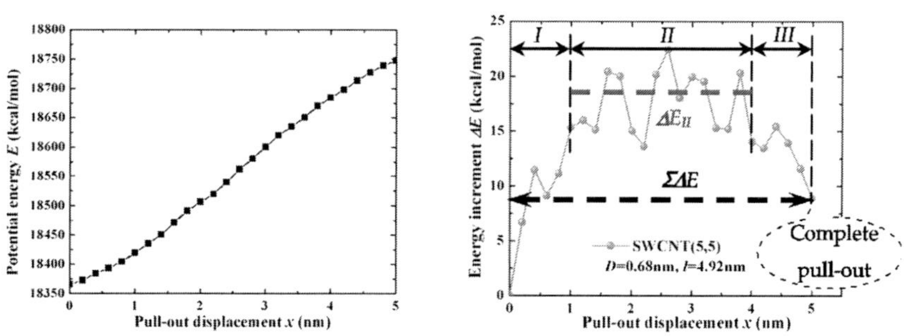

(a) Potential energy E (b) Energy increment $\Sigma\Delta E$

Figure 9. Variation of E and ΔE during the pull-out of SWCNT(5,5)

Generally, this variation of E can be divided into four parts, i.e. the variation of potential energy in polymer matrix; the variation of potential energy of CNT; the variation of interfacial bonding energy between CNT and polymer matrix; and possible thermal dissipation. Since the polymer was fixed during the pull-out process and the potential energy change of CNT was very small as confirmed in our computations, the variation of E in the present simulation can be mainly attributed to the variation of interfacial bonding energy between CNT and polymer matrix by neglecting thermal dissipation.

Taking an example of the above SWCNT(5,5) pull-out from PE matrix, the calculated energy increment ΔE versus pull-out displacement Δx is shown by pink balls in Fig.9b, where three successive stages can be discerned: in initial stage I, ΔE increases sharply; after that, ΔE goes through a long and platform stage II, followed by final descent stage III until the complete pull-out. As plotted, the total energy change during the pull-out (i.e., the pull-out energy) and the average energy increment in stage II are referred to as ΣΔE and ΔE_{II}, respectively. Moreover, stage I and stage III have the same approximate range of a=1.0nm which is very close to the cut-off distance of vdW interaction. This trend is surprisingly coincident with that observed sliding behaviour among nested walls in a MWCNT [Li et al., 2010]. The obvious severe fluctuation of energy increment may be attributed to the non-uniformity of polymer matrix in the length direction of CNT. Note that the pull-out energy may also be calculated by using the developed continuum theoretical model for short-fibre reinforced composites (e.g. [Fu & Lauke, 1997]), which plays important role in predicting the fracture toughness of the composites.

On this basis, the effects of CNTs' dimensions on interfacial properties of CNT/Polymer nanocomposites, were explored for the first time by comparing the energy increment of several CNTs with different nanotube length, diameter, or wall number.

3.1.4. Effects Of Cnts' Dimension

3.1.4.1. Effect Of Nanotube Length

The variation of energy increment corresponding to the pull-out of several SWCNTs(5,5) with different lengths, are plotted in Fig. 10. Just as that in Fig. 9b, three distinct stages are clearly observed for each case. Moreover, among these three curves, there is no obvious change in the magnitude of the platform stage II related to the stable CNT pull-out, which indicates that ΔE_{II} is independent of nanotube length.

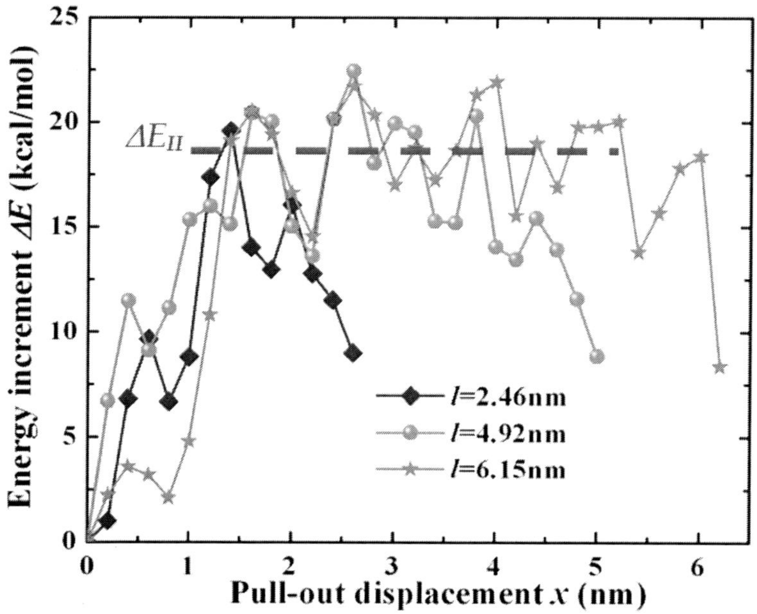

Figure 10. Effect of nanotube length on energy increment of SWCNT(5,5)

Due to this length-independent behaviour of ΔE_{II}, the length of l_0=3.44nm was employed for all CNTs in the following simulations.

3.1.4.2. Effect Of Nanotube Diameter

The energy increment ΔE corresponding to the pull-out of several SWCNTs with different diameters are illustrated in Fig. 11a, which increases with nanotube diameter. Both ΣΔE and ΔE_{II} were found to increase linearly with nanotube diameter D as shown in Fig. 11b, which can be fitted as follows:

$$\Sigma \Delta E = 277.79D + 62.39 \quad (1)$$

$$\Delta E_{II} = 19.29\, D + 4.27 \quad (2)$$

where D has the unit of nm, ΣΔE and ΔE_{II} have the unit of kcal/mol. This linear correlation can be explained by the increase of interfacial atoms accompanied with the increase of nanotube diameter. Note that the formula of Eq. (1) for calculating the pull-out energy is only applicable for the pull-out of CNT with the length of l_0=3.44nm, in contrast to the length-independent behaviour of ΔE_{II}. In view of that $\Sigma\Delta E$ increases with nanotube length, for a real SWCNT with length of l_r which is far longer than the present length l_0, $\Sigma\Delta E$ can be estimated as follows

$$\Sigma\Delta E^{*}=\Sigma\Delta E+\Delta E_{II}\ (l_r-l_0)/\Delta x\ (l_0<l_r) \tag{3}$$

where l_r is of the unit of nm, and the displacement increment here is Δx=0.2nm. Note that $\Sigma\Delta E$ represents the pull-out energy of SWCNT with length of l_0=3.44nm which can be calculated by Eq. (1).

In a word, the average increment in stage II (ΔE_{II}) during the pull-out process of the armchair SWCNT from polymer matrix, corresponding to the interfacial properties between CNT and polymer matrix, is independent of nanotube length, but proportional to nanotube diameter, which is similar to that for the sliding among nested walls in a MWCNT [Li et al., 2010]. Moreover, by using Eqs (1-3), the pull-out energy $\Sigma\Delta E$ and average energy increment in stage II $\Delta E_{II}$for the pull-out of any armchair SWCNT can be predicted.

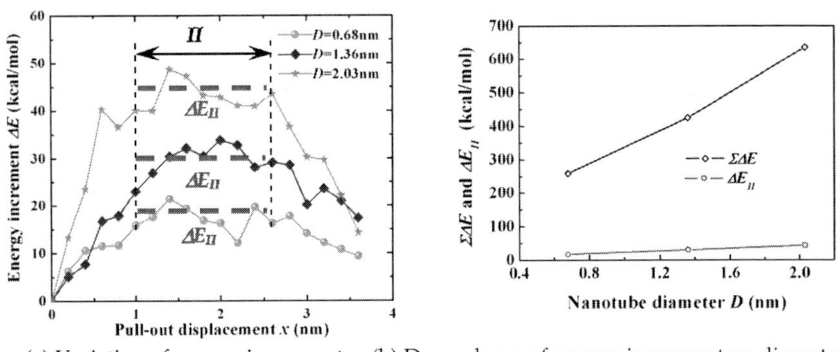

(a) Variation of energy increment (b) Dependence of energy increment on diameter

Figure 11. Effect of nanotube diameter on energy increment of SWCNT(5,5)

3.1.4.3. Effect Of Wall Number

To investigate the effect of wall number n, three MWCNTs with different wall number were embedded in PE matrix: SWCNT(15,15) with n=1, DWCNT(15,15)/(10,10) with n=2, and TWCNT(15,15)/(10,10)/(5,5) with n=3. The corresponding nanocomposites are referred to as: SWCNT/PE, DWCNT/PE,

TWCNT/PE, respectively. Note that the above three CNTs have the same outermost wall with diameter of D_0=2.03nm and length of l_0=3.44nm.

The corresponding energy increment ΔE are plotted in Fig. 12, whose average value in stage II (ΔE_{II}) are listed in Table 4. From this table, it can be found that the ΔE_{II} for DWCNT/PE is about 20% higher than that for SWCNT/PE. However, there is only a minor change of ΔE_{II} between DWCNT/PE and TWCNT/PE. The reason can be explained by the increase of distance between newly inserted inner walls and the interface. As the vdW interaction is mostly dependent on the distance, the longer the distance is, the weaker the induced vdW interaction is.

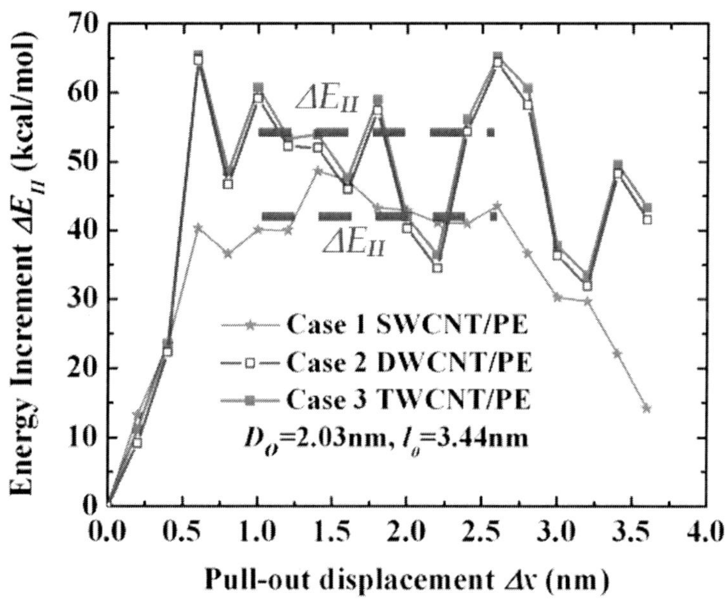

Figure 12. Effect of n on energy increment

Therefore, ΔE_{II} for the MWCNT (n≥2) pull-out can be approximately estimated as 1.2 times of that for corresponding SWCNT, which is actually the outermost wall of the MWCNT. To some extent, this finding is consistent with the reports of Schadler [Schadler et al., 1998] who concluded that only the outer walls are loaded in tension for CNT/Epoxy nanocomposites based on the observation of Raman spectrum.

Moreover, the calculated energy increment in the present simulation on the CNT pull-out from polymer matrix is compared with the reports [Li et al., 2010] on the pull-out of outermost wall in the same MWCNT as listed in Table 4. Obviously, the former is smaller than the latter. It may indicate that even for some CNTs with fractured outer walls in the CNT/PE nanocomposites, the CNT is easier to be pulled out from matrix instead of that the fractured outer walls are pulled out against the corresponding inner walls.

Table 4. Comparison of ΔE_{II} for two types of pull-out

	Model	ΔE_{II} (kcal/mol)
CNT Pull-out from PE	SWCNT/PE	43.07
	DWCNT/PE	51.13
	TWCNT/PE	52.71
Pull-out of the outermost wall	DWCNT	55.11
	TWCNT	59.32

3.1.5. Pull-Out Force

In practical CNT/Polymer nanocomposites, the real pull-out force can be contributed from the following factors [Bal & Samal, 2007; Wong et al., 2003]: vdW interaction between CNT and PE matrix, possible chemical bonding between CNT and PE matrix, mechanical interlocking resulted by local non-uniformity of nanocomposites, such as waviness of CNT, mismatch in coefficient of thermal expansion, statistical atomic defects, etc. Consequently, the pull-out force can be divided into two parts, i.e., $F=F_{vdW}+F_m$. Here, F_{vdW} is the component for overcoming the vdW interaction at the interface which can be calculated by the following Eq. (4); and F_m is the frictional sliding force caused by the other factors stated. The magnitudes of these two parts strongly depend on the interfacial state and CNT dimension. For almost perfect interface, F_{vdW} dominates the pull-out force. On the other hand, for the case of chemical bonding or mechanical interlocking, which in general occurs easily for large CNTs, F_m mainly contributes to the total pull-out force. In the present study, only F_{vdW} and the related ISS for perfect interface are considered as mentioned in the beforehand work.

According to that the work done by the pull-out force at each pull-out step is equal to the energy increment of nanocomposites, the corresponding pull-out force for the stable CNT pull-out stage should be also independent of nanotube length, but proportional to nanotube diameter, just as energy increment is.

From the obtained energy increment ΔE_{II} in Eq. (2) and pull-out displacement increment of Δx=0.2nm, we can get the pull-out force as follows:

$$F_{II} = \Delta E_{II} / \Delta x = \lambda(0.67D+0.15) \tag{4}$$

where F_{II} and D have the units of nN and nm, respectively. The value of λ represents the effect of wall number, which is 1.0 for SWCNT and 1.2 for MWCNT with consideration of the contribution of the inner walls.

3.1.6. Interfacial Shear Stress (Iss) And Surface Energy Density

Based on the above discussions, the corresponding ISS and surface energy density are analyzed in the following.

The pull-out force is equilibrated with the axial component of vdW interaction which induces the ISS. Conventionally, if we employ the common assumption of constant ISS with uniform distribution along the embedded CNT, the pull-out force F_{II} will vary with the embedded length of CNT, which is obviously contradict with the above length-independent reports of ΔE_{II}. For the extreme case of CNT with infinite length, the ISS tends to be zero, which is physically unreasonable. This indicates that the conventional assumption of ISS is improper for the perfect interface of CNT/PE nanocomposites. Therefore, in view of the above characteristic of the variation of energy increment ΔE_{II}, such as that the range of stage I or stage III is around 1.0nm, it is concluded that the ISS is distributed solely at each end of embedded CNT within the range of a=1.0nm at the beginning of stage II or the end of stage I. In view of the dependence of vdW force upon the distance between two atoms, the ISS at each end of the embedded CNT which is induced by the variation of vdW force, should at first increase sharply and then decrease slowly to zero after reaching the maximum. Here, by assuming its uniform distribution within two end regions for simplicity, the effective ISS can be derived as:

$$\tau_{II} = F_{II} / (2\pi D a) \tag{5}$$

By using Eqs. (1-5), the pull-out energy $\Sigma\Delta E$, the average pull-out force F_{II} and the ISS τ_{II} for CNT/PE nanocomposites can be calculated. As shown in Fig. 13a, the calculated ISS is found to decrease initially with nanotube diameter and saturate at the value of 106.7MPa for SWCNT/PE nanocomposites. For MWCNT/PE, the saturated value is 128MPa, which is about 1.2 times of that for SWCNT/PE, both of which has the same outermost wall.

On the other hand, in view of that two new surface regions are generated at the two ends of CNT after each pull-out step, the corresponding surface energy should be equal to the energy increment ΔE_{II} by neglecting thermal dissipation. Therefore, the surface energy density can be calculated as:

$$\gamma_{II} = \Delta E_{II}/(2\pi D\Delta x) = F_{II}/(2\pi D) \tag{6}$$

As shown in Fig. 13b, the surface energy density has the same trend as the ISS, which initially decreases slightly as nanotube diameter increases and finally converges at the value of 0.11N/m. This value is very close to the previous reports of 6-8meV/Å^2 (i.e., 0.09-0.12N/m) [Lordi & Yao, 2000] and 0.15 kcal/molÅ^2 (i.e., 0.1 N/m) [Al-Ostaz et al., 2008] for SWCNT/PE nanocomposites, which indicates the effectiveness of the present simulation.

3.1.7. Comparison With Previous Reports

The predicted ISS in the present study as given in Table 5 is obviously higher than that in the previous reports [Frankland et al., 2002; Zheng et al., 2009; Al-Ostaz et al.] which is calculated from

$$\tau^* = 2\Sigma\Delta E/(\pi D l^2) \tag{7}$$

The reason of this big difference is that the assumption of the constant ISS with uniform distribution along the total embedded length of CNT employed in previous numerical simulation, which has been verified to be unreasonable here.

For the pull-out force, the calculated values using the proposed formulae of Eqs.(1-4) are compared with that reported in direct CNT pull-out experiments at nano-scale from several different polymer matrices, as shown in Table 6. Obviously, the reported pull-out forces are much higher than the calculated values only when vdW interactions are considered at the interface between CNT and polymer matrix, although severe experimental data scattering has been observed which may be caused by manipulation process or force/displacement measurement. The reason can be attributed to the following factors.

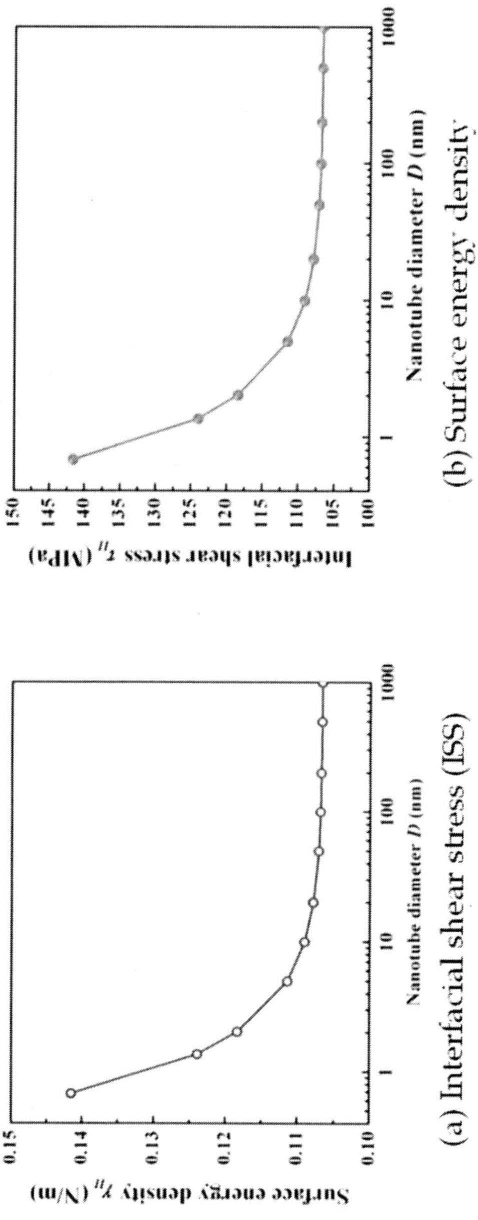

Figure 13. Dependence of ISS and surface energy density on nanotube diameter

Table 5. Prediction in previous simulations for SWCNT/PE nanocomposites

Ref.	SWCNT		Experimental reports		Present prediction	
	D (nm)	l (nm)	$\Sigma\Delta E^*$ (kcal/mol)	τ^* (MPa)	$\Sigma\Delta E^*$ (kcal/mol)	τ_{II} (MPa)
[Frankland et al., 2002]	1.36	5.3	×	2.7	707.4	123.9
[Zheng et al., 2009]	1.36	5.9	~500	33	798.8	123.9
[Al-Ostaz et al., 2008]	0.78	4.2	224	133	352.2	137

Firstly, if we take into account the curvature of CNT (Fig. 14a), the necessary pull-out force will increase. Considering a special case with the highest probability where an inclined angle θ between the axial direction of CNT and the pull-out direction is 45º(Fig. 14b), the corresponding pull-out force will increase about 40%. By observing the data in Table 6, this increase effect is still too small compared with the experimental data.

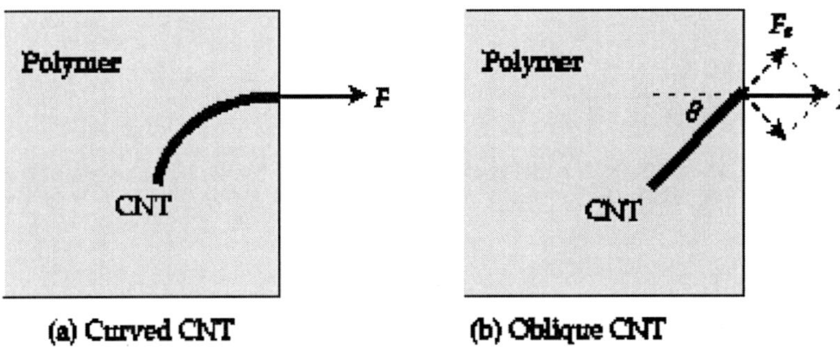

Figure 14. Simplified model for the pull-out of curved and oblique CNT

Table 6. Prediction of previous experiments for MWCNT/Polymer nanocomposites

Ref.	Matrix	MWCNT	Pull-out force (μN)	
		D (nm)	Exp.	Prediction
[Deng, 2008]	PEEK	49	1.04	0.04
		89	8.6	0.07
[Barber et al., 2003]	PE-butene	80	0.85	0.07
[Cooper et al., 2002]	Epoxy	8.2	3.8	0.01
		11	2.8	0.01
		13.4	0.6/2.3	0.01
		24	6.8/12.8	0.2

Secondly, for the effect of the pressure or residual stress in nanocomposites, a simple representative volume element (RVE) model (Fig. 15a) was constructed to perform FEM analysis. This continuum mechanics based computation is valid, at least qualitatively when the diameter of CNT is over several tens of nanometers. Assuming the symmetrical structure, the quarter part of interface region (i.e. the blue part in Fig. 15b) was employed.

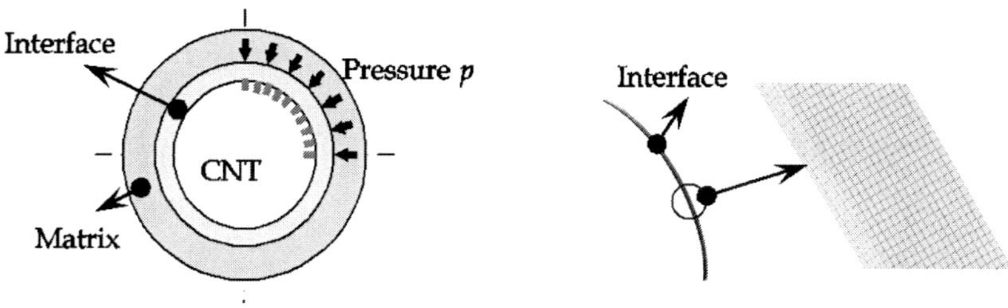

(a) Simplified composite model (b) FEM model of interface

Figure 15. Simplified FEM model of interface between CNT and matrix

The inner wall surface of the interface region was fixed as boundary condition, which represents the rigid CNT. The uniform static pressure was applied on the outer wall of interface region. Two values of Young's modulus E_i for the interface region were considered, i.e., 1GPa and 3GPa. The corresponding FEM model is shown in Fig. 15b, where the size of element size was taken as 0.05nm for convergence and accuracy.

The relationship of strain energy density γ and the applied pressure p is shown in Fig. 16, in which strain energy increases by the power of square with p. It can be found that a large pressure p can only cause very small increase of strain energy density γ, which indicates that the effect of pressure or residual stress is not as strong as we expected. In general, the residual stress in polymer nanocomposites ranges from 25.0MPa to 40.0MPa. For instance, for the case of E_i=1GPa of interface region, by applying for the pressure of 30.0MPa, the strain energy density is around $2.85\subseteq10^{-5}$N/m, which is still much smaller than the surface energy density of γ_{II}=0.11N/m caused by vdW interaction as stated previously. It means that the pressure or residual stress in polymer nanocomposite system is not a dominant factor.

The above discussion leads to an important conclusion, i.e., the interface properties between CNT and polymer matrix contributed by vdW interaction is quite minor for the real CNT pull-out from polymer matrix. Therefore, to accurately evaluate the interfacial properties for real case, it is necessary to incorporate the effects of frictional sliding caused by mechanical interlocking, atomic statistical defects, or chemical bonding. Moreover, for effectively improving the interfacial properties and therefore the mechanical properties of bulk nanocomposites, it is vital to incorporate into chemical bonding and mechanical interlocking, which will be the topic in the future.

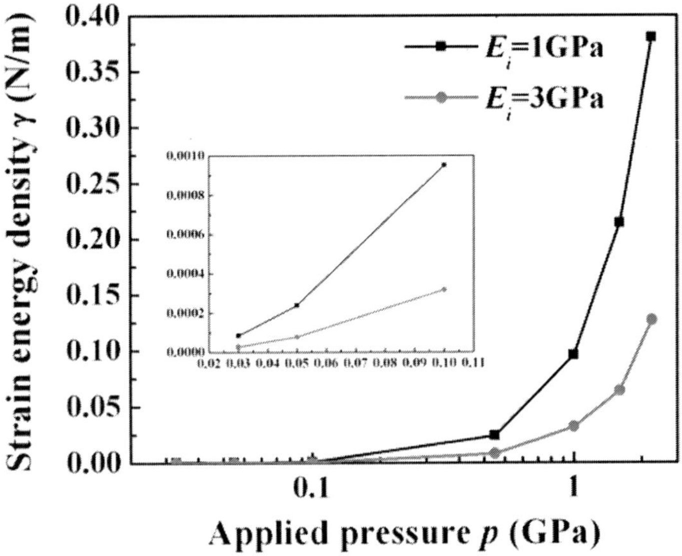

Figure 16. Relationship between applied pressure and strain energy density

Note that although the type of polymer will influence the value of ISS which is not discussed in the present work, the characteristics of pull-out force and the corresponding ISS will be similar to those discussed in CNT/PE nanocomposites here.

3.2. Characterization On Overall Mechanical Properties Of Cnt/Polymer Nanocomposites

To investigate the overall mechanical properties of CNT/Polymer nanocomposites, tensile tests and SENB tests were carried out, in which MWCNT-7 and VGCF® were used as reinforcement filler, respectively. Moreover, a sequential multi-scale model was developed by incorporating the interfacial properties between CNT and polymer matrix obtained from previous pull-out simulations into the conventional continuum theory, which connects the interfacial properties to overall mechanical properties of CNT/Polymer nanocomposites.

3.2.1. Mechanical Tests For Cnt/Epoxy Nanocomposites

Epoxy resin jER806 (Japan Epoxy Resins Co., Ltd., Japan) and the hardener Tohmide-245LP (Fuji Kasei Kogyou Co., Ltd., Japan) were used to prepare the polymer matrix with the weight ration of 100:62. According to the weight fraction W_f of reinforcements, two groups of nanocomposites were fabricated: MWCNT/Epoxy with W_f of MWCNT-7 varying at 2, 3, 4%, and VGCF/Epoxy with W_f of VGCF® varying at 2, 4, 6%. The fabrication process as shown in Fig. 17 is described below:

1 The epoxy resin was first heated to 60ºC in an oven to decrease the viscosity for its better miscibility with MWCNTs;

2 Then the CNTs were dispersed into the epoxy resin using the planetary centrifugal mixer at 2000rpm for 10min;

3 After adding the hardener, the mixture was agitated for another 10 min at 1000rpm;

4 Subsequently, the mixture was poured into the silicon mould for a pre-curing process with 12h at room temperature, followed by a post-curing process which was performed at 80º C in the oven for 6h.

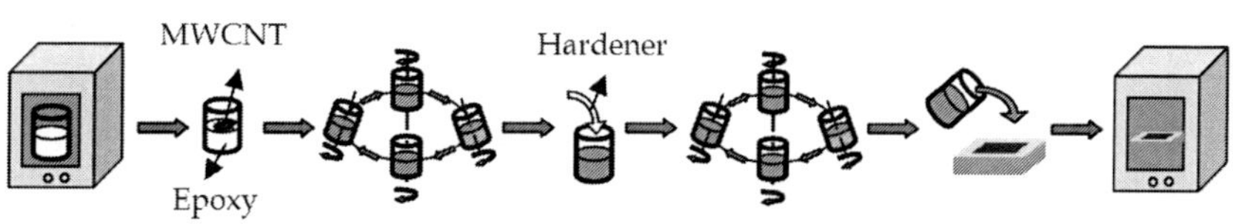

Figure 17. Fabrication of MWCNT/Epoxy nanocomposites for mechanical tests

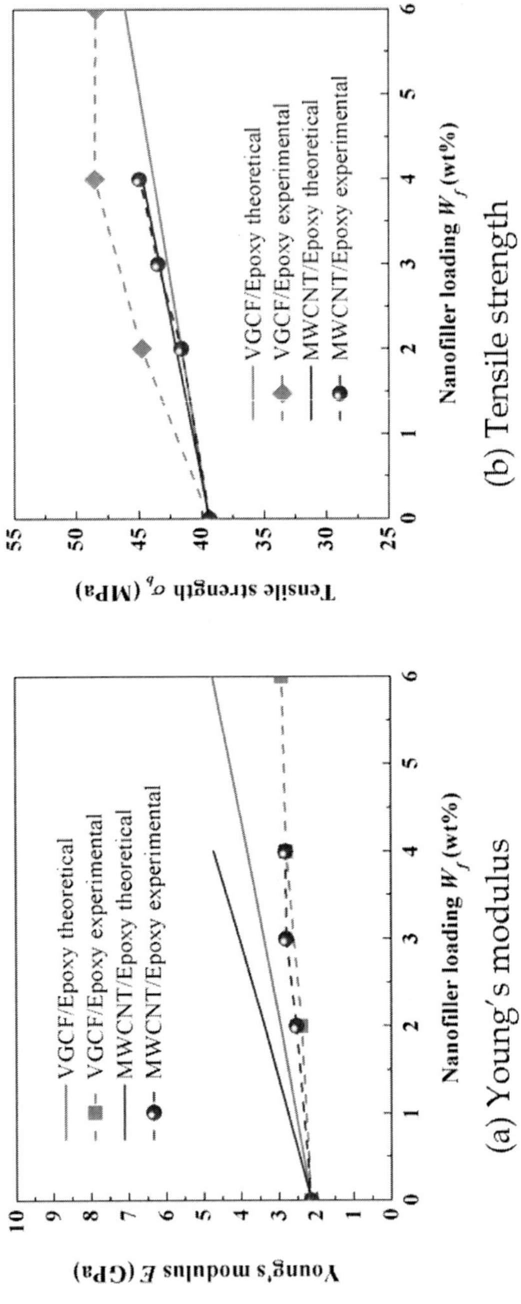

Figure 18. Comparison of theoretical and experimental Young's modulus and tensile strength

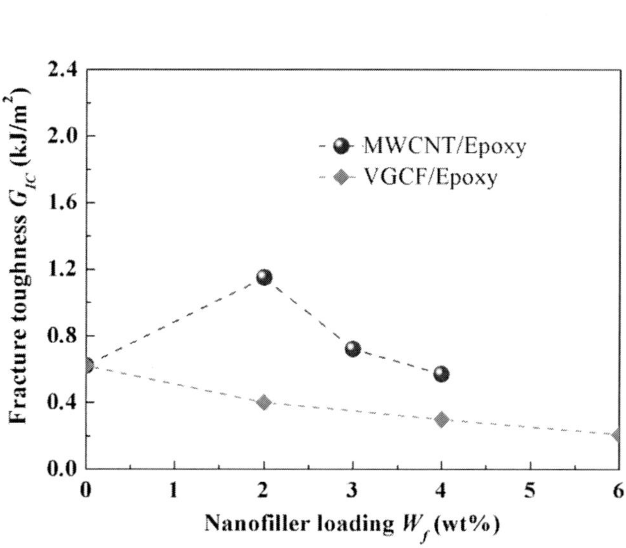

Figure 19. Experimental fracture toughness

According to ASTM D638 (Type V) and ASTM D5045 standards, tensile tests and SENB tests were preformed. Three specimens of the above fabricated composites at a specified CNT loading were prepared. A universal materials testing machine (Instron 5567) was used with a cross-head speed of 0.25mm/min. The Poisson's ratio of nanocomposites measured in tensile tests by using bi-axial strain gages was directly used in the evaluation of fracture toughness in SENB tests.

The obtained Young's modulus E_c, tensile strength σ_c of MWCNT/epoxy and VGCF/Epoxy nanocomposites are plotted in Fig. 18. As shown in Fig. 18a, the largest increase of 31% and 36% in E_c from 2.14GPa of neat epoxy occur at 4wt% loading for MWCNT/Epoxy and at 6wt% loading for VGCF/Epoxy, respectively. In Fig. 18b, σ_c increases obviously for nanocomposites. At 4wt% loading, the highest increases are 14.1% for MWCNT/Epoxy and 23.1% for VGCF/Epoxy from 39.4MPa of neat epoxy. The effect of VGCF in σ_c is slightly better than that of MWCNT. Moreover, σ_c decreases slightly when the addition is over 4wt% for VGCF/Epoxy. However, there is no obvious saturation trend in σ_c even at 4wt% loading of MWCNT, which means that σ_c may be enhanced with further addition of MWCNT. The corresponding Mode-I fracture toughness G_{IC} is given in Fig. 19, a remarkable increase, i.e. 85.5% can be identified for 2wt% MWCNT loading, although the effect of VGCF to G_{IC} is unpromising.

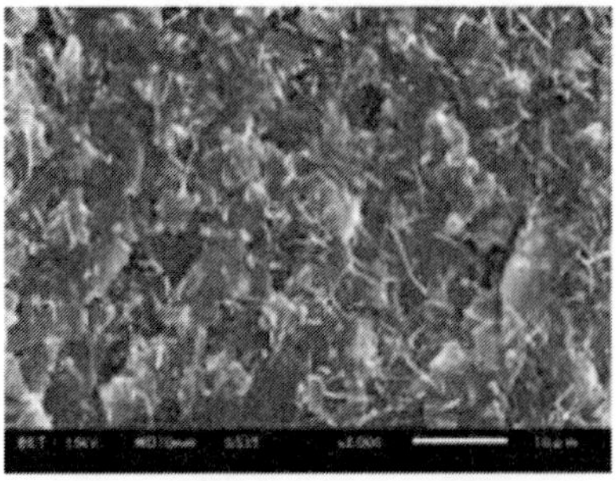

(a) MWCNT/Epoxy (3wt%)

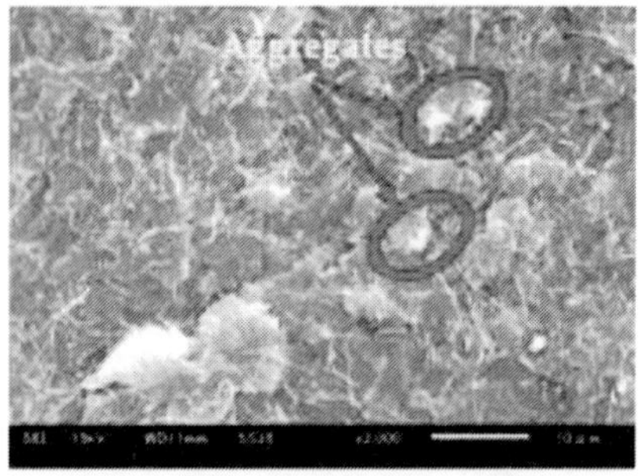

(b) MWCNT/Epoxy (4wt%)

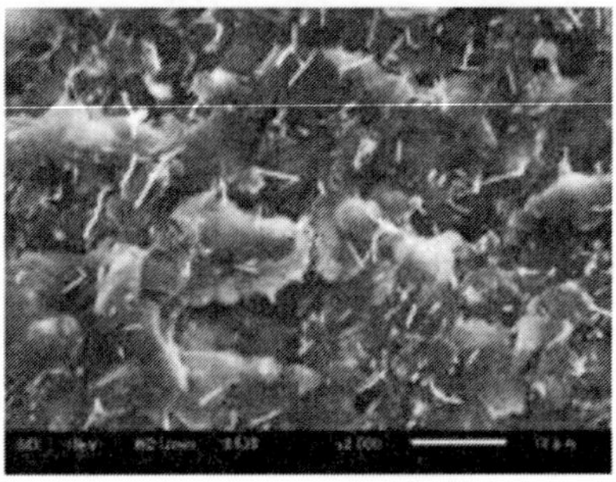

(c) VGCF/Epoxy (4wt%)

(d) VGCF/Epoxy (6wt%)

(e) MWCNT/Epoxy (2wt%)

Figure 20. SEM pictures of SENB specimens

In order to clarify the reinforcement mechanism, SEM observations were performed on the fracture surfaces of SENB specimens as shown in Fig. 20. Even there are some small aggregates at a high loading, e.g. 4wt% in Fig. 20(b), good dispersion of MWCNT or VGCF can be identified, which leads to the increase of E and σ_b in both nanocomposites. As shown in Fig. 19, 6wt% VGCF leads to the lowest G_{IC} while 2wt% MWCNT provides the highest G_{IC}. However, it is difficult to distinguish them through the fracture surfaces from Fig. 20 (d) and (e), although the former demonstrates a clear brittle fracture feature. One possible reason for the increase of G_{IC} in 2wt% MWCNT may be there are more pull-out holes of MWCNT, which improves the fracture toughness.

3.2.2. Theoretical Prediction On Tensile Properties Of Mwcnt/Polymer Nanocomposites

With consideration of fibre length and fibre orientation distribution, Cox's shear-lag model [Cox, 1952; Krenchel, 1964] predicts the longitudinal modulus of short-fiber reinforced composites as

$$E=\eta_o \eta_l E_f V_f + E_m(1-V_f) \tag{8}$$

where V_f is the volume fraction of fiber, E_f and E_m are Young's moduli of nanofiller and matrix, respectively. The orientation efficiency factor η_o is 1/5 for three-dimensional random distribution, and the nanofiller length efficiency factor η_l was given in Ref. [Cox, 1952]. The predicted Young's modulus is also plotted in Fig. 18(a), which is a bit higher than experimental ones, especially for higher nanofiller loadings, which may be caused by the difficult dispersion of nanofiller in experiments.

For short-fiber reinforced composites under the assumption of iso-strain state in the fibers and matrix,Fukuda and Chou [Fukuda & Chou, 1982] developed a probabilistic theory as follows

$$\sigma_{ultc} = \begin{cases} \sigma_{ultf} V_f \left(1-\dfrac{l_c}{2l_f}\right) C_0 + \sigma_m{}'(1-V_f) & (l > l_c) \\ \sigma_{ultf} V_f \left(\dfrac{l_f}{2l_c}\right) C_0 + \sigma_m{}'(1-V_f) & (l \le l_c) \end{cases} \tag{9}$$

where l_f is the length of fiber, σ_{ultf} the tensile strength of fiber, σ'mσm' the matrix stress at the failure of composites, and C_0 the orientation factor which is 1/8 in the case of three-dimensional random array model. The critical length l_c is defined as [Kelly & Tyson, 1965]

$$l_c=\sigma_{ultl} r_f V_f / \tau_y \tag{10}$$

where r_f is the fiber radius, and τ_y is the shear strength of fiber/matrix interface.

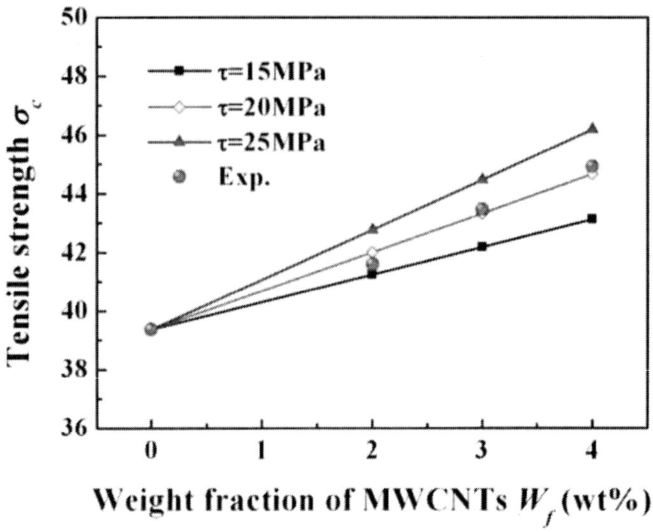

Figure 21. Prediction of σ_c for MWCNT/Epoxy

Here, the key point is how to define the shear strength τ_y by using the previous MM simulation results. With consideration of its distribution which is only at each end of embedded CNT within 1.0nm, the shear strength should be around 106.7MPa for SWCNT, and 128.04MPa for MWCNT which is about 1.2 times of that for SWCNT. On the other hand, if we define the shear strength based on the whole embedded length of CNT (e.g., 5μm for present experiments), the corresponding shear strength can be calculated as 0.09MPa for SWCNT, and 0.1MPa for MWCNT. Naturally, these values defined from the conventional conception of shear strength seem to be very small from common sense. It further confirms the previous analysis, which indicates that the contribution from vdW interaction to the pull-out force may be comparatively small in practical nanocomposites compared with that from frictional sliding caused by possible chemical bonding or mechanical interlocking. From the above statements, no matter what kind of definition for the shear strength is used, the interfacial shear strength can be assumed to range from 0.1MPa to 128.04MPa for MWCNT. Naturally, by considering the obliqueness of CNT (Fig.14) to the direction of pull-out force, this shear strength range can modified as 0.14~179.26MPa. With consideration of the effects of residual stress in polymer systems, and especially possible chemical bonding or mechanical interlocking, the above range may be increased significantly. Taking the possible range of 0.5~200MPa, the corresponding critical length is about 9750 ~24.38μm, all of which are much longer than that of the MWCNTs used, i.e. $l_f \ll l_c$. It means that the reinforcement effect predicted theoretical in Eq. (7) is unpromising. Replacing σ'mσm' by the measured tensile strength of neat epoxy σ_m, the tensile strength of MWCNT/Epoxy nanocomposites can be predicted. As shown in Fig. 21, the interfacial shear strength of 15~25MPa provides good consistence between experimental measurements and theoretical prediction. Therefore, the previously stated wide range for ISS can rightly cover this narrow band, which indicates that the present multi-scale method is meaningful. Moreover, if we employ the conventional definition of interfacial shear strength which leads to a value lower than 1MPa, it is obviously lower than 15~25MPa. This indicates that the contribution from frictional sliding dominates the interfacial shear strength, which is at least 10 times higher than that from vdW interaction. Taking the value of 20MPa for interfacial shear strength, the predicted value is plotted in Fig. 18b, which indicates good consistence with experimental results.

4. CONCLUSION

The above research results clearly clarify the different reinforcement effects of two kinds of CNTs (i.e., MWCNT-7 and VGCF®) in two types of nanocomposites (i.e., three-phase hybrid CFRP laminates and two-phase CNT/Polymer nanocomposites).

For three-phase hybrid CFRP laminates, a new simple fabrication method powder method was developed to disperse CNTs at the interface between CFRP plies. DCB experiments and FEM analysis were carried out which verifies the significant improvement of interlaminar mechanical properties.

For two-phase CNT/Polymer nanocomposites, the interfacial properties between CNT and polymer matrix was clarified, where the effects of nanotube length, diameter and wall number were explored for the first time. Moreover, the obtained interfacial mechanical properties were incorporated into the continuum mechanics to develop a sequential multi-scale model, which predicts the overall mechanical properties of nanocomposites and indicates the relationship between the interfacial properties and overall mechanical properties.

ACKNOWLEDGEMENTS

This work is partly supported by two Grand-in-Aids for Scientific Research (No. 19360045 and No. 22360044) from the Japanese Ministry of Education, Culture, Sports, Science and Technology. The authors acknowledge Prof. C. B.

Fan (Beijing Institute of Technology, China) for kindly providing the computational resources.

REFERENCES

1. P. M. Ajayan, O. Stephan, C. Colliex, D. Trauth, 1994 Aligned Carbon Nanotube Arrays Formed by Cutting a Polymer Resin-nanotube Composite. Science, 265 1212

2. Al-Ostaz, G. Pal, P. R. Mantena, A. Cheng, 2008 Molecular Dynamics Simulation of SWCNT-Polymer Nanocomposite and its Constituents. Journal of Materials Science, 43 164 173 .

3. R. Andrews, D. Jacques, M. Minot, T. Rantell, 2002 Fabrication of Carbon Multiwall Nanotube/Polymer Composites by Shear Mixing. Macromolecular Materials and Engineering, 287 395 403 .

4. M. Arai, Y. Noro, K. Sugimoto, M. Endo, 2008 Mode I and Mode II Interlaminar Fracture Toughness of CFRP Laminates Toughened by Carbon Nanofiber Interlayer. Composites Science and Technology, 68 516 525 .

5. S. Bal, S. S. Samal, 2007 Carbon Nanotube Reinforced Polymer Composites-A State of Art. Bulletin of Materials Science, 30 4 379 386 .

6. D. S. Bangarusampath, H. Ruckdaschel, V. Altstadat, J. K. W. Sandler, D. Garray, M. S. P. Shaffer, 2009 Rheology and Properties of Melt-Processed Poly (ether ether ketone)/ Multi-Wall Carbon Nanotube Composites. Polymer, 50 5803 5811 .

7. H. Barber, S. R. Cohen, S. Kenig, H. D. Wagner, 2003 Measurement of Carbon Nanotube-Polymer Interfacial Strength. Applied Physics Letters, 82 23 4140 4142 .

8. H. Barber, S. R. Cohen, S. Kenig, H. D. Wagner, 2004 Interfacial Fracture Energy Measurements for Multi-walled Carbon Nanotubes Pulled from a Polymer Matrix. Composite Science and Technology, 64 2283 2289 .

9. R. Bhattacharyya, T. V. Sreekumar, T. Liu, S. Kumar, L. M. Ericson, R. H. Hauge, R. E. Smalley, 2003 Crystallization and Orientation Studies in Polypropylene/Single Wall Carbon Nanotube Composite. Polymer, 44 2373 2377 .

10. P. P. Camanho, C. G. Davila, 2002 Mixed-mode Decohesion Finite Elements for the Simulation of Delamination in Composite Materials. NASA/TM, 211737

11. Y. Cao, N. Hu, J. Lu, H. Fukunaga, Z. Yao, 2002 A 3D Brick Element based on Hu-Washizu Variational Principle for Mesh Distortion. International Journal for Numerical Methods in Engineering, 53 2529 2548 .

12. S. C. Chowdhury, T. Okabe, 2007 Computer Simulation of Carbon Nanotube Pull-out from Polymer by the Molecular Dynamics Method. Composites Part A, 38 747 754 .

13. T. E. Chang, L. R. Jensen, A. Kisliuk, R. B. Pipes, R. Pyrz, A. P. Sokolov, 2005 Microscopic Mechanism of Reinforcement in Single-Wall Carbon Nanotube/Polypropylene Nanocomposite. Polymer, 46 439 444 .

14. T. E. Chang, A. Kisliuk, S. M. Rhodes, W. J. Brittain, A. P. Sokolov, 2006 Conductivity and Mechanical Properties of Well-Dispersed Single-Wall Carbon Nanotube/ Polystyrene Composite. Polymer, 47 7740 7746 .

15. L. Ci, J. Bai, 2006 The Reinforcement Role of Carbon Nanotues in Epoxy Composites with Different Matrix Stiffness. Composite Science and Technology, 66 599 603 .

16. A. Cooper, D. Ravich, D. Lips, J. Mayer, H. D. Wagner, 2002 Distribution and Alignment of Carbon Nanotubes and Nanofibrils in a Polymer Matrix. Composite Science and Technology, 62 1105 1112 .

17. A. Cooper, S. R. Cohen, A. H. Barber, H. D. Wagner, 2002 Detachment of Nanotubes from a Polymer Matrix. Applied Physics Letters, 81 20 3873 3875 .

18. H. L. Cox, 1952 The Elasticity and Strength of Paper and Other fibrous Materials. British Journal of Applied Physics, 3 72 79 .

19. S. Curran, A. P. Davey, J. Coleman, A. Dalton, B. Mc Carthy, S. Maier, A. Drury, D. Gray, M. Brennan, K. Ryder, M. L. Chapelle, C. Journet, P. Bernier, H. J. Byrne, D. Carroll, P. M. Ajayan, S. Lefrant, W. Blau, 1999 Evolution and Evalution of the Polymer/Nanotube Composite. Synthetic Metals, 103 2559 2562 .

20. G. Demczyk, Y. M. Wang, J. Cumings, M. Hetman, W. Han, A. Zettl, R. O. Ritchie, 2002 Direct Mechanical Measurement of the Tensile Strength and Elastic Modulus of Multiwalled Carbon Nanotubes. Materials Science and Engineering A, 334 173 178 .

21. F. Deng, T. Ogasawara, N. Takeda, 2007 Experimental Characterization of Poly (Ether Ether Ketone)/Multi-Wall Carbon Nanotube Composites. Key Engineering Materials, 334-335 , 721 724 .

22. F. Deng, 2008 Investigation of the Interfacial Bonding and Deformation Mechanism of the Nano Composites Containing Carbon Nanotubes. Tokyo University. PhD Dissertation.

23. W. Ding, A. Eitan, F. T. Fisher, X. Chen, D. A. Dikin, R. Andrews, L. C. Brinson, L. S. Schadler, R. S. Ruoff, 2003 Direct Observation of Polymer Sheathing in Carbon Nanotube-Polycarbonate Composites. Nano Letters, 3 11 1593 1597 .

24. S. J. V. Frankland, A. Caglar, D. W. Brenner, M. Griebel, 2002 Molecular Simulation of the Influence of Chemical Cross-links on the Shear Strength of Carbon Nanotube-Polymer Interfaces. Journal of Physical Chemistry B, 106 3046 3048 .

25. S. Y. Fu, B. Lauke, 1997 The Fiber Pull-out Energy of Misaligned Short Fibre Composites. Journal of Materials Science, 32 1985 1993

26. X. L. Gao, K. Li, 2005 A Shear-lag Model for Carbon Nanotube-Reinforced Polymer Composites. International Journal of Solids and Structures, 42 1649 1667 .

27. J. Garcia, B. L. Wardle, A. J. Hart, 2008 Joining Prepreg Composite Interfaces with Aligned Carbon Nanotubes. Composites Part A, 39 1065 1070 .

28. H. Gojny, J. Nastalczyk, Z. Roslaniec, K. Schulte, 2003 Surface Modified Multi-Walled Carbon Nanotubes in CNT/Epoxy-Composites. Chemical Physics Letters, 370 820 824 .

29. H. Gojny, M. H. G. Wichmann, B. Fiedler, K. Schulte, 2005 Influence of Different Carbon Nanotubes on the Mechanical Properties of Epoxy Matrix Composites- A Comparative Study. Composite Science and Technology, 65 2300 2313 .

30. H. Gojny, M. H. G. Wichmann, U. Kopke, B. Fiedler, K. Schulte, 2004 Carbon Nanotube-Reinforced Epoxy-Composites: Enhanced Stiffness and Fracture Toughness at Low Nanotube Content. Composite Science and Technology, 64 2363 2371 .

31. J. Gou, B. Minaie, B. Wang, Z. Liang, C. Zhang, 2004 Computational and Experimental Study of Interfacial Bonding of Single-walled Nanotube Reinforced Composites. Computational Materials Science, 31 225 236 .

32. R. Haggenmueller, H. H. Gommans, A. G. Rinzler, J. E. Fischer, K. I. Winey, 2000 Aligned Single-wall Carbon Nanotubes in Composites by Melt Processing methods. Chemical Physics Letters, 330 10 219 225 .

33. Y. Han, J. Elliott, 2007 Molecular Dynamics Simulations of the Elastic Properties of Polymer/Carbon Nanotube Composites. Computational Materials Science, 39 2 315 323 .

34. X. He, F. Zhang, R. Wang, W. Liu, 2007 Preparation of a Carbon Nanotube/Carbon Fiber Multi-Scale Reinforcement by Grafting Multi-walled Carbon Nanotubes onto the Fibers. Carbon, 45 2559 2563 .

35. F. Inam, D. W. Y. Wong, M. Kuwata, T. Peijs, 2010 Multiscale Hybrid Micro-Nanocomposites based on Carbon Nanotubes and Carbon Fibers. Journal of Nanomaterials. (In press)

36. X. Jiang, Y. Bin, M. Matsuo, 2005 Electrical and Mechanical Properties of Polyimide-Carbon Nanotubes Composites Fabricated by In Situ Polymerization. Polymer, 46 7418 7424 .

37. L. Y. Jiang, Y. Huang, H. Jiang, G. Ravichandran, H. Gao, K. C. Hwang, B. Liu, 2006 A Cohesive Law for Carbon Nanotube/Polymer Interfaces based on the van der Waals Force. Journal of the Mechanics and Physics of Solids, 54 2436 2452 .

38. L. Jin, C. Bower, O. Zhou, 1998 Alignment of Carbon Nanotube in a Polymer Matrix by Mechanical Stretching. Applied Physics Letters, 73 9 1197 1193 .

39. P. Karapappas, A. Vavouliotis, P. Tsotra, V. Kostopoulos, 2009 Enhanced Fracture Properties of Carbon Reinforced Composites by the Addition of Multi-Wall Carbon Nanotubes. Journal of Composite Materials, 43 9 977 985 .

40. Kelly, W. R. Tyson, 1965 Tensile Properties of Fiber-reinforced Metals: Copper/Tungsten and Copper/Molybdenum. Journal of the Mechanics and Physics of Solids, 13 329 350 .

41. K. L. Kepple, G. P. Sanborn, P. A. Lacasse, K. M. Gruenberg, W. J. Ready, 2008 Improved Fracture Toughness of Carbon Fiber Composite Functionalized with Multi Walled Carbon Nanotubes. Carbon, 46 2026 2033 .

42. Fukuda, T. W. Chou, 1982 A Probabilistic Theory of the Strength of Short-fiber Composites with Variable Fiber Length and Orientation. Journal of Materials Science, 17 1003

43. T. Kimura, H. Ago, M. Tobita, S. Ohshima, M. Kyotani, M. Yumura, 2003 Polymer Composites of Carbon Nanotubes Aligned by a Magnetic Field. Advanced Materials, 14 19 1380 1383 .

44. Krenchel, 1964 Fibre Reinforcement. Akademisk Forlag. Copenhagen.

45. Lau, 2003 Interfacial Bonding Characteristics of Nanotube/Polymer Composites. Chemical Physics Letters, 370 399 405 .

46. W. Lee, S. Lee, C. Kim, 2006 The Mechanical Properties of MWNT/PMMA Nanocomposites Fabricated by Modified Injection Molding. Composite Structures, 76 406 410 .

47. Y. Li, N. Hu, G. Yamamoto, Z. Wang, T. Hashida, H. Asanuma, C. Dong, T. Okabe, M. Arai, H. Fukunaga, 2010 Molecular Mechanics Simulation of the Sliding Behavior between Nested Walls in a Multi-walled Carbon Nanotube. Carbon, 48 2934 2940 .

48. Liao, S. Li, 2001 Interfacial Characteristics of a Carbon Nanotube-Polystyrene Composite System. Applied Physics Letters, 79 25 4225 4227 .

49. V. Lordi, N. Yao, 2000 Molecular Mechanics of Binding in Carbon-Nanotube-Polymer Composites. Journal of Materials Research, 15 12 2770 2779 .

50. O. Marietta-Tondin, 2006 Molecular Modelling of Nanotube Composite Materials: Interface Formation, Interfacial Strength, and Thermal Expansion. The Florida State University. PhD Dissertation.

51. T. Natsuki, F. Wang, Q. Q. Ni, M. Endo, 2007 Interfacial Stress Transfer of Fiber Pullout for Carbon Nanotubes with a Composite Coating. Journal of Materials Science, 42 4191 4196 .

52. T. Ogasawara, Y. Ishida, T. Ishikawa, R. Yokota, 2004 Characterization of Multi-walled Carbon Nanotube/Phenylethynyl Termainated Polyimide Composites. Composite: Part A, 35 67 74 .

53. Qian, E. C. Dickey, R. Andrews, T. Rantell, 2000 Load Transfer and Deformation Mechanisms in Carbon Nanotube-Polystyrene Composites. Applied Physics Letters, 76 20 2868 2870 .

54. D. Rosca, S. V. Hoa, 2009 Highly Conductive Multiwall Carbon Nanotube and Epoxy Composites Produced by Three-roll Milling. Carbon, 47 1958 1968 .

55. B. Safadi, R. Andrews, E. A. Grulke, 2002 Multiwalled Carbon Nanotube Polymer Composites: Synthesis and Characterization of Thin Films. Journal of Applied Polymer Science, 84 2660 2669 .

56. S. Schadler, S. C. Giannaris, P. M. Ajayan, 1998 Load Transfer in Carbon Nanotube Epoxy Composites. Applied Physics Letters, 73 26 3842 3844 .

57. R. Sen, B. Zhao, D. Perea, M. E. Itkis, H. Hu, J. Love, E. Bekyarova, R. C. Haddon, 2004 Preparation of Single-walled Carbon Nanotube Reinforced Polystyrene and Polyurethane Nanofibers and Membranes by Electrospinning. Nano Letters, 4 3 459 464 .

58. H. Sun, 1998 COMPASS: An ab Initio Force-Field Optimized for Condensed-Phase Applications- Overview with Details on Alkane and Benzene Compounds. Journal of Physical Chemistry B, 102 38 7338 7364 .

59. Tai, M. Yeh, J. Liu, 2004 Enhancement of the Mechanical Properties of Carbon Nanotube/Phenolic Composites using a Carbon Nanotube Network as the Reiforcement. Carbon, 42 2735 2777 .

60. T. Thostenson, W. Z. Li, D. Z. Wang, Z. F. Ren, T. W. Chou, 2002 Carbon Nanotube/Carbon Fiber Hybrid Multiscale Composites. Journal of Applied Physics, 91 9 6034 6037 .

61. T. Thostenson, T. W. Chou, 2002 Aligned Multi-walled Carbon Nanotube-Reinforced Composites: Processing and Mechanical Characterization. Journal of Physics D, 35 L77 L80 .

62. Tsai, T. Lu, 2009 Investigating the Load Transfer Efficiency in Carbon Nanotubes Reinforced Nanocomposites. Composite Structures, 90 172 179.

63. V. P. Veedu, A. Cao, X. Li, K. , C. Soldano, S. Kar, P. M. Ajayan, M. N. Ghasemi-Nejhad, 2006 Multifunctional Composites using Reinforced Laminae with Carbon-nanotube Forests. Nature Materials, 5 457 462 .

64. B. Vigolo, A. Penicaud, C. Coulon, C. Sauder, R. Pailler, C. Journet, P. Bernier, P. Poulin, 2000 Macroscopic Fibers and Ribbons of Oriented Carbon Nanotubes. Science, 290 17 1331 1334 .

65. Wong, M. Paramsothy, X. J. Xu, Y. Ren, S. Li, K. Liao, 2003 Physical Interactions at Carbon Nanotube-Polymer Interface. Polymer, 44 7757 7764 .

66. Q. Xiao, L. C. Zhang, 2004 The Stress Transfer Efficiency of a Single-walled Carbon Nanotube in Epoxy Matrix. Journal of Materials Science, 39 4481 4486 .

67. T. Yokozeki, Y. Iwahori, S. Ishiwata, 2007 Matrix Cracking Behaviors in Carbon Fiber/Epoxy Laminates Filled with Cup-stacked Carbon Nanotubes (CSCNTs). Composites Part A, 38 917 924 .

68. Q. Zheng, D. Xia, Q. Xue, K. Yan, X. Gao, Q. Li, 2009 Computational Analysis of Effect of Modification on the Interfacial Characteristics of a Carbon Nanotube-Polyethylene Composites System. Applied Surface Science, 255 3524 3543 .

Chapter 10

Improving Electrical Conductivity, Thermal Stability, and Solubility of Polyaniline-Polypyrrole Nanocomposite by Doping with Anionic Spherical Polyelectrolyte Brushes

Na Su

School of Printing and Packaging Engineering, Shanghai Publishing and Printing College, Shanghai 200093, China

ABSTRACT

The extent to which anionic spherical polyelectrolyte brushes (ASPB) as dopant improved the performance of polyaniline-polypyrrole (PANI-PPy) nanocomposite was investigated. Different characterization and analytical methods including Fourier transform infrared spectroscopy (FTIR), thermo-gravimetric analysis (TGA), scanning electron microscopy (SEM), and X-ray diffraction (XRD) confirmed that ASPB serving as dopant could improve the comprehensive properties of PANI-PPy nanocomposite. It was different from dopants such as SiO_2, poly(sodium-*p*-styrenesulfonate) (PSS), and canonic spherical polyelectrolyte brushes (CSPB) which only enhanced the performance of PANI-PPy nanocomposite on one or two sides. The electrical conductivity of (PANI-PPy)/ASPB nanocomposite at room temperature was 8.3 S/cm, which was higher than that of PANI-PPy (2.1 S/cm), (PANI-PPy)/PSS (6.8 S/cm), (PANI-

PPy)/SiO_2 (7.2 S/cm), and (PANI-PPy)/CSPB (2.2 S/cm). Meanwhile, (PANI-PPy)/ASPB nanocomposite possessed enhanced thermal stability and good solubility. In addition, the effects of polymerization temperature, the molecular weight of grafted polyelectrolyte brushes, and storage time on electrical conductivity were discussed.

KEYWORDS

Nanocomposite, Polyaniline-polypyrrole, Anionic spherical polyelectrolyte brushes, Doping, Electrical conductivity, Thermal stability, Solubility

1. BACKGROUND

During the last several decades, conducting polymers have been the subject of numerous investigations due to their excellent physical and chemical properties originating from their unique π-conjugated system [1–3]. Among the conducting polymers studied, polyaniline (PANI) and polypyrrole (PPy) are of particular interest because of their promising electrical conductivity, high environmental stability, interesting redox properties, magnetoresistance (MR) behaviors, and electrochemical performances [4–10]. In comparison with numerous reports about PANI and PPy, researches on copolymers of aniline and pyrrole are still far from enough. Because the copolymer may overcome the shortcomings of a single π-electron in the homopolymer, obtaining composites with excellent property [11], studies on copolymerization of aniline and pyrrole gradually attract people's attention. Electrochemical [12, 13] and chemical oxidative polymerization methods [14, 15] are the most common methods used in the synthesis of conducting copolymers. However, the large-scale application of PANI-PPy composite is sometimes limited by the difficulty of insolubility and infusibility of the material which can lead to poor electronic conductivity and mechanical properties. Therefore, the improvement of the comprehensive properties of PANI-PPy nanocomposite is of significance.

To date, most of the published research on this topic has been developed to improve the performance of conducting polymers on certain aspects. Xin et al. [16] prepared poly(aniline-*co*-pyrrole) nanocomposite by chemical oxidation polymerization using iron(III) chloride hexahydrate ($FeCl_3{\cdot}H_2O$) as an oxidant and sodium dodecylbenzenesulfonate (SDBS) as a surfactant. They found that the nanocomposite had high electrical conductivity by selecting proper conditions for the synthesis process. Li et al. [17] proposed a method to prepare poly(pyrrole-*co*-aniline) nanofibrils using a template by chemical copolymerization technique. It was reported that the length, diameter, and thickness of copolymer nanofibers were controlled by using AAO as a template, and the copolymer nanofibers had good thermal stability. Moreover, as for PPy, enhanced mechanical properties and reduced flammability were obtained by doping with epoxy resin [18]. However, since the properties of conducting polymers are mutually restraining, the improvement of their comprehensive performance is an important and difficult task.

In recent years, conducting polymers doped with polyelectrolyte have achieved outstanding progress [19]. Wu and coworkers [20] developed PPy, which exhibited excellent electrical conductivity and solubility by using different concentrations of water-soluble polystyrene sulfonate (PSS). The reason for this may be that de-doping does not easily happen for doped ions due to the large size of the polyelectrolyte, so the electrical conductivity of conducting polymers is more stable. Meanwhile, the entanglement effect of the long flexible chains of the polyelectrolyte can effectively hinder the growth of copolymer chains, helping to enhance its solubility. It is undoubtedly a good reference for the development of conducting polymers with excellent performance. In addition, in order to improve the thermal stability, magnetoresistance, and processing performance of conductive nanocomposites, besides organic materials just like MWNTs/PANI [21], inorganic materials were used, such as PPy/Fe_3O_4 [22,23], PPy/SiO_2 [24, 25], and PPy/Co_3O_4 [26]. In view of this, highly branched anionic spherical polyelectrolyte brushes (ASPB), consisting of polyelectrolyte chains affixing to the surface of spheres, may be a novel dopant of conducting polymers which can improve the performance of conducting polymers by introducing the brush polymers with certain functional groups.

This paper presented a facile method for the synthesis of (PANI-PPy)/ASPB nanocomposite by chemical oxidative polymerization. On the basis of a previous work [27], the advantage of ASPB serving as dopant in the synthesis of (PANI-PPy)/ASPB nanocomposite was evaluated. Compared to the PANI-PPy nanocomposite doped with PSS, SiO_2, and canonic spherical polyelectrolyte brushes (CSPB), (PANI-PPy)/ASPB nanocomposite exhibited good performances of electrical conductivity, thermal stability, and solubility. Moreover, since electrical conductivity was an important performance for conducting polymers, the effects of polymerization temperature, the molecular weight of grafted polyelectrolyte brushes, and storage time on electrical conductivity were also studied.

2. METHODS

2.1 Materials

Aniline and pyrrole (Sinopharm Chemical Reagents Co., Ltd, Shanghai, China) were distilled under reduced pressure before use. Ammonium persulfate (APS, 98 %) was purchased from Sinopharm Chemical Reagents Co., Ltd, Shanghai, China. Other chemicals and solvents including hydrochloric acid (HCl, 36–38 %) and ethanol were analytical reagents and were used without further purification. The ASPB ($D_z \approx 100$ nm, $M_w = 500–2000$ g/mol) consisting of modified SiO_2 cores and PSS brushes were prepared by surface-initiated polymerization [28]. CSPB ($D_z \approx 100$ nm, $M_w = 2000$ g/mol) were composed of SiO_2 cores and poly(diallyldimethylammonium chloride) (p-DMMPAC) brushes.

2.2 Synthesis of (PANI-PPy)/ASPB Composite

In this paper, we presented a facile method for the synthesis of (PANI-PPy)/ASPB nanocomposite by chemical oxidative polymerization, and the synthesis process is shown in Fig. 1. In a typical procedure, 92 mg of ASPB was firstly added into 25.2 mL of 2 M HCl (aq) with ultrasonic dispersion for 20 min, and then 1.8 mL of aniline and 1.4 mL of pyrrole were added. After the mixture was cooled to 5 °C and degassed under a flow of N_2 for 20 min, copolymerization was initiated by dissolving 4.5 g of APS in 12.6 mL of 2 M HCl (aq). After 6 h, the products were collected via filtration and washed with ethanol and distilled water for three times to remove small molecular compounds and unreacted monomers. The resulting products were vacuum dried at 60 °C for 12 h. In order to compare the doping effect of different dopants involving ASPB, PSS, SiO_2, and CSPB, (PANI-PPy)/PSS, (PANI-PPy)/SiO_2, (PANI-PPy)/CSPB, and PANI-PPy nanocomposites were synthesized.

2.3 Characterization

Fourier transform infrared spectroscopy (FTIR) was obtained using a Nicolet AVATAR 360FT spectrometer (USA). The chemical composition of (PANI-PPy)/ASPB nanocomposite was inspected by energy-dispersive X-ray diffraction (EDX) spectroscopy attached to a scanning electron microscope (SEM) which was used to investigate the morphology of samples. It was recorded on a Quanta 200 microscope (FEI, Netherlands) operated at 30 kV. X-ray diffraction (XRD) measurements were performed on a Shimadzu “XRD-6000” instrument (Japan) operating at a voltage of 40 kV and a current of 40 mA with

CuK$_\alpha$ radiation, λ = 1.54060 Å. The samples were measured in a continuous scan mode at 5°–50° (2ϑ) with a scanning rate of 5°/min. Thermo-gravimetric analysis (TGA) was carried out on a SETSYS-1750 instrument at a heating rate of 10 °C/min under nitrogen atmosphere.

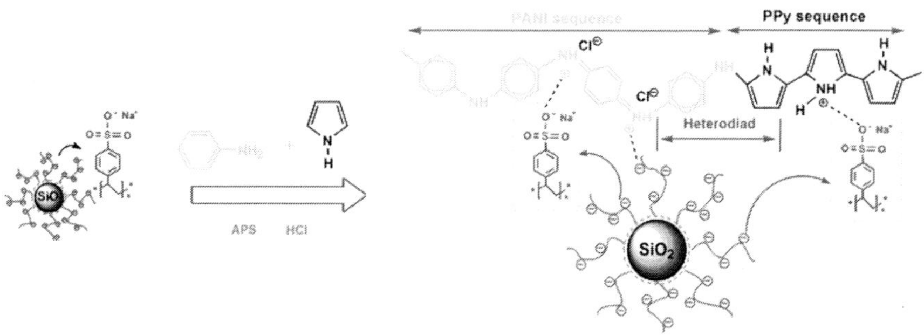

Figure 1. Schematic representation of the synthesis of (PANI-PPy)/ASPB nanocomposite

2.4 Measurements

The electrical conductivity of the samples was measured in a four-point probe apparatus (RTS-4, China) at room temperature. The samples were firstly pressed into a circular tablet with D = 13 mm at 20 MPa, and the measurement of the thickness W of each tablet was then performed. In order to calculate the source current I according to formula (1) [29], $F(W/S)$ and $F(D/S)$ (S = 1) which denoted the width correction coefficient and the diameter correction coefficient, respectively, were looked up from a table. The resistivity ρ at room temperature could be obtained from the four-point probe apparatus. Since conductivity (σ; S/cm) = 1/ρ, the electrical conductivity of the samples was calculated.

$$I = F(W/S) \times F(D/S) \times W \times 0.1 \quad (1)$$

The solubility of the samples can be reflected by the conductivity of the saturated solution (T = 25 °C, pH = 6). It is measured with a DDS-12A digital conductivity meter (Hubei Provincial Institute of Measurement and Testing). The specific process is as follows: 10 mg of samples was dissolved in 4 mL of ethanol with stirring and heating to boiling, then the supernatant and soluble impurities were removed. The process was repeated twice. After 10 mL of ethanol was added and heated to boiling for the purpose of fully dissolving the samples, the samples were placed in a bath at constant temperature for 20 min to precipitate the solid. To avoid the effect of the solid particles suspended in the electrode on experimental results, 3 or 4 mL of supernatant was added into the beaker. Additionally, the conductivity of the saturated solution of ethanol used as reference was required to be measured.

3. RESULTS AND DISCUSSION

3.1 FTIR Spectra

FTIR spectroscopy is used to study the chemical bonding of samples (see Fig. 2). The characteristic peaks of PANI-PPy, represented by the absorption bands at 1551 and 1476 cm^{-1}, can be attributed to the stretching vibration of the quinonoid and benzenoid rings. The peaks at 1293 and 1203 cm^{-1} are assigned to C–N and C=N stretching vibrations, respectively, which are consistent with the literature [30]. While adding a variety of different dopants, no new peaks appear. However, the ratio of the integrated absorption areas at approximately 1476 and 1551 cm^{-1} (A1476/A1551) for the composites is different, indicating that they have different conjugation lengths. The higher ratio represents the longer conjugation length [31]. The conjugation length follows the order (PANI-PPy)/ASPB > (PANI-PPy)/SiO_2 > (PANI-PPy)/PSS > (PANI-PPy)/CSPB, which is also the order of the electrical conductivity of nanocomposites.

In order to demonstrate that the conducting composites are composed of copolymers of aniline and pyrrole, instead of a simple blend of the homopolymers, the contrast FTIR spectra of the copolymers and blends are also shown in Fig. 2b. As can be seen from the figure, most characteristic absorption peaks for the blends also appear in the spectrum of the copolymers. However, compared with the blends which have only one C–N stretching vibration at 1130 cm^{-1} (Fig. 2b (a')), two C–N stretching vibrations (1203 and 1097 cm^{-1}) and a carbonyl group (1701 cm^{-1}) are displayed (Fig. 2b (a)), which is consistent with the literature [32].

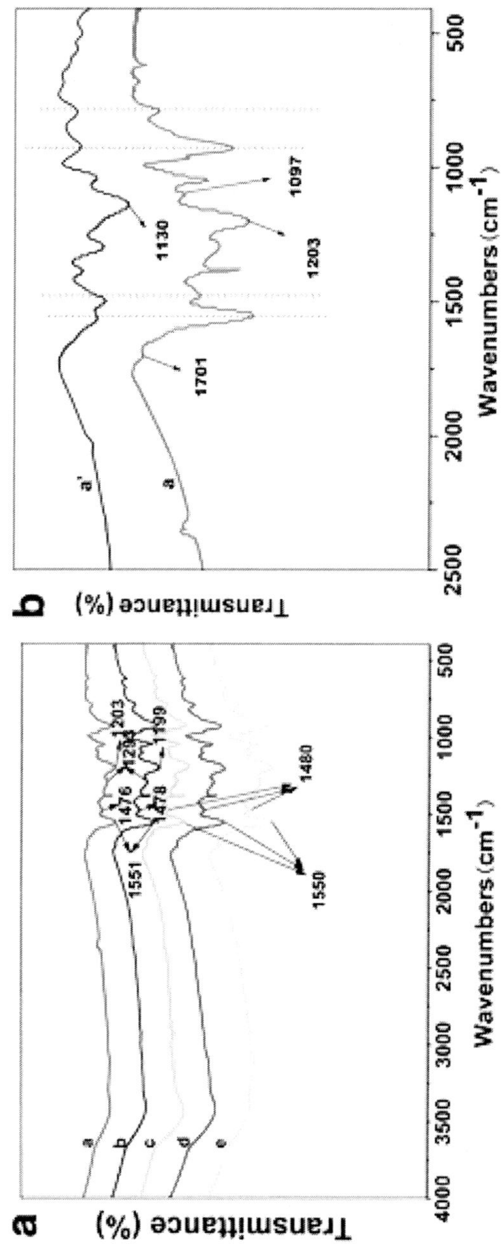

Figure 2. **a** FTIR spectra of (*a*) PANI-PPy, (*b*) (PANI-PPy)/ASPB, (*c*) (PANI-PPy)/PSS, (*d*) (PANI-PPy)/CSPB, and (*e*) (PANI-PPy)/SiO_2nanocomposites. **b** FTIR spectra of (*a*) PANI-PPy nanocomposites and (*a′*) the blends of the two homopolymers

3.2 XRD Patterns

The crystallographic structure of the samples is probed by X-ray diffraction. As shown in Fig. 3, the broader peak in the XRD pattern of PANI-PPy nanocomposite is observed at $2\vartheta = 23.5°$, implying an amorphous structure [31]. For the (PANI-PPy)/ASPB nanocomposite, the characteristic peak has shifted to $2\vartheta = 22.4°$. Similarly, for (PANI-PPy)/ASPB, (PANI-PPy)/SiO_2, (PANI-PPy)/PSS, and (PANI-PPy)/CSPB nanocomposites, only a broad peak presents in their XRD spectra, but these peak positions have shifted to $2\vartheta = 22.4°$, 21.8°, 22.6°, and 22.8°, respectively. Accordingly, the interplanar distances (*d*) corresponding to these diffraction peaks are 0.391, 0.402, 0.389, and 0.385 nm separately calculated by the Prague formula ($2d\sin\vartheta = n\lambda$) [33]. Compared to PANI-PPy (0.373 nm), the interplanar distance can be increased by doping. The reason for this may be that the addition of dopants changes the original lattice structure of the conductive composite, and the morphological changes are expected, which will be observed in SEM images.

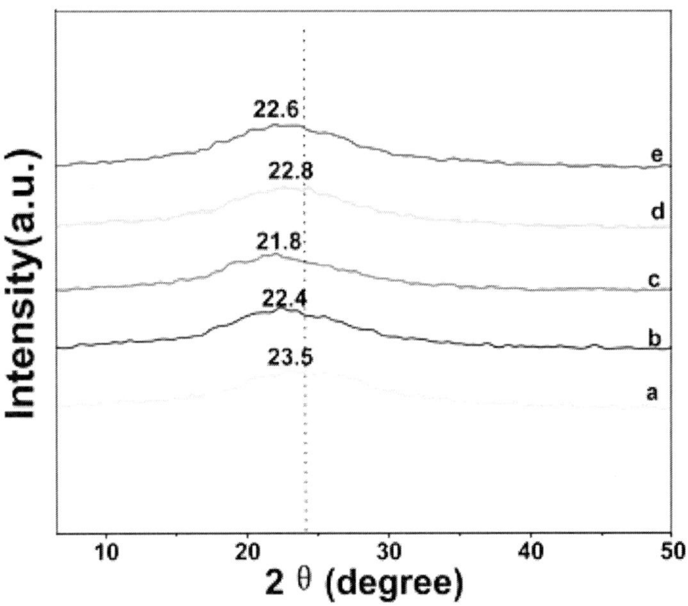

Figure 3. XRD patterns of (*a*) PANI-PPy, (*b*) (PANI-PPy)/ASPB, (*c*) (PANI-PPy)/SiO_2, (*d*) (PANI-PPy)/CSPB, and (*e*) (PANI-PPy)/PSS nanocomposite

3.3 Electrical Conductivity

Electrical conductivity is determined using a RTS-4 four-point probe which is a simple apparatus for measuring the resistivity of conductive composites. Results indicate that the room-temperature electrical conductivities of PANI-PPy, (PANI-PPy)/PSS, (PANI-PPy)/SiO_2, and (PANI-PPy)/CSPB nanocomposites are 2.1, 6.8, 7.2, and 2.2 S/cm, respectively, while (PANI-PPy)/ASPB nanocomposite shows a high value of electrical conductivity (8.3 S/cm). The increase in magnitude of the electrical conductivity is consistent with the FTIR results.

3.4 Thermal Stability

Figure 4 displays the thermo-gravimetric analysis of PANI-PPy, (PANI-PPy)/ ASPB, (PANI-PPy)/CSPB, (PANI-PPy)/SiO_2, and (PANI-PPy)/PSS nanocomposites under N_2 atmospheres at 60 % RH. As observed from the TGA curve of PANI-PPy, the decreased water content is because of the hydrophobic property of PPy and PANI [34]. A large amount of mass loss of copolymers begins at 200 °C, and the mass loss of copolymers in the temperature range of 200–700 °C is 31.6 %, which is mainly due to the degradation of the polymer [35]. When doped with ASPB, the mass loss of (PANI-PPy)/ASPB nanocomposite in the same temperature range was 23.6 % with a 7 % decrease, indicating that the addition of ASPB can improve the thermal stability of the conductive composite. It may be the reason that the generated direction of polymer chains for polypyrrole and polyaniline is promoted, which will be proved by the SEM images. It also can be seen from the TGA curves of the PANI-PPy nanocomposite doped with CSPB, PSS, and SiO_2 that the mass losses of (PANI-PPy)/ASPB, (PANI-PPy)/CSPB, (PANI-PPy)/PSS, and (PANI-PPy)/SiO_2 at 200–700 °C were 23.6, 30.9, 31.1, and 29.3 %, respectively, which are lower than that of PANI-PPy (31.6 %). Results show that the addition of PSS and CSPB can improve the thermal stability of PANI-PPy nanocomposite, but the effect is not obvious. The DSC curves of PANI-PPy nanocomposite with different dopants are also shown in Fig. 4. Compared with PANI-PPy nanocomposite, nearly 20 °C (from 233 to 252 °C) decomposition temperature is increased by the addition of ASPB.Since (PANI-PPy)/ASPB nanocomposite also has a higher decomposition temperature than (PANI-PPy)/CSPB, (PANI-PPy)/PSS, and (PANI-PPy)/SiO_2, better thermal stability is obtained.

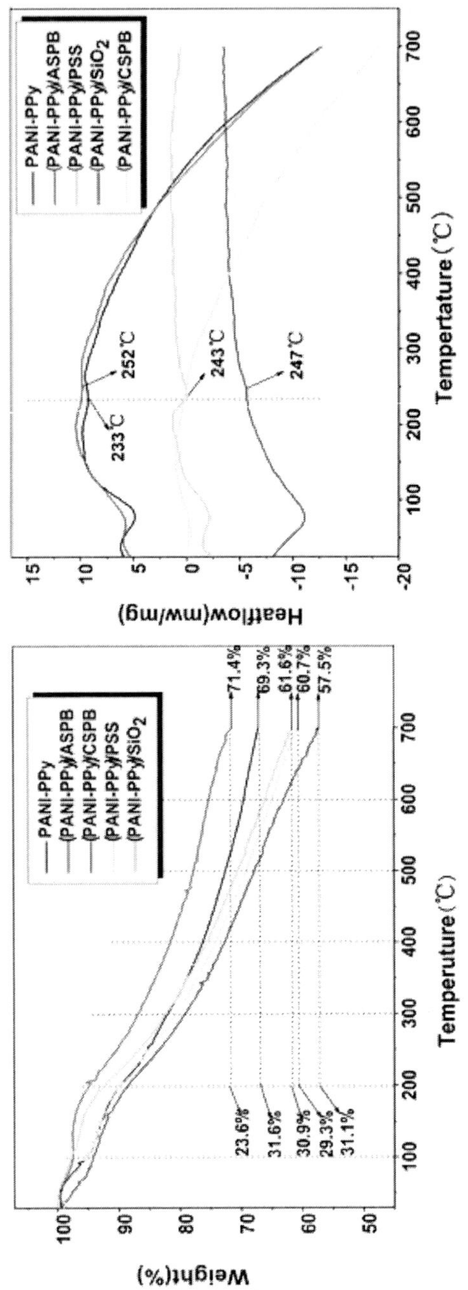

Figure 4. Effect of polymerization temperature and the molecular weight of grafted polyelectrolyte brushes on electrical conductivity

3.5 Solubility

Figure 5 represents the solubility of the samples which is reflected by the conductivity of the saturated solution (T = 25 °C, pH = 6). As shown in Fig. 5a, the PANI-PPy nanocomposite with different dopants (0.01 g) are dissolved in ethanol (4 mL) with ultrasonic dispersion for 10 min and then placed for 30 h. If the conductive nanocomposite has good solubility, its solution should be green. For the PANI-PPy nanocomposite doped with 5 wt% SiO_2, PSS, CSPB, and ASPB, different from the (PANI-PPy)/SiO_2 and (PANI-PPy)/CSPB nanocomposites whose solutions are clear, the solutions show a dark green color for (PANI-PPy)/PSS and (PANI-PPy)/ASPB nanocomposites, indicating that the solubility for these two composites is good. The main reason is that the resulting copolymer chains become short and small due to the flexible long chains of PSS, which makes the nanocomposite well dissolved in the solvent. In addition, the entry of PSS in the PANI-PPy chains may make copolymer chains grow in a direction such that the crosslinking degree is minimal. The addition of ASPB can improve the solubility of conductive nanocomposites which is mainly because of the grafting of PSS brushes. The crystallinity of conductive composites increases with the addition of SiO_2, thus leading to a larger grain size sample, and a poor solubility is obtained. This result is consistent with the results of XRD. The quantitative analysis method is also shown in Fig. 5b. After testing the conductivity of the saturated solution of ethanol (0.2 μS/cm), the conductivities of the saturated solution of PANI-PPy, (PANI-PPy)/ASPB, (PANI-PPy)/CSPB, (PANI-PPy)/SiO_2, and (PANI-PPy)/PSS are 2.8, 7, 5.4, 3.8, and 6.4 μS/cm, respectively. Therefore, the contributions of different dopants on the solubility of conductive composites follow the order ASPB > PSS > CSPB > SiO_2.

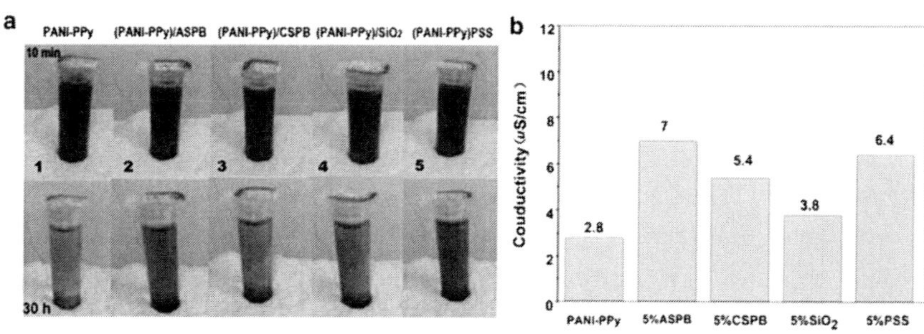

Figure 5. **a** Qualitative and **b** quantitative analysis of the conductivity of saturated solution of PANI-PPy nanocomposite with different dopants (T = 25 °C, pH = 6)

3.6 Evaluation of the Effects of Dopants

In summary, a variety of characterization methods prove that the effects of dopant on PANI-PPy nanocomposite are different. The performances of conductivity, solubility, and thermal stability for conductive nanocomposites when the amount of dopant is 5 wt% are shown in Table 1.

It can be observed from the table that the contributions of each dopant on electrical conductivity, solubility, and thermal stability of PANI-PPy nanocomposite follow the order ASPB > SiO_2 > PSS > CSPB, ASPB > PSS > CSPB > SiO_2, and ASPB > SiO_2 > PSS > CSPB, respectively. When doped with CSPB, the performances of conductive nanocomposites are not significantly improved except their solubility, indicating CSPB are not suitable as the dopant of conducting polymers. It may be due to the repulsive interaction between their cationic charge of brush layers and radical cation of conducting copolymers, so that CSPB do not enter into the conducting copolymer chains. For (PANI-PPy)/SiO_2 nanocomposite, the addition of SiO_2 may increase the structure regularity of conducting polymers, so the electrical conductivity and thermal stability of conductive composites improve. However, the insolubility of SiO_2 in most solvents also makes it powerless in improving the solubility of conducting copolymers. Furthermore, the main advantage of PSS is that it can increase the solubility of PANI-PPy nanocomposite. On the one hand, its flexible polymer chains in the copolymer system hinder the growth of the copolymer chain, making it smaller and shorter. On the other hand, the hydrophilicity of PSS also promotes the water solubility of conducting copolymers. But studies have shown that it has no obvious effect on improving the thermal stability. For ASPB consisting of a SiO_2 core and PSS brushes, just a combination of the advantages of SiO_2 and PSS is conducted. The thermal stability and solubility of copolymers are enhanced by SiO_2particles and PSS chains, respectively. Therefore, ASPB are an excellent dopant which can improve the comprehensive performances of PANI-PPy nanocomposite.

Table 1. Effect of dopant species on the performance of conductive nanocomposites

Performance	Sample				
	PANI-PPy	(PANI-PPy)/ASPB	(PANI-PPy)/SiO_2	(PANI-PPy)/PSS	(PANI-PPy)/CSPB
Electrical conductivity (S/cm)	2.1	8.3	7.2	6.8	2.2
Conductivity of saturated solution (μS/cm)	2.8	7	3.8	6.4	5.4
Exothermic peak (°C)	233	252	247	243	Unconspicuous

3.7 Morphology

In order to explore the doping mechanism of ASPB, the morphologies of PANI-PPy (a), (PANI-PPy)/ASPB nanocomposite (b), and ASPB (b′) are studied using SEM images (see Fig. 6), and the chemical composition of (PANI-PPy)/ASPB nanocomposite is determined by the simultaneous EDX spectroscopy (c). As shown in Fig. 6a, PANI-PPy displays a closely stacked granular structure, while the micrograph of (PANI-PPy)/ASPB nanocomposite (Fig. 6b) shows a spherical-like structure. It may be because the addition of ASPB with uniform spherical structure (Fig. 6b (b′)) provides the space factors for copolymer orderly growth, which is consistent with the XRD and TGA results. ASPB serve as template for the synthesis of (PANI-PPy)/ASPB nanocomposite. The EDX analysis shows that the signals corresponding to silicon and sulfur appear on the spectrum, suggesting that the doping of ASPB in the PANI-PPy matrix has occurred.

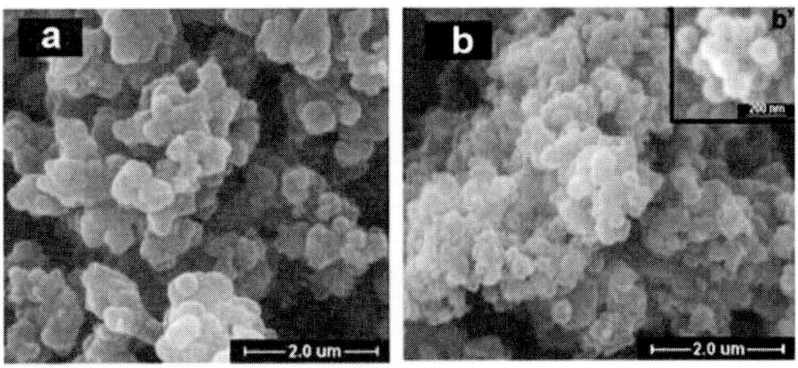

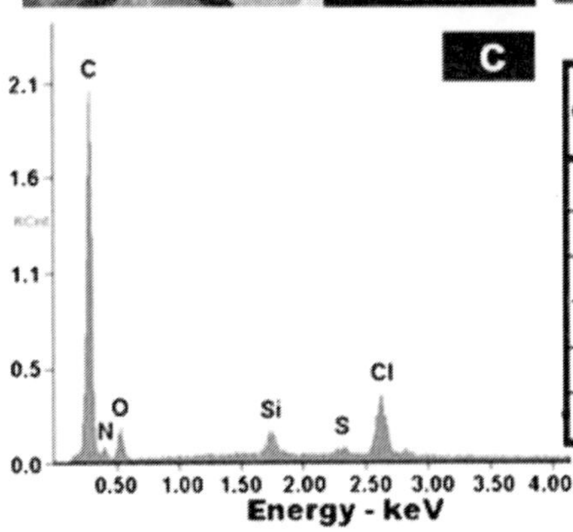

elements	contents	
	Wt%	At%
CK	73.16	78.17
OK	13.09	11.94
NK	10.95	08.80
SiK	00.63	00.27
SK	00.26	00.12
ClK	01.90	00.69

Figure 6. SEM images of **a** PANI-PPy, **b** (PANI-PPy)/ASPB composite, and (*b′*) ASPB. **c** EDX analysis of (PANI-PPy)/ASPB nanocomposite

3.8 Doping Mechanism

Figure 7 describes the doping mechanism and specific distribution in the ASPB within a charged reagent. The copolymerization reaction of aniline and pyrrole monomers in the PSS chains happens. *R* denotes the distance between the two copolymer chains. The amount of negative charge per unit length in PSS chains is much more than the amount of positive charge per unit length in PANI-PPy chains, so part of SO^{3-} is balanced by Na^{+}. Due to the effects of density gradient formed by densely grafted PSS chains, copolymerization reaction occurs preferentially in high-concentration aniline and pyrrole monomers [36], which are located at the cores of ASPB. According to the doping mechanism of protonic acid, the copolymerization of aniline and pyrrole monomers forms a polymeric cation, which raises the function of multivalent counter ions in brush layers. Therefore, the complexation between PANI-PPy and PSS is irreversible. As a result, the copolymerization of aniline and pyrrole monomers is fixed in the brush layers, and ASPB play a role as template in the synthesis of (PANI-PPy)/ASPB nanocomposite.

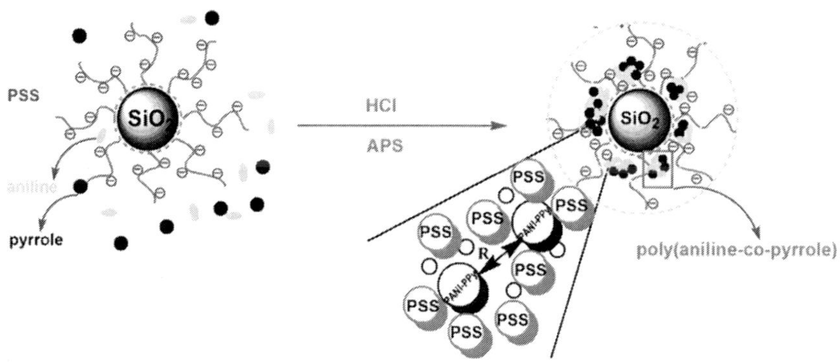

Figure7. Doping mechanism of ASPB

3.9 Effect of Polymerization Temperature and the Molecular Weight of Grafted Polyelectrolyte Brushes on Electrical Conductivity

The influences of polymerization reaction temperature and the molecular weight of PSS brushes on electrical conductivity are investigated, with 5 wt% ASPB used and 6 h reaction time. As shown in Fig. 8, the electrical conductivity of PANI-PPy composite at room temperature is lower than that at low temperatures (5 °C). The reason for this is that the polymerization process is slow at low temperatures, the content of defect structure in the molecular structure of copolymers decreases along with orderly improvement, which helps to increase the electrical conductivity. While at high reaction temperature, the formation of increased defects in the copolymer makes the surface morphology rough, thereby affecting electrical conductivity [37]. In addition, when other conditions remain the same, ASPB with different molecular weights of grafted brushes (500, 1000, and 2000 g/mol) are selected for the doping experiment, and the results are also shown in Fig. 8. The electrical conductivities of (PANI-PPy)/ASPB nanocomposites increase with the increase of the molecular weight of grafted polyelectrolyte brushes. The reason for this may be that the long chains of the polyelectrolyte make pyrrole and aniline monomers grow better in the ASPB system. The crosslinking degree of conducting copolymer chains reduces, resulting in generating a polymer chain with more extended and extended conjugation length, so the electrical conductivity increases [19].

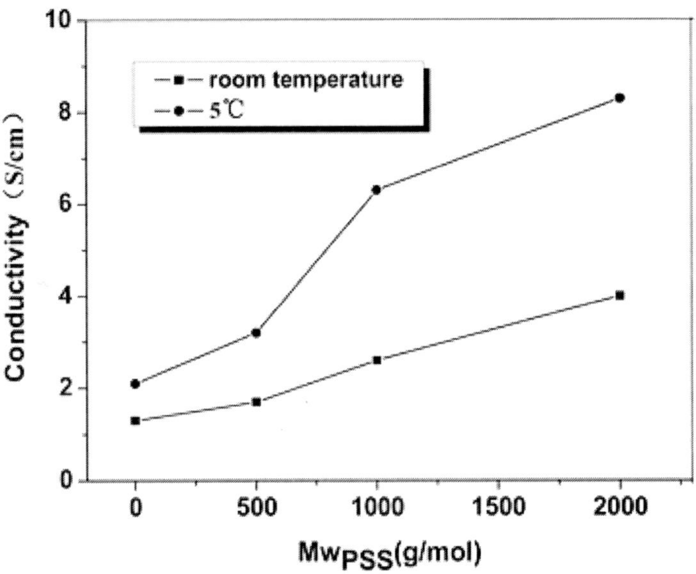

Figure 8. Thermal gravimetric analysis of PANI-PPy, (PANI-PPy)/ASPB, (PANI-PPy)/SiO_2, (PANI-PPy)/CSPB, and (PANI-PPy)/PSS nanocomposites

3.10 Effect of Storage Time on Electrical Conductivity

The PANI-PPy nanocomposite doped with proton acid and ASPB is placed in the air for a period of time at room temperature to study the stability of electrical conductivity. As shown in Fig. 9, the electrical conductivity of PANI-PPy nanocomposite doped with ASPB is almost unchanged after a few months, while the electrical conductivity of the small ion-doped nanocomposite declines. It is mainly because that the de-doping phenomenon is difficult to happen when macromolecular doping is conducted, and the structure of copolymer chains is more stable.

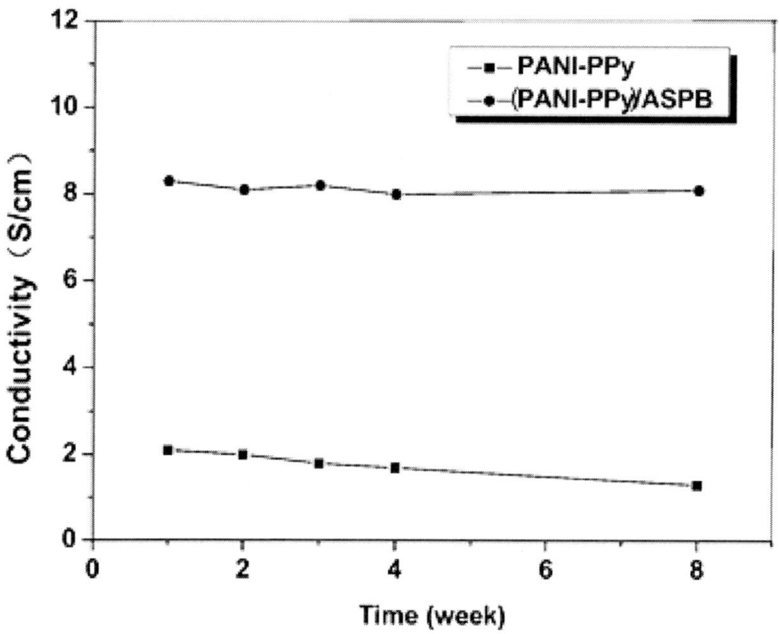

Figure 9. Effect of storage time on the electrical conductivity

4. CONCLUSIONS

This paper proposes a simple and effective method to improve the comprehensive performance of conductive composites. By doping the ASPB, the electrical conductivity, thermal stability, and solubility of PANI-PPy composite are enhanced. The electrical conductivity of (PANI-PPy)/ASPB nanocomposite is 8.3 S/cm, which is higher than that of (PANI-PPy)/CSPB

(2.2 S/cm), (PANI-PPy)/PSS (6.8 S/cm), and (PANI-PPy)/SiO_2 (7.2 S/cm). Furthermore, the influences of polymerization temperature, the molecular weight of grafted polyelectrolyte brushes, and storage time on the electrical conductivity of (PANI-PPy)/ASPB nanocomposite are investigated. Results show that the long grafted chains and low reaction temperature help to improve the electrical conductivity of conductive composites.

ACKNOWLEDGEMENTS

This work is supported by Innovation Program of Shanghai Municipal Education Commission (15ZZ112).

COMPETING INTERESTS

The authors declare that they have no competing interests.

REFERENCES

1. Gospodinova N, Terlemezyan L. Conducting polymers prepared by oxidative polymerization: polyaniline. Prog Polym Sci. 1998;23:1443.
2. Carrasco PM, Grande HJ, Cortazar M, Alberdi JM, Areizaga J, Pomposo JA. Structure-conductivity relationships in chemical polypyrroles of low, medium and high conductivity. Synthetic Met. 2006;156:420.
3. Lim VWL, Kang ET, Neoh KG, Ma ZH, Tan KL. Determination of pyrrole–aniline copolymer compositions by X-ray photoelectron spectroscopy. Appl Surf Sci. 2001;181:317.
4. Saravanan C, Shekhar RC, Palaniappan S. Synthesis of polypyrrole using benzoyl peroxide as a novel oxidizing agent. Macromol Chem Phys. 2006;207:342.
5. Ding HJ, Wan MX, Wei Y. Controlling the diameter of polyaniline nanofibers by adjusting the oxidant redox potential. Adv Mater. 2007;19:465.
6. Gu H, Guo J, Wei H, Zhang X, Zhu J, Shao L, et al. Magnetoresistive conductive polymer-tungsten trioxide nanocomposites with ultrahigh sensitivity at low magnetic field. Polymer 2014;55:944–950

7. Gu H, Guo J, Yan X, Wei H, Zhang X, Liu J, et al. Electrical transport and magnetoresistance in advanced polyaniline nanostructures and nanocomposites. Polymer. 2014;55:4405–19.

8. Gu H, Wei H, Guo J, Haldolaarachige N, Young DP, Wei S, et al. Hexavalent chromium synthesized polyaniline nanostructures: magnetoresistance and electrochemical energy storage behaviors. Polymer. 2013;54:5974–85.

9. Gu H, Huang Y, Zhang X, Wang Q, Zhu J, Shao L, et al. Magnetoresistive polyaniline-magnetite nanocomposites with negative dielectrical properties. Polymer. 2012;53(3):801–9.

10. Zhao Y, Wei H, Arowo M, Yan X, Wu W, Chen J, et al. Electrochemical energy storage by polyaniline nanofibers: high gravity assisted oxidative polymerization vs. rapid mixing chemical oxidative polymerization. Phys Chem Chem Phys. 2015;17(2):1498–502.

11. Fusalba F, Bélanger D. Electropolymerization of polypyrrole and polyaniline-polypyrrole from organic acidic medium. J Phys Chem B. 1999;103:9044–54.

12. Ouyang J, Li Y. Great improvement of polypyrrole films prepared electrochemically from aqueous solutions by adding nonaphenol polyethyleneoxy (10) ether. Polymer. 1997;38:3997.

13. Wei H, Wang Y, Guo J, Yan X, O'Connor R, Zhang X, et al. Electropolymerized polypyrrole nanocoatings on carbon paper for electrochemical energy storage. ChemElectroChem. 2015;12(1):119–26.

14. Chakrabarti S, Banerjee D, Bhattacharyya R. Enhancement of room temperature electrical conductivity of polypyrrole by chemical modification. J Phys Chem B. 2002;106:3061.

15. Qiu B, Xu C, Sun D, Yi H, Guo J, Zhang X, et al. Polyaniline coated ethyl cellulose with improved hexavalent chromium removal. ACS Sustainable Chemistry &Engineering. 2014;2(8):2070–80.

16. Xin S, Zhao C, Zhou T, et al. Preparation and characterization of polyaniline-polypyrrole composite from polyaniline dispersions. J Appl Polym Sci. 2007;104:3523–9.

17. Li XH, Zhang XG, Li HL. Preparation and characterization of pyrrole/aniline copolymer nano fibrils using the template-synthesis method. J Appl Polym Sci. 2001;81:3002.

18. Zhang X, Yan X, Guo J, Liu Z, Jiang Z, Jiang D, et al. Polypyrrole doped epoxy resin nanocomposites with enhanced mechanical properties and reduced flammability. Journal of Materials Chemistry C. 2015;3:162–76.

19. Errachid A. Electropolymerization of nano-dimensioned polypyrrole micro-ring arrays on gold substrates prepared using submerged micro-contact printing. Nanotechnology. 2007;18:485301.

20. Wu TM, Chang HL, Lin YW. Synthesis and characterization of conductive polypyrrole with improved conductivity and processability. Soc Chem Ind. 2009;58:1065–70.

21. Gu H, Guo J, He Q, Jiang Y, Huang Y, Haldolaarachige N, et al. Magnetoresistive polyaniline/multi-walled carbon nanotube nanocomposites with negative permittivity. Nanoscale. 2014;6:181–9.

22. Chen W, Li XW, Xue G. Magnetic and conducting particles: preparation of polypyrrole layer on Fe_3O_4 nanospheres. Appl Surf Sci. 2003;218(1–4):215–21.

23. Qiu B, Xu C, Sun D, Wang Q, Gu H, Zhang X, et al. Polyaniline coating with various substrates for hexavalent chromium removal. Appl Surf Sci. 2015;334:7–14.

24. Tuyen LTT, Kamloth K, Liess HD. Electrical properties of doped polypyrrole/silicon heterojunction diodes and their response to NO_x gas. Thin Solid Films. 1997;292:293–8.

25. Liu XH, Wu HY, Ren FL. Controllable fabrication of SiO(2)/polypyrrole core-shell particles and polypyrrole hollow spheres. Mater Chem Phys. 2008;109(1):5–9.

26. Wei H, He C, Liu J, Gu H, Wang Y, Yan X, et al. Electropolymerized polypyrrole nanocomposites with cobalt oxide coated on carbon paper for electrochemical energy storage. Polymer. 2015;67:192–9.

27. Huang Y, Su N, Zhang X, Zhao J, Li H, Liu X, et al. Controllable synthesis and characterization of poly(aniline-co-pyrrole) using anionic spherical polyelectrolyte brushes as dopant and template. Polym Composits. 2014;5:1858.

28. Su N, Li H, Zheng H, Yi S, Liu X. Synthesis and characterization of poly(sodium-p-styrenesulfonate)/modified SiO_2 spherical brushes. Express Polym Lett. 2012;6:680.

29. Su N, Li H, Yuan S, Yi S, Yin E. Synthesis and characterization of polypyrrole doped with anionic spherical polyelectrolyte brushes. Express Polym Lett. 2012;6:697.

30. Li G, Zhang Z. Synthesis of dendritic polyaniline nanofibers in a surfactant gel. Macromolecules. 2004;37:2683–5.

31. Menon VP, Lei J, Martin CR. Investigation of molecular and supermolecular structure in template-synthesized polypyrrole tubules and fibrils. Chem Mater. 1996;8:2382.

32. Xu P, Han XJ, Wang C, Zhang B, Wang HL. Morphology and physico-electrochemical properties of poly(aniline-co-pyrrole). Synth Met. 2009;159:430.

33. Zhang X, Zhu J, Haldolaarachchige N, Ryu J, Young DP, Wei S, et al. Synthetic process engineered polyaniline nanostructures with tunable morphology and physical properties. Polymer. 2012;53(10):2109–20.

34. Qiu B, Xu C, Sun D, Wei H, Zhang X, Guo J, et al. Polyaniline coating on carbon fiber fabrics for improved hexavalent chromium removal. RSC Advances. 2014;4:29855–65.

35. Bose S, Kuila T, Uddin ME. In-situ synthesis and characterization of electrically conductive polypyrrole/graphene nanocomposites. Polymer. 2010;51:5921.

36. Korovin AN, Sergeyev VG, Pyshkina OA. Nanoreactor-assisted polymerization toward stable dispersions of conductive composite particles. Macromol Rapid Commun. 2011;32:462.

37. Cairns DB. Synthesis and characterization of submicrometer-sized polypyrrole–polystyrene composite particles. Langmuir. 1999;15:8052–8.

Index